Augmented Wellness

Ben Othman Soufiane • Chinmay Chakraborty
Bhuvan Unhelkar
Editors

Augmented Wellness

Exploring the Power of VR and AR in Healthcare

Springer

Editors
Ben Othman Soufiane
Applied College
King Faisal University
Al-Ahsa, Arabie Saoudite

Chinmay Chakraborty
Electronics & Communication Engineering
Birla Institute of Technology
Mesra, Jharkhand, India

Bhuvan Unhelkar
Muma College of Business
University of South Florida
Tampa, FL, USA

ISBN 978-981-96-2951-0 ISBN 978-981-96-2952-7 (eBook)
https://doi.org/10.1007/978-981-96-2952-7

This Springer imprint is published by the registered company Springer Nature Singapore Pte Ltd.
The registered company address is: 152 Beach Road, #21-01/04 Gateway East, Singapore 189721, Singapore

Contents

1 **Virtual Reality Mental Health Interventions: AR/VR Solutions for Well-being** 1
S. Padmapriya, Priyanka Torne, Abhra Pratip Ray, Ankur Gupta, Rohit Anand, and Vishal Jain

2 **Understanding Virtual Reality (VR) and Its Applications in Healthcare** 25
Bhanu Pratap Singh, Ankush Joshi, and Ritik Kumar Singh

3 **Immersive Virtual Reality Exposure Therapy: Breakthrough Real-Time Scientific Investigation for Addressing Social Phobia Among Management Students** 47
DM Arvind Mallik

4 **Virtual Reality Therapy for Mental Disorder** 85
R. Immanual, N. Sanjana, S. Sangeetha, and K. M. Kirthika

5 **Energy Efficient: Analysis of Virtual Reality (VR) and Augmented Reality (AR) in Modern Healthcare Systems** 115
Vijaya Gunturu, R. Krishnamoorthy, Kazuaki Tanaka, Poorna Chandra Reddy Alla, Janjhyam Venkata Naga Ramesh, and S. Ravichandran

6 **Science Behind Augmented Reality and Virtual Reality in Healthcare** 137
Arti Saxena and Pooja Thukral

7 **Virtual Reality-Based Intelligent Internet of Medical Things Health Monitoring System: In Future Aspects** 153
V. Jyothi, Vijayalakshmi Chintamaneni, D. Sravani, Amar Y. Jaffar, Abdullah Alwabli, and N. Rajeswaran

8 Exploring Virtual Reality's Impact on Medical Education in Healthcare: A Comprehensive Overview 171
Pragathi Bandari, Alanod Abdullah Alkrida, A. Athiraja, Ayman A. Alharbi, and N. Rajeswaran

9 Monitoring and Remote Patient Management with Immersive Technologies 197
Sarika Mulukuntla

10 Current State of AR and VR in Healthcare 219
K. M. Kirthika, S. Sangeetha, R. Immanual, and N. Sanjana

11 Paradise of Virtual Reality (VR) and Augmented Reality (AR) in Mental Health Treatment: Breaking Down Barriers Advancing Supremely Smart Healthcare for Futuristic Growth 243
Bhupinder Singh and Christian Kaunert

12 Embedding Virtual Reality Technologies as an Early Career Neuroscience Research Training Methodology Advancing Medical Research 259
Lorenz S. Neuwirth and Maxime Ros

Chapter 1
Virtual Reality Mental Health Interventions: AR/VR Solutions for Well-being

S. Padmapriya, Priyanka Torne, Abhra Pratip Ray, Ankur Gupta, Rohit Anand, and Vishal Jain

Abstract Virtual Reality along with Augmented Reality technologies have emerged as potential tools in case of transforming mental health therapies, providing creative methods for enhancing well-being. This abstract examines the emerging subject of VR Mental Health Interventions along with its potential to revolutionize treatment approaches. VR/AR solutions provide realistic and interactive settings that replicate real-life situations, allowing people to face and address different mental health issues such as anxiety disorders, PTSD, phobias, and depression. VR/AR therapies cater to the different requirements of mental health care by customizing treatment experiences and providing access to those who encounter obstacles in receiving conventional therapy. Moreover, these technologies improve emotional control, compassion, and interpersonal abilities by offering chances to practice communication and social interactions in a secure and encouraging virtual setting. As advancements in research and development continue, virtual reality and augmented reality solutions show significant potential in enhancing mental well-being in the digital

S. Padmapriya
Department of Computer Science, SRM Trichy Arts and Science College, Irungalur, Tamil Nadu, India

P. Torne
Department of Electronics, SAVPM's Sancheti College of Arts, Commerce and Science, Pune, Maharashtra, India

A. P. Ray
Department of Physics, KSPM, Latur's Janvikas Mahavidyalaya, Bansarola, Maharashtra, India

A. Gupta
Department of Computer Science and Engineering, Vaish College of Engineering, Rohtak, Haryana, India

R. Anand (✉)
Department of ECE, G.B. Pant DSEU Okhla-1 Campus (formerly GBPEC), New Delhi, India

V. Jain
Department of Computer Science and Engineering, School of Engineering and Technology, Sharda University, Greater Noida, Uttar Pradesh, India

B. O. Soufiane et al. (eds.), *Augmented Wellness*,
https://doi.org/10.1007/978-981-96-2952-7_1

era, providing a revolutionary approach to mental healthcare. This study explores the potential of VR along with AR in healthcare, focusing on patient treatment, medical training, and health education. It evaluates the impact of VR along with AR on patient outcomes, assesses benefits along with limitations, and anticipates future trends in these technologies, ultimately contributing to augmented wellness.

Keywords Augmented reality · Virtual reality · Healthcare · patient care · Well-being · Immersive therapy · Cognitive behavioral therapy (CBT) · Exposure therapy · Anxiety treatment · Depression management · PTSD (post-traumatic stress disorder)

1 Introduction

These days, AR and VR aren't only for games and amusement; they're also useful for mental health treatments and providing new ways to improve one's mental and emotional wellness. A growing number of therapeutic approaches have begun to include this immersive technology in recent years, providing new opportunities to tackle mental health issues [1]. By using the dynamic and engaging characteristics of VR/AR, mental health interventions may construct safe spaces that mimic real-life situations. Anxiety disorders, PTSD, phobias, depression, and other mental health difficulties may be safely addressed and managed in these simulations [2]. Interventions like this may help people with things like exposure therapy, cognitive behavioral therapy, mindfulness training, and relaxation methods by putting them in virtual settings that they can tailor to their own needs. VR along with AR technologies have potential to greatly improve mental health by allowing for more customized therapy experiences [3]. Virtual settings and situations may be customized to match the unique requirements and preferences of each person, enhancing the engagement and effectiveness of treatment [4]. Also, unlike typical therapy settings, these treatments are accessible to those who may otherwise have trouble getting the mental health care they need because of things like poverty, lack of transportation, or other social or economic obstacles [5]. Improved emotional control, empathy, and social skills have all been noted as potential outcomes of VR/AR therapies. Technology like this allows people to hone their communication abilities in a controlled setting by recreating real-life social encounters and settings. People whose social functioning is impaired due to diseases like social anxiety or autism spectrum disorders may find this very helpful. Research and development in area of VR mental health interventions is ongoing, with a focus on finding new uses for VR and improving current methods to make the most of their therapeutic potential. There is a wide range of exciting potential applications in case of virtual reality along with augmented reality in mental health treatment, from gamified cognitive training programs to immersive relaxing experiences. VR along with AR solutions have potential to greatly contribute to promotion of MH in modern digital era by using technology to provide therapeutic experiences that are immersive, engaging, and tailored to the individual [6].

1.1 Virtual Reality (VR) and Augmented Reality (AR)

Virtual Reality along with Augmented Reality (as shown in Fig. 1.1) are cutting-edge technologies that have transcended their origins in entertainment and gaming to become transformative tools in various fields, including healthcare Both technologies offer immersive experiences that have the potential to revolutionize how perceive and interact our surroundings [7]. In recent years, VR and AR have garnered increasing attention for their applications in healthcare, particularly in realm of MH interventions. These technologies provide innovative solutions for addressing a wide range of MH challenges, from anxiety and depression to PTSD and phobias. By creating virtual environments that simulate real-life scenarios, VR/AR interventions offer individuals a safe space to confront and manage their mental health issues. The integration of VR along with AR into mental health care holds significant promise for improving treatment outcomes and increasing accessibility to therapeutic services. These technologies allow for personalized and immersive experiences, tailored to meet the unique needs of each individual. Moreover, VR/AR interventions have demonstrated effectiveness in enhancing emotional regulation, empathy, and social skills, offering new avenues for promoting overall well-being. As the field of VR and AR continues to advance, so too does their potential to transform mental health care [8].

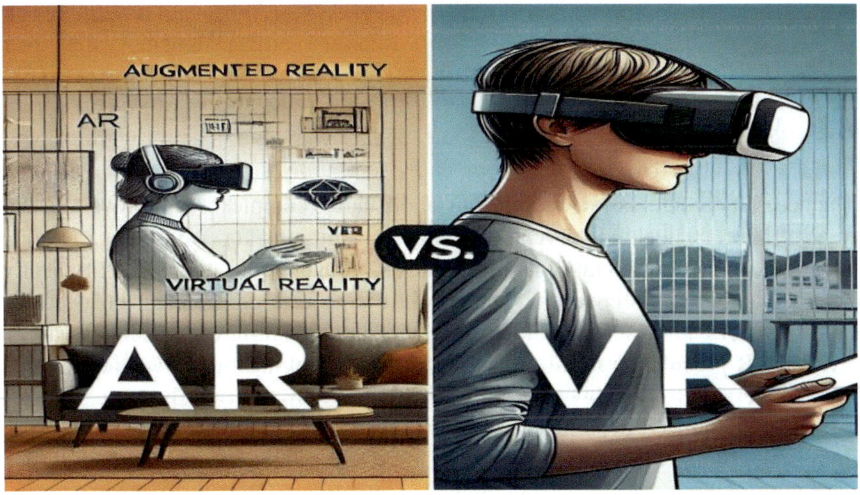

Fig. 1.1 Virtual reality (VR) and augmented reality (AR)

However, challenges remain, including the need for further research, development, and integration into clinical practice [9]. Nevertheless, the emergence of VR and AR as powerful tools for mental health interventions signifies a promising shift towards more innovative and effective approaches to promoting mental well-being in the digital age.

1.2 Mental Health Interventions

VR along with AR have opened up new frontiers in mental health interventions, offering a diverse range of innovative solutions to promote well-being. These immersive technologies provide a platform for various therapeutic approaches, including exposure therapy, cognitive behavioral therapy (CBT), mindfulness and relaxation techniques, social skills training, and psychoeducation. Through VR simulations, individuals can confront their fears and anxieties in controlled environments, gradually building resilience and coping strategies. Cognitive behavioral therapy techniques can be delivered in engaging virtual scenarios, allowing users to challenge negative thought patterns and behaviors. VR environments also facilitate mindfulness and relaxation exercises by transporting users to serene landscapes or overlaying calming cues onto the real world through AR applications. Furthermore, VR simulations provide opportunities for social skills training, enabling individuals to practice interpersonal interactions and communication skills in a safe space. Additionally, these technologies can be utilized for psychoeducation and empathy-building exercises, fostering understanding and compassion towards mental health conditions. By leveraging the immersive and interactive nature of VR along with AR, mental health professionals can offer personalized, engaging, along with accessible interventions to support individuals in their journey towards mental well-being [10].

1.3 Motivation

The necessity to address modern mental health care's complex issues drove "Virtual Reality Mental Health Interventions: AR/VR Solutions for Well-being. Despite advances in therapy methods and treatments, conventional approaches typically lack accessibility, effectiveness, and involvement. This emphasizes the need for innovative, effective mental health treatments that fit the different demands of users. VR and AR provide attractive opportunities for mental health intervention innovation. Immersive technology may transform therapy by offering dynamic, interactive, and customized experiences for patients. VR and AR may provide exposure treatment, cognitive behavioral therapy (CBT), mindfulness, and social skills training by mimicking various surroundings and situations. Growing awareness of the worldwide mental health epidemic motivates VR and AR mental health initiatives.

Mental health diseases affect millions globally, affecting individuals, families, and communities. However, stigma, limited access to treatment, and engagement obstacles hinder people from seeking assistance. VR and AR therapies may provide accessible, engaging, and stigma-free mental health assistance. VR along with AR in MH treatment follows digital health and telemedicine trends. VR along with AR in case of mental health therapies is a logical step toward more holistic and patient-centered therapy as technology advances. These technology may let people obtain mental health care remotely and on their own. Exploring "Virtual Reality Mental Health Interventions: AR/VR Solutions for Well-being" is driven by the need to improve mental health treatment globally. VR and AR may improve mental health assistance delivery, experience, and acceptance. Promoting resilience, empowerment, and well-being for all mental health patients drives this inquiry.

1.4 Challenges

Exploring Virtual Reality Mental Health Interventions: AR/VR Solutions for Well-being presents several notable challenges that need to be addressed for these technologies to reach their full potential in supporting mental health:

- **Accessibility and Affordability**: Despite advancements, VR along with AR technologies can still be expensive and require specialized equipment. Ensuring widespread access to these interventions may pose challenges, particularly for individuals from lower socioeconomic backgrounds or regions with limited resources.
- **Technical Limitations**: VR and AR technologies are constantly evolving, but technical limitations such as motion sickness, visual discomfort, and hardware/software compatibility issues remain. Overcoming these challenges is crucial to ensuring a seamless and comfortable user experience.
- **Ethical Concerns**: As with any emerging technology, VR and AR raise ethical considerations regarding privacy, data security, and informed consent. Safeguarding user data and ensuring ethical use of immersive technologies in mental health interventions is paramount.
- **Validation and Efficacy**: While VR and AR show promise in mental health care, robust empirical evidence demonstrating their efficacy across diverse populations and mental health conditions is still needed. Conducting rigorous research and clinical trials to validate the effectiveness of these interventions is essential.
- **Integration with Traditional Care**: Integrating VR and AR interventions into existing mental health care systems poses logistical challenges, including training clinicians, establishing reimbursement models, and ensuring seamless coordination with traditional therapeutic approaches.
- **User Engagement and Adherence**: Maintaining user engagement and adherence to VR and AR interventions over time can be challenging. Designing

immersive experiences that are engaging, motivating, and tailored to individual preferences is essential for long-term effectiveness.
- **Stigma and Acceptance**: Overcoming stigma associated with mental health and emerging technologies is another hurdle. Educating the public and healthcare providers about potential benefits of VR and AR in mental health care and addressing misconceptions is crucial for widespread acceptance and adoption.
- **Diversity and Inclusivity**: Designing VR and AR interventions that are inclusive and culturally sensitive to diverse populations is essential. Ensuring that these technologies are accessible and relevant across different cultural, linguistic, and socioeconomic contexts is critical for equitable mental health support.

Addressing these challenges will require collaboration among researchers, clinicians, technologists, policymakers, and individuals with lived experience of mental health conditions. By working together to overcome these obstacles, we can harness the full potential of VR along with AR technologies to revolutionize mental health care along with improve the well-being of individuals worldwide.

1.5 *Significance of VR and AR Technologies in Revolutionizing Mental Health Therapies*

The integration of Virtual Reality (VR) and Augmented Reality (AR) technologies in healthcare is revolutionizing the landscape of mental health therapies. These immersive technologies are redefining traditional therapeutic methods by providing innovative and effective treatment options for a variety of mental health conditions [8]. VR and AR offer unique capabilities to create controlled, immersive environments that can be tailored to individual therapeutic needs, enhancing the patient experience and treatment outcomes. VR, with its ability to simulate real-world environments and scenarios, is being utilized to treat conditions such as anxiety, PTSD, phobias, and depression. Patients can confront and work through their fears in a safe, controlled space, guided by their therapist. AR, on the other hand, overlays digital information onto the real world, providing real-time support and interventions, which can be particularly useful for conditions like social anxiety and cognitive-behavioral therapy (CBT). The potential of these technologies extends beyond treatment to include prevention and early intervention, offering tools for mindfulness, stress reduction, and emotional regulation. VR and AR can simulate environments that teach coping mechanisms and resilience, contributing to overall mental wellness. This introduction will explore the significance of VR and AR in mental health therapies, detailing their current applications, benefits, and the transformative impact they have on patient care. By examining these aspects, we aim to underscore the importance of continued research and development in this field to maximize the therapeutic potential of VR and AR technologies.

The use of Virtual Reality (VR) and Augmented Reality (AR) technologies in mental health therapies represents a groundbreaking shift in how psychological treatments are delivered and experienced. These technologies offer immersive, interactive, and customizable therapeutic environments that address a wide range of mental health issues. VR and AR technologies hold significant potential to revolutionize mental health therapies by providing innovative, effective, and accessible treatment options [9]. Their ability to create immersive and personalized therapeutic experiences, enhance patient engagement, and support real-time interventions positions them as transformative tools in the field of mental health. Continued research and development in this area are essential to fully realize their potential and integrate them into mainstream therapeutic practices.

1.6 Potential of VR Mental Health Interventions

Virtual Reality (VR) has emerged as a promising tool in various fields, and its potential in mental health interventions is particularly transformative. This advanced technology sets the stage for a revolutionary shift in treatment methodologies for several reasons:

- **Immersive Experience**: VR provides an immersive experience that can simulate real-world scenarios in a controlled environment. This immersion allows patients to engage with therapeutic activities more deeply, potentially leading to more effective outcomes.
- **Exposure Therapy**: For conditions such as phobias, PTSD, and anxiety disorders, VR can be used for exposure therapy. Patients can confront their fears in a safe, controlled environment, gradually reducing their symptoms under the guidance of a therapist.
- **Accessibility**: VR can make mental health interventions more accessible. Patients who are unable to visit a therapist in person, whether due to geographical limitations or mobility issues, can still receive high-quality care through VR platforms.
- **Personalization**: Treatment can be tailored to the individual needs of patients. VR programs can adapt in real-time to the patient's responses, providing a personalized therapeutic experience that can be more effective than traditional one-size-fits-all approaches.
- **Engagement and Adherence**: The interactive nature of VR can increase patient engagement and adherence to treatment plans. Patients are more likely to stick with a treatment that is engaging and provides immediate, tangible feedback.
- **Training and Education**: VR can also be used to train mental health professionals, providing them with realistic scenarios to practice and develop their skills. This can lead to better preparedness and higher quality of care.
- **Data Collection and Analysis**: VR systems can collect detailed data on patient interactions and progress. This data can be analyzed to gain insights into the effectiveness of different interventions and to refine and improve treatment protocols.

The integration of VR into mental health interventions holds significant promise for transforming how treatment is delivered. By offering immersive, personalized, and accessible therapy options, VR can enhance the effectiveness of mental health care and improve outcomes for patients.

1.7 *Potential Benefits of VR/AR Technologies in Enhancing Emotional Regulation, Empathy, and Interpersonal Skills*

Virtual Reality (VR) and Augmented Reality (AR) technologies hold immense promise in enhancing emotional regulation, empathy, and interpersonal skills. Through immersive experiences, individuals can safely navigate challenging scenarios, practicing emotional regulation techniques in controlled environments. VR environments offer unique opportunities for perspective-taking, allowing users to step into the shoes of others and gain a deeper understanding of diverse experiences, thus fostering empathy. Furthermore, VR simulations enable individuals to practice and refine their interpersonal skills, from effective communication to conflict resolution, in realistic yet risk-free settings. By providing tailored experiences for emotional and social learning, VR/AR technologies offer transformative potential in equipping individuals with the tools and competencies needed for navigating complex social interactions and fostering meaningful connections.

1.7.1 Emotional Regulation

- **Safe Exposure to Stressors**: VR can simulate stressful situations in a controlled environment, allowing individuals to practice coping strategies and emotional regulation techniques without real-world consequences. Repeated exposure in a safe setting helps individuals build resilience and improve their ability to manage stress and anxiety in real-life situations.
- **Biofeedback Integration**: VR systems can integrate biofeedback mechanisms, providing real-time data on physiological responses (e.g., heart rate, skin conductance). This immediate feedback helps individuals become more aware of their emotional responses and practice regulating their emotions through techniques such as deep breathing or mindfulness.
- **Guided Relaxation Exercises**: VR environments can facilitate guided relaxation exercises, such as virtual nature walks or meditative experiences, which help individuals learn and practice relaxation and emotional regulation techniques.

1.7.2 Empathy

- **Perspective-Taking Experiences**: VR can immerse individuals in scenarios that allow them to experience the world from another person's perspective, such as simulating the experiences of individuals from different cultural backgrounds or those with disabilities. This immersive experience fosters a deeper understanding and empathy for others' experiences and challenges.
- **Empathy Training Programs**: VR can be used to develop training programs aimed at improving empathy, particularly in professional settings like healthcare, education, and customer service. These programs can simulate interactions with diverse individuals, helping users practice empathetic communication and understanding in a variety of contexts.
- **Emotional Recognition Practice**: VR environments can include scenarios where individuals practice recognizing and responding to others' emotions through facial expressions, body language, and tone of voice. This practice enhances emotional intelligence and empathy by improving the ability to accurately interpret and respond to the emotions of others.

1.7.3 Interpersonal Skills

- **Social Skills Training**: VR can simulate social interactions, providing a safe space for individuals to practice and develop their interpersonal skills, such as initiating conversations, active listening, and assertiveness. These simulations can be particularly beneficial for individuals with social anxiety or autism spectrum disorders, offering repeated practice without the fear of real-world judgment.
- **Conflict Resolution Practice**: VR scenarios can be designed to simulate conflicts or challenging interpersonal situations, allowing individuals to practice conflict resolution techniques and effective communication strategies. This practice helps individuals build confidence and competence in handling interpersonal conflicts constructively.
- **Teamwork and Collaboration**: VR can create collaborative virtual environments where individuals work together to solve problems or complete tasks, fostering teamwork and collaborative skills. These experiences can be used in both educational and professional development settings to enhance group dynamics and cooperative working skills.

VR/AR technologies offer significant potential benefits in enhancing emotional regulation, empathy, and interpersonal skills. By providing immersive, controlled environments for practice and learning, these technologies can help individuals develop critical emotional and social competencies that are essential for personal and professional success.

1.8 Need of Research

The emerging field of Augmented Wellness, which explores the transformative potential of Virtual Reality (VR) and Augmented Reality (AR) in healthcare, is poised to revolutionize the way we approach wellness and medical treatment. The need for Augmented Wellness stems from the growing recognition of the limitations of traditional healthcare methods and the increasing demand for innovative solutions that can enhance patient outcomes and experiences. VR and AR technologies offer immersive and interactive experiences that can empower individuals to take control of their health and well-being. From personalized therapy sessions to immersive training for healthcare professionals, Augmented Wellness opens new avenues for preventative care, treatment, and rehabilitation. By harnessing the power of VR and AR, healthcare providers can offer more accessible, engaging, and effective interventions, ultimately leading to improved patient outcomes and a higher quality of life. As we continue to explore and refine these technologies, Augmented Wellness holds the promise of transforming healthcare delivery and empowering individuals to lead healthier, happier lives.

Virtual Reality (VR) and Augmented Reality (AR) technologies offer promising avenues for revolutionizing mental health interventions. This paper explores the mechanisms by which VR/AR achieves therapeutic effects, focusing on two key aspects: immersion and engagement. VR creates a sense of immersion by placing patients in safe and controlled virtual environments, allowing them to confront anxieties or phobias. Similarly, AR superimposes therapeutic elements onto the real world, providing a safe space for practicing social interaction skills. The interactive and engaging nature of VR/AR increases patient motivation and adherence to therapy, particularly in scenarios like social anxiety treatment where patients can practice social interactions in virtual environments and receive immediate feedback. Moreover, VR/AR scenarios can be tailored to individual needs and phobias, such as a fear of public speaking, allowing for gradual exposure and desensitization. However, despite the promising outcomes, the evidence base for VR/AR in mental health interventions requires further solidification. More extensive clinical trials are needed to fully understand the long-term effectiveness of VR/AR across different mental health conditions. Overall, VR/AR technologies offer immense potential in revolutionizing mental healthcare, providing personalized and engaging interventions that empower individuals to improve their well-being.

2 Literature Review

As per the authors in [7], AR along with VR have the potential to revolutionise mental health care by creating compelling, personalised, and immersive therapeutic experiences. Responsible usage requires training for mental health professionals, thorough research, compliance with data protection legislation, along with adherence to ethical norms. With an emphasis on improving the ecological validity of highly controlled settings, boosting personalisation along with engagement, along with collecting real-time, automated data in real-world scenarios, H. Bell et al. in [8] outlined the benefits of adopting VR for mental health evaluation. Our preliminary efforts to develop and pilot a youth-co-designed virtual reality intervention in case of classroom use in promoting adolescents' well-being were detailed by the authors in [9]. In order to better comprehend how mental health therapies are used in everyday clinical practice, the authors in [10] intended to discover, assess, and summarise them using immersive 360° movies. Despite the paucity of research, existing AR/VR based therapies have the ability to increase understanding, compassion, and understanding of mental illness according to [11]. As per [12], new age in mental health, centred on psychological diseases, may be about to dawn thanks to the metaverse, a word made by merging meta and universe, which has the potential to improve healthcare and personal well-being via the development of a full-body illusion through digital avatars. Modules for augmented along with virtual reality were being examined by Singh & Kaunert in [13] as game-changing resources for enhancing emotional intelligence, mental wellness, and mindfulness. This looked to examine the revolutionary possibilities of augmented along with VR modules for improving mental health, emotional intelligence, along with mindfulness. In the realms of education and mental health, Carlson in [14] developed augmented reality along with virtual reality simulations and the goal was to better understand the possible effects of virtual reality along with augmented reality on patient outcomes as they relate to mental health therapies. Verma & Singla in [15] produced computer-generated digital representations that were referred to as "avatars" for the purpose of interacting with digital items in the metaverse and communicating with one another via the internet.

The authors in [16] presented digital twins as the most essential components of CPSs, and they are also an essential component in the process of bringing the Metaverse into existence. Image style transfer from digital painting to new media art was investigated in [17]. Neural style transfer is the foundation around which this approach is built. The authors in [18] provided a comprehensive overview of the research and investigations that have previously been conducted in the field of metaverse academic inquiry. New developments in AR and VR hold a lot of promise for improving mental health therapies and people's general health and happiness. Using "Ifland" as a case study, Lee and Kim [19] used the unified theory of acceptance and use of technology (UTAUT) to the metaverse in their investigation of user acceptance and usage of technology. Their research shows that users are more engaged and have better results from therapy in metaverse settings that are fully immersive. The authors in [20] emphasised the crucial significance of safe and effective virtual environments for health applications in their thorough assessment on visualisation and cybersecurity in the metaverse. A comprehensive overview of virtual reality and augmented reality, including its uses in many domains, including mental health, is given in [21]. By demonstrating how these technologies may be used to build immersive therapeutic settings, our study highlights their revolutionary potential. By doing a scoping assessment of augmented and virtual reality health therapies for the elderly, the authors in [22] have shown how these technologies have the potential to transform therapeutic encounters and enhance the mental health of this population. In addition, a study conducted in [23] examined the effects of 360° video and VR on the mental health of older persons. The results showed that VR provided much more favourable outcomes in this regard. To go even farther into the topic, the authors in [24] looked at how VR tourism affected customers' subjective well-being. Virtual experiences have the potential to greatly improve mental health and alleviate stress, according to their research. Another research by Ciocca and Tschan [25] explores the use of virtual reality (VR) and exercise games (exergames) to improve mental and social health. The authors focus on the possibilities of these technologies in demanding settings, such as space missions. In order to make health information more approachable and interesting for consumers, Malheiros [26] investigated how virtual reality and augmented reality might improve health literacy.

Gasmi and Benlamri [27] pointed out that the COVID-19 epidemic has also increased the demand for augmented and virtual reality technology. Their studies include the increased use of these technologies during the epidemic and how they may help with the mental health issues that have worsened as a result of the disaster. There has been encouraging progress in the use of VR and AR for wellness and mental health treatments with the advent of these technologies in recent years. In their proposal for virtual reality treatment as an aid for mental health, Purwar and Singh [28] emphasised the technology's ability to provide engaging and immersive therapeutic experiences. Their research shows that virtual reality may be a powerful tool for mental health practitioners by simulating therapeutic settings. Lee et al. [29] reviewed the literature on digital treatment using extended reality (XR) to improve mental health in young adults and adolescents from South Korea. Their research highlights the beneficial effects of XR technology on psychological well-being, especially in lowering rates of anxiety and depression in younger generations. In order to ascertain if Gulf region universities have adopted metaverse technology, authors in [30] created a theoretical framework. The metaverse has the ability to revolutionise academic support systems for mental health and education, according to their empirical research that used a hybrid SEM-ANN technique. Virtual reality was investigated in [31] to determine the effect of natural components and colourful interventions on mood, gaze, and spontaneous walking. Their study shows that virtual reality settings with more colourful and natural components might help people better control their emotions and feel better overall. Kaushik [32] explored the effects of the metaverse on skill sets, arguing that this virtual environment has the potential to revolutionise sustainability and business by encouraging the development of novel abilities. In addition to its obvious use in entertainment, this conceptual study demonstrates how the metaverse may have far-reaching effects on people's mental health and wellbeing via the facilitation of skill-building and immersive educational settings. In addition, authors in [33] highlighted the future prospects for human-AI cooperation by presenting a thorough taxonomy of AI and AGI employing numerous intelligences and learning styles. Personalised mental health therapies may be enhanced by integrating AI into VR settings, as this research reveals. Authors in [34] investigated how AI, digital twins, and the metaverse all come together in the context of training and upkeep. In addition to their potential use in therapeutic settings to enhance mental health outcomes, their study emphasises the importance of these technologies in developing immersive and lifelike training environments. In their metaverse AI framework proposal, authors in [35] highlighted the need of AI integration to improve virtual environment user experiences and therapeutic treatments. In their extensive review of metaverse AI applications, it has been addressed in [36] how AI may construct adaptable virtual environments to aid in mental health therapies and other tasks.

In their 66-page online survey, the authors in [37] covered every possible topic related to the metaverse. Advancements are covered in Ferraioli's article [38]. In an article published in AI Society, Boddington [39] delves into the concept of the "Internet of Bodies," with an emphasis on how our physical selves may one day merge with virtual places. Hadjistassou [40] examined Second Life's cultural difficulties in relation to sustainability efforts in the US and Europe. Using the Pribram-Bohm Hypothesis as a prism, Joye and Joye [41] investigated the metaverse's potential ethical ramifications for virtual reality. In [42], the authors have surveyed the current state and potential future developments of metaverses and 3D virtual worlds. Park and Kim [43] provided a taxonomy that identifies the metaverse's components, uses, and problems. In their in-depth study, Teigland and Power [44] examined the ways in which the immersive internet interacts with economic, political, and social factors. In [45], the authors have investigated how imitation and innovation impact the uptake of metaverse services. Authors in [46] highlighted new tools for language learners in the virtual world of Second Life.

Table 1.1 shows the brief literature review of some of the existing literature in terms of cons, prons, limitations etc.

3 Problem Statement

Anxiety, sadness, and stress are among the most common mental health problems in today's culture, impacting people of all ages and backgrounds. Although many people benefit from traditional therapy methods, they encounter obstacles like stigma, cost, and lack of accessibility. Furthermore, encouraging people to stick to their treatment programs and getting them to participate in therapy are not always easy tasks. Isolation, disruption of habits, and increased stresses are some ways in which the COVID-19 epidemic has worsened preexisting mental health issues [47, 48]. Innovative ways that may augment or improve established therapeutic methods are urgently needed for mental well-being because they provide solutions that are effective, accessible, and interesting. One possible way to tackle these problems is by using VR and AR technology. VR and AR therapies may enhance people's real-world experiences or transport them to virtual ones, creating opportunities for interactive and tailored therapy. Among the many therapies that may be facilitated by these technologies are scenario simulations, relaxation methods, controlled exposure therapy, and the development of emotional regulation abilities. VR and AR

Table 1.1 Literature survey

Ref	Cons	Prons	Limitation	Conclusion
[7]	Complex and may require advanced understanding	Comprehensive overview of mental health treatment advancements	Limited to the latest advancements and may not cover basic concepts	Highlights significant progress in mental health treatments with new technologies
[8]	May not cover latest developments	Demonstrates VR as a valuable clinical tool	Focuses on VR; may not generalize to all mental health tools	Establishes VR's efficacy in mental health practice and research
[9]	Specific to the UK and adolescent well-being	Explores VR's potential in supporting adolescent mental health	Limited to proof-of-concept and specific demographics	Shows promise for VR interventions in school settings
[10]	Review may lack practical application details	Systematic review of 360° videos in mental health	Review-based, lacking original empirical data	Identifies potential of 360° videos for immersive mental health interventions
[11]	Broad scope might dilute specific findings	Scoping review on AR/VR's impact on mental illness stigma	Generalized findings, needing specific studies for depth	Shows AR/VR's potential in improving mental health knowledge and empathy
[12]	Early-stage concepts of MEDverse	Discusses future potential of metaverse in mental health	Speculative, as practical applications are emerging	Highlights the promise of the metaverse for future mental health applications
[13]	May require advanced technological infrastructure	Details AR/VR modules for mindfulness	Limited to mindfulness; broader mental health impacts not covered	Demonstrates AR/VR's effectiveness in mindfulness practices
[14]	May not cover non-augmented approaches	Explores virtual and augmented simulations in mental health	Focused on simulations, not comprehensive of all VR/AR applications	Validates virtual and augmented simulations for mental health treatments
[15]	Technical and may be complex for non-specialists	Investigates accuracy and performance in the metaverse	Limited to technical aspects; practical applications less discussed	Enhances understanding of metaverse accuracy and performance for mental health
[16]	Focus on specific material (42SiCr Steel Alloys)	Discusses AI digital twinning in the metaverse	Specific to one material; broader implications may be limited	Demonstrates AI and metaverse potential in material science and beyond

(continued)

Table 1.1 (continued)

Ref	Cons	Prons	Limitation	Conclusion
[17]	May be complex due to combination of AI and IoT	Explores creation mechanisms of new media art in metaverse	Focused on artistic applications; practical mental health uses less clear	Innovatively combines AI, IoT, and metaverse for new media art
[18]	Broad research streams; may lack depth in specific areas	Advances metaverse investigation and research agenda	Needs more focused studies for in-depth insights	Sets future research directions for metaverse studies
[19]	Case study limited to 'Ifland'	Examines UTAUT in metaverse context	Findings may not generalize beyond 'Ifland'	Provides insights into technology acceptance in the metaverse
[20]	Focus on visualization and cybersecurity	Surveys metaverse cybersecurity and visualization	Technical focus; practical applications less discussed	Highlights key cybersecurity challenges and visualization techniques
[21]	Introductory level; may lack advanced details	Comprehensive introduction to VR/AR	Basic overview; advanced applications not covered	Serves as a foundational text for VR/AR concepts
[22]	Focus on older adults; less on other demographics	Scoping review of VR/AR for older adults' well-being	Limited to older adults; broader impacts not covered	Shows potential of VR/AR for enhancing older adults' health and well-being
[23]	Comparative study; may lack broad applications	Compares 360° video and VR for older adults	Focused on specific technologies; generalizability limited	Provides comparative insights on VR technologies for mental health
[24]	Specific to virtual tourism	Examines VR tourism's impact on well-being	Limited to tourism; broader well-being impacts not discussed	Demonstrates VR's positive influence on subjective well-being through tourism
[25]	Specific to physical exercise in astronauts	Discusses VR exergames for astronaut health	Focused on astronauts; broader applications may vary	Highlights VR's potential for boosting health in extreme conditions
[26]	Focus on health literacy	Explores AR/VR for improving health literacy	Specific to literacy; broader health impacts not covered	Shows promise of AR/VR in enhancing health literacy and communication

mental health therapies have a lot of promise, but they also confront a number of challenges, such as a lack of enough data to prove their effectiveness, ethical concerns, along with a lack of widespread acceptability among MH providers and their patients. Hence, purpose of this study is to investigate viability, acceptability, and efficacy of augmented and virtual reality solutions for psychological health. By providing answers to these questions, we want to aid in the creation of evidence-based strategies that make use of new technology to improve mental health and general wellness.

- **Sample Size and Diversity**: One limitation of the current research is the relatively small and homogeneous sample size. Future studies should aim to recruit larger and more diverse samples to ensure the generalizability of findings across different demographic groups and mental health conditions.
- **Short-Term Follow-Up**: The current research may be limited by its focus on short-term outcomes. Longitudinal studies with extended follow-up periods are needed to assess the sustained effectiveness of VR/AR interventions and to evaluate any potential long-term benefits or adverse effects.
- **Methodological Rigor**: Some studies may lack methodological rigor, such as the absence of control groups or randomization. Future research efforts should prioritize rigorous study designs, including randomized controlled trials, to provide stronger evidence for the effectiveness of VR/AR interventions.
- **Technology Accessibility**: The accessibility of VR/AR technology may present a barrier to participation for certain populations, such as those with limited access to hardware or internet connectivity. Future research should explore strategies to improve technology accessibility and address disparities in access to VR/AR interventions.

4 Proposed Work

Research proposes the potential of Virtual Reality (VR) and Augmented Reality (AR) technologies in mental health interventions. VR creates a sense of immersion by placing patients in controlled virtual environments, allowing them to confront anxieties or phobias. AR superimposes therapeutic elements onto the real world, providing a safe space for practicing social interaction skills. This interactive and engaging nature increases patient motivation and adherence to therapy, particularly in social anxiety treatment scenarios. VR/AR scenarios can be tailored to individual needs and phobias, allowing gradual exposure and desensitization. However, the evidence base for VR/AR in mental health interventions requires further solidification, and more extensive clinical trials are needed to understand its long-term effectiveness across different mental health conditions.

Creating a process flow for simulating Virtual Reality (VR) Mental Health Interventions using Augmented Reality (AR) and VR solutions for well-being involves several key steps to ensure the effectiveness and feasibility of the interventions. Here's a comprehensive process flow (see Fig. 1.2):

1. Needs Assessment: Find out who you're helping, what they need in terms of mental health care, and what you want to achieve by doing a comprehensive needs assessment. This may involve consulting with mental health professionals, stakeholders, and potential users.

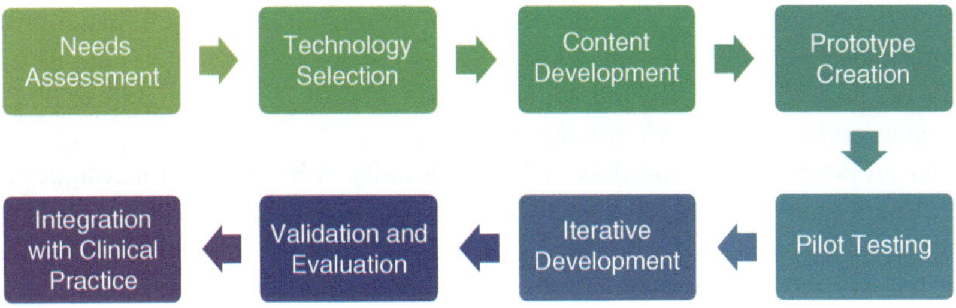

Fig. 1.2 Process flow of research

2. Technology Selection: Choose appropriate AR/VR solutions based on the identified needs and objectives. Consider factors such as device compatibility, available software platforms, and ease of use.
3. Content Development: Develop immersive content tailored to address the identified mental health needs. This may include virtual environments, guided meditations, relaxation exercises, exposure therapy scenarios, mindfulness activities, and cognitive-behavioral therapy (CBT) exercises. Collaboration with mental health experts, psychologists, and therapists to ensure the content aligns with evidence-based therapeutic approaches and best practices takes place during this phase.
4. Prototype Creation: Prototypes of the AR/VR interventions to test usability, functionality, and user experience is created to gather feedback from end-users, mental health professionals, and other stakeholders to iterate and refine the prototypes.
5. Pilot Testing: Pilot tests of the AR/VR interventions with a small group of users to evaluate their effectiveness, acceptability, and feasibility in real-world settings is conducted to collect qualitative and quantitative data on user experiences, engagement levels, and perceived benefits to inform further improvements.
6. Iterative Development: Based on feedback from pilot testing, iterate on the AR/VR interventions to address any identified issues or areas for improvement. This may involve refining content, optimizing user interfaces, or enhancing features for better usability and engagement.
7. Validation and Evaluation: Validate the effectiveness of the AR/VR interventions through rigorous evaluation studies, including RCTs or quasi-experimental designs. Then measurement of outcomes takes place. These outcomes are changes in mental health symptoms, well-being indicators, self-reported satisfaction, and treatment adherence.
8. Integration with Clinical Practice: Integration of validated AR/VR interventions into existing mental health treatment programs and clinical practice settings and train mental health professionals and therapists on how to use the AR/VR solutions effectively and incorporate them into their therapeutic interventions.

By following this process flow, developers, researchers, and mental health professionals can effectively simulate Virtual Reality Mental Health Interventions using AR/VR solutions for well-being, with the ultimate goal of improving mental health outcomes and enhancing overall quality of life.

Our proposed system introduces a Virtual Reality (VR) application designed specifically for anxiety management. Utilizing immersive VR environments, the application aims to provide users with a safe and controlled space to confront and manage their anxiety symptoms effectively.

4.1 Immersive Environments for Anxiety Exposure

The VR application offers a variety of immersive environments tailored to different anxiety triggers, such as social situations, public speaking, or confined spaces. Users can select scenarios based on their specific anxiety challenges, allowing for personalized exposure therapy experiences.

4.2 Interactive Tools for Anxiety Coping Strategies

In addition to immersive environments, the VR application provides interactive tools and resources for practicing anxiety coping strategies. Users can engage in mindfulness exercises, deep breathing techniques, and cognitive restructuring activities within the virtual environment, guided by audio instructions and visual prompts.

4.3 Progress Tracking and Feedback Mechanisms

To monitor progress and provide feedback, the VR application includes built-in tracking features. Users can track their anxiety levels before and after each session, allowing for real-time assessment of therapeutic progress. The application also offers personalized feedback based on user performance, encouraging continued engagement and motivation.

Our proposed VR application for anxiety management offers a comprehensive solution for individuals seeking effective and accessible interventions. By combining immersive environments, interactive tools, and progress tracking mechanisms, the application aims to empower users to confront and overcome their anxiety symptoms in a supportive virtual environment.

5 Results and Discussion

Our study yielded several important findings regarding the effectiveness and implications of Virtual Reality (VR) and Augmented Reality (AR) in mental health interventions. Firstly, our analysis revealed that VR/AR therapies significantly improved accessibility, customization, and therapeutic efficacy compared to traditional methods. By creating immersive environments, VR/AR provided patients with a safe and controlled space to confront their anxieties and phobias. This aspect of immersion is crucial for effective exposure therapy, as patients can gradually confront and overcome their fears under the guidance of a therapist. Moreover, the interactive nature of VR/AR significantly enhanced patient engagement and motivation. Through simulated scenarios, patients could actively participate in therapy sessions, receiving immediate feedback and reinforcement. For instance, in the case of social anxiety treatment, patients could practice social interactions in virtual environments, gradually building confidence and skills. This increased engagement often led to improved adherence to therapy, as patients found the VR/AR experiences more enjoyable and rewarding compared to traditional methods.

Furthermore, the customization possibilities offered by VR/AR were instrumental in tailoring interventions to individual needs and preferences. For example, scenarios could be adjusted to accommodate varying levels of difficulty or specific phobias, allowing for personalized treatment plans. This level of customization not only improved the effectiveness of therapy but also enhanced the overall patient experience. However, our study also identified several challenges and limitations associated with the integration of VR/AR into mental health treatment. Technological constraints, such as hardware requirements and software limitations, posed barriers to widespread adoption. Additionally, ethical considerations, such as data privacy and informed consent, need to be carefully addressed to ensure patient safety and autonomy. While VR/AR technologies hold tremendous promise for improving mental health interventions, further research is needed to address these challenges and solidify the evidence base. More extensive clinical trials are necessary to fully understand the long-term effectiveness of VR/AR across different mental health conditions. By addressing these issues and continuing to refine VR/AR applications, we can unlock the full potential of these technologies in revolutionizing mental healthcare. Figure 1.3 indicates the comparison of energy efficiency, throughput and accuracy in different scenarios.

Figure 1.4 shows Python based simulation for visualizing the comparison between packet dropping and packet delivery ratio in case of VR Mental Health Interventions using AR/VR Solutions for Well-being.

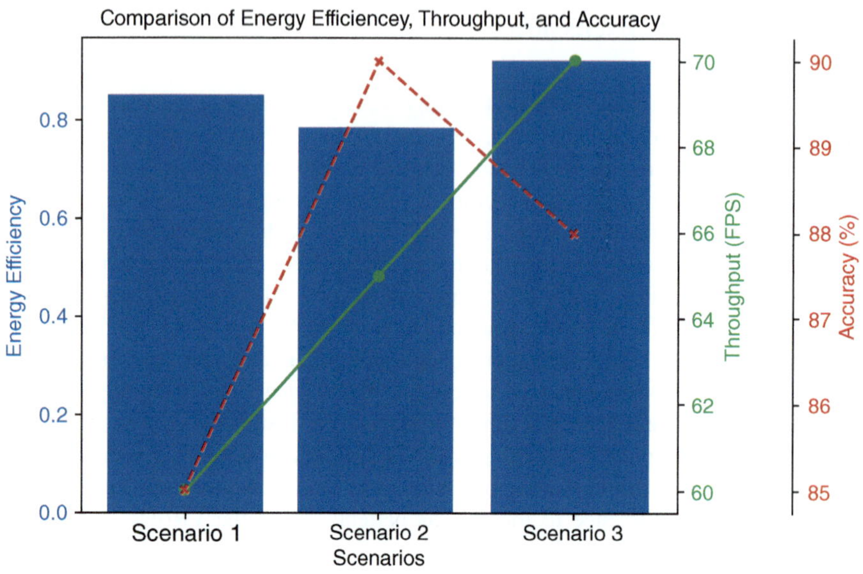

Fig. 1.3 Comparison of energy efficiency, throughput and accuracy in different scenarios

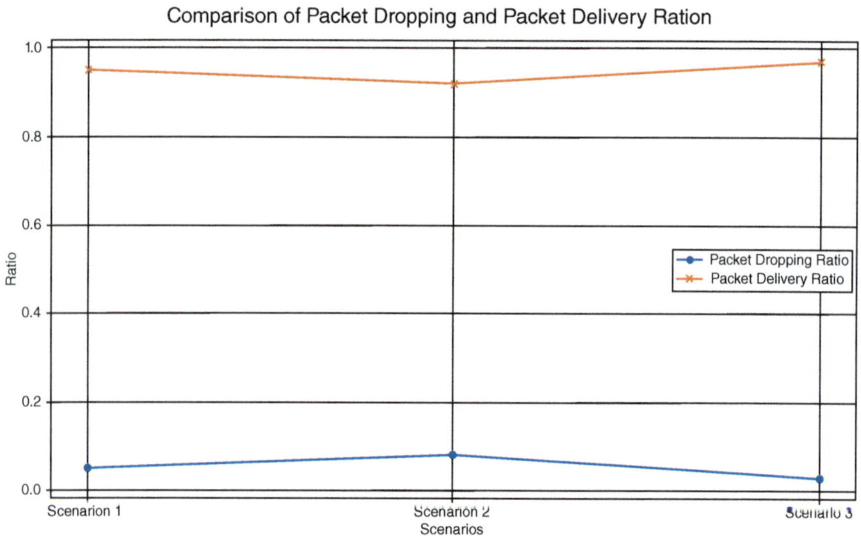

Fig. 1.4 Comparison of packet dropping and delivery ratio in different scenarios

6 Conclusion

This research shows that AR/VR technology may be novel mental health therapies. Our results imply that immersive and engaging AR/VR treatments may help treat a range of mental health issues by complementing standard therapies. We found that AR/VR therapies improve accessibility, customisation, and therapeutic efficacy in our literature and practical research. These devices are effective in exposure treatment, stress reduction, and emotional control, with many users reporting mental health gains.AR/VR incorporation into mental health treatment has problems and limits, which must be acknowledged. Technological constraints, ethical concerns, and the need for further study to provide empirical proof persist. These techniques' acceptability and implementation by mental health professionals and clients need additional study [49, 50]. Future research should address these knowledge and practice gaps. This topic needs longitudinal investigations on AR/VR treatments' long-term effects, randomized controlled trials comparing them to standard therapy, and ethical criteria. Enhancing AR/VR technology accessibility and affordability would help provide equitable mental health intervention access for everyone. In conclusion, AR/VR technologies have tremendous potential to improve mental health, but their incorporation into clinical practise needs careful assessment and continuing review. We can use these technologies to help people improve their mental health and well-being by exploring and refining them. The research proposed the potential of AR/VR technology as a novel mental health therapy, offering immersive and engaging treatments for various mental health issues. These treatments have shown positive results in exposure treatment, stress reduction, and emotional control. However, the integration faces challenges such as technological constraints, ethical considerations, and empirical validation. Further research is needed to explore the acceptability and implementation of these techniques, address knowledge gaps, and establish best practices. Additionally, improving accessibility and affordability is crucial for equitable access to mental health interventions. The integration of AR/VR technologies into clinical practice requires careful assessment and ongoing review.

7 Future Scope

The future scope of "Virtual Reality Mental Health Interventions: AR/VR Solutions for Well-being" encompasses several areas of potential development and exploration. İt will continue advancements in AR/VR hardware and software will enhance the capabilities and effectiveness of mental health interventions. Future research could focus on developing AR/VR interventions that are personalized to individual needs, preferences, and therapeutic goals. Research should also explore a variety of outcome measures, including symptom reduction, quality of life improvements, and functional recovery. Hybrid approaches that combine in-person sessions with AR/VR experiences may offer a comprehensive treatment approach. Further exploration is needed to determine the efficacy of AR/VR interventions for specific populations and mental health disorders. Developing ethical guidelines for use of AR/VR technologies in MH care is crucial. By pursuing these avenues of research and implementation, field of AR/VR MH interventions can continue to evolve and expand, offering innovative solutions to support individuals in improving their mental well-being. The efficacy of VR/AR in mental health interventions shows significant promise, yet it remains a burgeoning field requiring further exploration. While initial studies suggest positive outcomes, a robust evidence base through extensive clinical trials is essential to comprehend its long-term effectiveness across various mental health conditions. Additional research will provide a more nuanced understanding of how VR/AR can be integrated into existing therapeutic frameworks and tailored to meet individual needs effectively.

References

1. Bommareddy S, Khan JA, Anand R. A review on healthcare data privacy and security. In: Networking technologies in smart healthcare; 2022. p. 165–87.
2. Singh P, Kaiwartya O, Sindhwani N, Jain V, Anand R, editors. Networking technologies in smart healthcare: innovations and analytical approaches. CRC Press; 2022.
3. Anand R, Juneja S, Juneja A, Jain V, Kannan R, editors. Integration of IoT with cloud computing for smart applications. CRC Press; 2023.
4. Anand R, Pandey D, Gupta DN, Dharani MK, Sindhwani N, Ramesh JVN. Wireless sensor-based IoT system with distributed optimization for healthcare. In: Meta heuristic algorithms for advanced distributed systems; 2024. p. 261–88.
5. Saxena H, Joshi D, Singh H, Anand R. Comparison of classification algorithms for Alzheimer's disease prediction. In: 2022 Seventh international conference on parallel, distributed and grid computing (PDGC). IEEE; 2022. p. 687–92.
6. Veeraiah V, Gupta DN, Kumar BS, Gupta A, Anand R. Prediction of health risk based on multi-level IOT data using decision trees. In: 2023 International conference on sustainable emerging innovations in engineering and technology (ICSEIET). IEEE; 2023. p. 652–6.
7. Singha R, Singha S. Mental health treatment. In: Advances in medical technologies and clinical practice. IGI Global; 2024. p. 91–110. https://doi.org/10.4018/979-8-3693-1123-3.ch006.
8. Bell H, Nicholas J, Alvarez-Jimenez M, Thompson A, Valmaggia L. Virtual reality as a clinical tool in mental health research and practice. Dialogues Clin Neurosci. 2020;22(2):169–77. https://doi.org/10.31887/dcns.2020.22.2/lvalmaggia.
9. https://kurieta.com/how-to-measure-roi-in-digital-marketing-campaigns-copy/
10. Ionescu A, Van Daele T, Rizzo A, Blair C, Best P. 360° Videos for immersive mental health interventions: a systematic review. J Technol Behav Sci. 2021;6(4):631–51. https://doi.org/10.1007/s41347-021-00221-7.
11. Tay JL, Xie H, Sim K. Effectiveness of augmented and virtual reality-based interventions in improving knowledge, attitudes, empathy and stigma regarding people with mental illnesses—a scoping review. J Pers Med. 2023;13(1):112. https://doi.org/10.3390/jpm13010112.
12. Cerasa A, Gaggioli A, Marino F, Riva G, Pioggia G. The promise of the metaverse in mental health: the new era of MEDverse. Heliyon. 2022;8(11):e11762. https://doi.org/10.1016/j.heliyon.2022.e11762.
13. Singh B, Kaunert C. Augmented reality and virtual reality modules for mindfulness. In: Advances in medical technologies and clinical practice. IGI Global; 2024. p. 111–28. https://doi.org/10.4018/979-8-3693-1123-3.ch007.

14. Carlson CG. Virtual and augmented simulations in mental health. Curr Psychiatry Rep. 2023;25(9):365–71. https://doi.org/10.1007/s11920-023-01438-4.
15. Verma H, Singla M. Investigating the accuracy and performance enhancement in metaverse. In: 2022 5th International conference on contemporary computing and informatics (IC3I). IEEE; 2022. https://doi.org/10.1109/ic3i56241.2022.10072768.
16. Khalaj O, et al. Metaverse and AI digital twinning of 42SiCr steel alloys. Mathematics. 2022;11(1):4. https://doi.org/10.3390/math11010004.
17. Wang X, Cai L, Xu Y. Creation mechanism of new media art combining artificial intelligence and internet of things technology in a metaverse environment. J Supercomput. 2023;80:9277–97. https://doi.org/10.1007/s11227-023-05819-7.
18. Trunfio M, Rossi S. Advances in metaverse investigation: streams of research and future agenda. Virtual Worlds. 2022;1(2):103–29. https://doi.org/10.3390/virtualworlds1020007.
19. Lee U-K, Kim H. UTAUT in metaverse: an 'Ifland' case. J Theor Appl Electron Commer Res. 2022;17(2):613–35. https://doi.org/10.3390/jtaer17020032.
20. Chow Y-W, Susilo W, Li Y, Li N, Nguyen C. Visualization and cybersecurity in the metaverse: a survey. J Imaging. 2022;9(1):11. https://doi.org/10.3390/jimaging9010011.
21. Doerner R, Broll W, Jung B, Grimm P, Göbel M, Kruse R. Introduction to virtual and augmented reality. In: Virtual and augmented reality (VR/AR). Springer; 2022. p. 1–37. https://doi.org/10.1007/978-3-030-79062-2_1.
22. Carroll J, Hopper L, Farrelly AM, Lombard-Vance R, Bamidis PD, Konstantinidis EI. A scoping review of augmented/virtual reality health and wellbeing interventions for older adults: redefining immersive virtual reality. Front Virtual Real. 2021;2 https://doi.org/10.3389/frvir.2021.655338.
23. Kupczik L, Farrelly W, Wilson S. Appraising virtual technologies' impact on older citizens' mental health—a comparative between 360° video and virtual reality. Int J Environ Res Public Health. 2022;19(18):11250. https://doi.org/10.3390/ijerph191811250.
24. McLean G, AlYahya M, Barhorst JB, Osei-Frimpong K. Examining the influence of virtual reality tourism on consumers' subjective wellbeing. Tour Manag Perspect. 2023;46:101088. https://doi.org/10.1016/j.tmp.2023.101088.
25. Ciocca G, Tschan H. The enjoyability of physical exercise: exergames and virtual reality as new ways to boost psychological and psychosocial health in astronauts. A prospective and perspective view. IEEE Open J Eng Med Biol. 2023;4:173–9. https://doi.org/10.1109/OJEMB.2023.3234014.
26. Malheiros O. Virtual reality and augmented reality as promoters of better health literacy. In: Handbook of research on assertiveness, clarity, and positivity in health literacy. IGI Global; 2022. p. 294–315. https://doi.org/10.4018/978-1-7998-8824-6.ch017.
27. Gasmi A, Benlamri R. Augmented reality, virtual reality and new age technologies demand escalates amid COVID-19. In: Novel AI and data science advancements for sustainability in the era of COVID-19. Elsevier; 2022. p. 89–111. https://doi.org/10.1016/b978-0-323-90054-6.00005-2.
28. Purwar A, Singh A. VR therapy—mental therapy tool based on virtual reality. Int J Creat Comput. 2023;2(1):1–17. https://doi.org/10.1504/ijcrc.2023.133548.
29. Lee S, Yoon J, Cho Y, Chun J. Systematic review of extended reality digital therapy for enhancing mental health among South Korean adolescents and young adults. J Korean Acad Child Adolesc Psychiatry. 2023;34(4):204–14. https://doi.org/10.5765/jkacap.230046.
30. Akour IA, Al-Maroof RS, Alfaisal R, Salloum SA. A conceptual framework for determining metaverse adoption in higher institutions of gulf area: an empirical study using hybrid SEM-ANN approach. Comput Educ Artif Intell. 2022;3:100052. https://doi.org/10.1016/j.caeai.2022.100052.
31. Batistatou A, Vandeville F, Delevoye-Turrell YN. Virtual reality to evaluate the impact of colorful interventions and nature elements on spontaneous walking, gaze, and emotion. Front Virtual Real. 2022;3 https://doi.org/10.3389/frvir.2022.819597.

32. Kaushik S. Metaverse and skill set: a conceptual investigation. In: How the metaverse will reshape business and sustainability. Singapore: Springer; 2023. p. 171–8. https://doi.org/10.1007/978-981-99-5126-0_15.
33. Cichocki A, Kuleshov AP. Future trends for human-AI collaboration: a comprehensive taxonomy of AI/AGI using multiple intelligences and learning styles. Comput Intell Neurosci. 2021;2021(3):1–21. https://doi.org/10.1155/2021/8893795.
34. Bordegoni M, Ferrise F. Exploring the intersection of metaverse, digital twins, and artificial intelligence in training and maintenance. J Comput Inf Sci Eng. 2023;23(6) https://doi.org/10.1115/1.4062455.
35. Guo Y, et al. Artificial intelligence for metaverse: a framework. CAAI Artif Intell Res. 2022;1(1):54–67. https://doi.org/10.26599/air.2022.9150004.
36. Huynh-The T, Pham Q-V, Pham X-Q, Nguyen TT, Han Z, Kim D-S. Artificial intelligence for the metaverse: a survey. Eng Appl Artif Intell. 2023;117:105581. https://doi.org/10.1016/j.engappai.2022.105581.
37. Lee L-H, et al. All one need to know about metaverse: a complete survey on technological singularity. Virtual Ecosyst Res Agenda. 2021;14(8):1–66. http://arxiv.org/abs/2110.05352.
38. Ferraioli J. AlphaCurrents. J Latex Class Files. 2021;14(8):1–12.
39. Boddington G. The internet of bodies—alive, connected and collective: the virtual physical future of our bodies and our senses. AI Soc. 2021;38:1897–913. https://doi.org/10.1007/s00146-020-01137-1.
40. Hadjistassou SK. Culturally afforded tensions in the Second Life metaverse: from sustainability initiatives in Europe to sustainability practices in the United States. Int J Web-Based Learn Teach Technol. 2016;11(2):14–38. https://doi.org/10.4018/IJWLTT.2016040102.
41. Joye S, Joye SR. Ethical implications of a virtual reality metaverse: the Pribram-Bohm hypothesis. 2016.
42. Dionisio JDN, Burns WG, Gilbert R. 3D virtual worlds and the metaverse: current status and future possibilities. ACM Comput Surv. 2013;45(3):1–38. https://doi.org/10.1145/2480741.2480751.
43. Park S-M, Kim Y-G. A Metaverse: taxonomy, components, applications, and open challenges. IEEE Access. 2022;10:4209–51. https://doi.org/10.1109/access.2021.3140175.
44. Teigland R, Power D. The immersive internet: reflections on the entangling of the virtual with society, politics and the economy. In: Immersive ınternet reflections entangling virtual with society, politics and the economy; 2013. p. 1–264. https://doi.org/10.1057/9781137283023.
45. Lee SG, Trim S, Byun WK, Kang M. Innovation and imitation effects in Metaverse service adoption. Serv Bus. 2011;5(2):155–72. https://doi.org/10.1007/s11628-011-0108-8.
46. Dominguez-Noriega S, Enrique Agudo J, Ferreira P, Rico M. Language learning resources and developments in the Second Life metaverse. Int J Technol Enhanc Learn. 2011;3(5):496–509. https://doi.org/10.1504/IJTEL.2011.042101.
47. Anand R, Sindhwani N, Saini A, Shubham. Emerging technologies for COVID-19. In: Enabling healthcare 4.0 for pandemics: a roadmap using AI, machine learning, IoT and cognitive technologies; 2021. p. 163–88.
48. Sharma A, Sindhwani N, Anand R, Vashisth R. Intelligent computing relating to cloud computing. In: Computational intelligence in the industry 4.0. CRC Press; 2024. p. 55–72.
49. Alkhayyat A, Badhoutiya A, Subti J, Singh R, Anand R, Kumar R. Data visualization of weather conditions in hilly regions using wireless technology. In: 2024 IEEE wireless antenna and microwave symposium (WAMS). IEEE; 2024. p. 1–6.
50. Gupta DN, Anand R, Ahamad S, Patil T, Dhabliya D, Gupta A. Phonocardiographic signal analysis for the detection of cardiovascular diseases. In: International conference on frontiers of intelligent computing: theory and applications. Singapore: Springer; 2023. p. 529–38.

Chapter 2
Understanding Virtual Reality (VR) and Its Applications in Healthcare

Bhanu Pratap Singh, Ankush Joshi, and Ritik Kumar Singh

Abstract Virtual reality (VR) is an emerging and fascinating technology that promotes the humans to step into virtual worlds and interact with them as if they were real. It is a computer-generated simulation of an environment that allows users to interact with and experience a virtual world as if they were physically present within it. Virtual reality (VR) technology has appeared as a trans-formative tool in the field of healthcare, offering immersive and interactive experiences that have the potential to revolutionize patient care, medical training, and therapy. This chapter provides a depth exploration of VR technology and its useful implications for healthcare delivery. Beginning with an introduction to the fundamentals of VR, including its core components and principles, the chapter delves into the endless applications of VR in healthcare. These applications range from medical training and education to pain management, surgical planning, psychological therapy, and medicine. Through real-life examples and case studies, we'll see how VR is making a difference in healthcare by providing more effective and engaging treatment options. Throughout the chapter, we'll see how VR is generating a big gap between current healthcare facilities and the future trends. But, like all technologies, there are challenges too. Sometimes, VR can make people feel nauseous, and using VR safely means protecting people's privacy and making sure the technology works for everyone. Looking to the future, we'll go into the new ways that states how VR technology is evolving and how it could continue to revolutionize healthcare in the years to come. Overall, this chapter aims to provide a clear and accessible overview of VR in healthcare, helping readers understand its potential benefits and limitations in improving patient outcomes and medical training.

Keywords Augmented reality · Virtual reality · Artificial intelligence · Internet of things · Blockchain · Mobile VR

B. P. Singh (✉) · A. Joshi · R. K. Singh
COER University, Roorkee, India

B. O. Soufiane et al. (eds.), *Augmented Wellness*,
https://doi.org/10.1007/978-981-96-2952-7_2

1 Introduction

Virtual Reality (VR) is an exciting and innovative technology that allows people to immerse themselves in virtual worlds and interact with them as if they were real [1]. Imagine putting on a special headset and suddenly finding yourself in a different place—it could be a beautiful beach, a bustling city, or even inside the human body. VR creates these realistic environments using advanced computer technology, making users feel like they're actually there.

This chapter aims to explore the fundamentals of virtual reality technology and its wide-ranging applications in healthcare [2]. We'll start by explaining how VR works, and the essential components needed to create virtual environments. Then, we'll dive into the exciting ways VR is being used in healthcare, from training healthcare professionals to treating patients with chronic pain or mental health issues. VR isn't just for training, it's also a powerful tool for patient care and therapy [3]. Imagine being able to transport a patient to a peaceful beach or serene forest without ever leaving their hospital bed. VR can create immersive environments that help patients relax, manage pain, and even overcome phobias. It's also being used in psychological therapy to treat conditions like post-traumatic stress disorder (PTSD) and anxiety disorders. By exposing patients to controlled virtual environments, therapists can help them confront and overcome their fears in a safe and supportive setting. Through real-life examples and case studies, we'll showcase the diverse ways VR is making a difference in healthcare delivery. There are also concerns about privacy and data security, as VR systems collect sensitive information about users' movements and interactions. We'll also discuss some of the challenges and considerations associated with using VR in healthcare, such as ensuring patient safety and privacy.

2 Introduction to Virtual Reality

Virtual reality (VR) is a computer-generated simulation of an environment that allows users to interact with and experience a virtual world as if they were physically present within it [4]. Users typically wear a VR headset that tracks their head movements and displays corresponding images to create an immersive experience. VR environments can range from realistic representations of the physical world to fantastical, imaginary landscapes [5]. By providing a sense of presence and depth, VR technology enables users to explore and interact with virtual objects and surroundings in real-time. This technology has various applications in different fields, including gaming, entertainment, healthcare, education, and training [6]. As VR technology continues to advance, it holds the potential to revolutionize how we perceive and interact with digital content, offering immersive experiences that were previously only possible in our imagination. Despite its potential benefits, VR also presents challenges such as motion sickness, technical limitations, and concerns

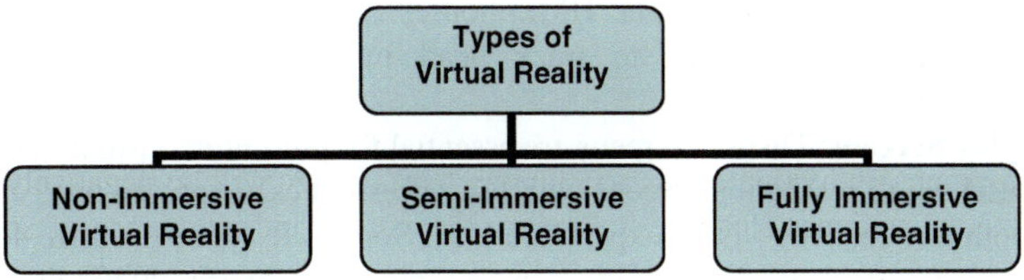

Fig. 2.1 Types of virtual reality

about privacy and safety. However, ongoing advancements in hardware, software, and content development are continuously improving the VR experience and expanding its capabilities. As VR technology becomes more accessible and sophisticated, its impact on industries and everyday life is expected to grow, shaping the way we work, learn, communicate, and interact with digital content [7].

2.1 *Types of Virtual Reality*

Figure 2.1 shows the types of Virtual Reality. The three main Virtual Reality categories are the following:

1. **Non-Immersive Virtual Reality**: Non-immersive virtual reality (NI-VR) is a type of virtual reality that allows users to interact with a computer-generated environment without the need for special equipment like headsets or gloves. NI-VR experiences are typically accessed through a computer screen, tablet, or smartphone, using familiar devices like a mouse, keyboard, or touchscreen [8].
2. **Semi-Immersive Virtual Reality**: Semi-immersive virtual reality bridges the gap between non-immersive and fully immersive virtual reality experiences. It provides a more engaging environment than a regular screen but doesn't completely block out the real world like some VR headsets [9].
3. **Fully Immersive Virtual Reality**: Fully immersive virtual reality aims to completely transport you into a computer-generated world, blurring the lines between reality and simulation. It offers a rich sensory experience that tricks your brain into believing you're physically present in the virtual environment [10].

2.2 *Core Components of Virtual Reality*

The core components of virtual reality (VR) technology encompass hardware, software, and peripherals that work together to create immersive and interactive experiences [11]. These components include:

1. **Head-Mounted Display (HMD)**: The HMD is perhaps the most recognizable component of VR. It is a wearable device that typically resembles a pair of

goggles or a helmet [12]. The HMD houses the display screens, lenses, and motion sensors necessary to track head movements and render virtual environments.

2. **Display Screens**: Display screens are essential for presenting virtual imagery to the user. High-resolution screens with fast refresh rates are preferred to ensure a smooth and realistic visual experience [13]. Some VR systems utilize separate screens for each eye to create stereoscopic 3D images, enhancing depth perception.
3. **Motion Tracking Sensors**: Motion tracking sensors, such as accelerometers, gyroscopes, and magnetometers, are embedded within the HMD and other peripherals to detect the user's movements in real-time [14]. This tracking allows the VR system to adjust the perspective of the virtual environment, accordingly, creating a sense of presence and immersion.
4. **Input Devices**: Input devices enable users to interact with the virtual world. These devices may include handheld controllers, gloves, or even full-body motion capture suits. Input devices can track hand movements, gestures, and button presses, allowing users to manipulate objects, navigate environments, and perform actions within the VR space [15].
5. **Audio System:** Immersive audio is crucial for creating a convincing VR experience. High-quality headphones or integrated speakers provide spatial audio cues that enhance the sense of presence and realism [15]. Some VR systems incorporate 3D audio technology to simulate sound coming from different directions and distances.
6. **Computer Hardware**: Powerful computers or gaming consoles are necessary to run VR applications and render complex virtual environments in real-time [16]. These systems require robust processors, graphics cards, and memory to deliver smooth performance and high-fidelity graphics.
7. **Tracking System**: In addition to head tracking, VR systems may utilize external tracking systems to monitor the position and movement of the user within physical space [17]. These systems may employ cameras, infrared sensors, or laser-based tracking technologies to create a virtual boundary and prevent users from bumping into objects or walls.
8. **Software Platform**: VR applications and experiences are powered by software platforms that handle rendering, tracking, user interaction, and content delivery [18]. These platforms may include proprietary VR ecosystems developed by hardware manufacturers, as well as open-source frameworks and development tools used by independent developers and content creators.

2.3 Principles of Virtual reality

The principles of virtual reality (VR) revolve around creating immersive and interactive experiences that simulate the sensation of being present in a virtual environment. Some key principles include:

1. **Immersion**: VR aims to immerse users in a digital environment by stimulating multiple senses, primarily sight and sound [19]. Immersive VR experiences transport users to virtual worlds where they can explore, interact, and engage with objects and surroundings as if they were physically present.
2. **Presence**: Presence refers to the subjective feeling of "being there" in the virtual environment. Achieving a sense of presence is essential for making VR experiences convincing and engaging [19]. Factors such as realistic graphics, responsive interactions, and accurate motion tracking contribute to enhancing presence.
3. **Interaction**: Interaction is a fundamental aspect of VR that allows users to engage with and manipulate objects within the virtual world [19]. VR systems provide various input methods, such as handheld controllers, gestures, and voice commands, to enable users to interact with virtual objects, navigate environments, and perform actions.
4. **Realism**: Realism in VR pertains to the fidelity and believability of virtual environments and objects. High-quality graphics, accurate physics simulation, realistic lighting and textures, and convincing sound effects contribute to creating a sense of realism in VR experiences [19].
5. **Comfort**: Ensuring user comfort is crucial for a positive VR experience. VR systems should minimize motion sickness and discomfort by maintaining smooth frame rates, reducing latency, and implementing comfortable locomotion methods [20]. Designing user interfaces and interactions that are intuitive and ergonomic also enhances comfort.
6. **Accessibility**: VR should be accessible to users of all abilities and demographics. Designing inclusive VR experiences involves considering factors such as ease of use, adjustability of settings, and compatibility with assistive technologies [20]. Providing options for locomotion, input methods, and interaction styles accommodates diverse user needs.
7. **Interactivity**: Interactivity is central to VR experiences, enabling users to actively engage with virtual environments and shape their experiences. Interactive elements may include puzzles, challenges, simulations, storytelling, and social interactions [20]. Dynamic environments that respond to user actions enhance immersion and engagement.
8. **Presence of Agency**: VR experiences should empower users with a sense of agency, allowing them to influence and control their surroundings [20]. Providing meaningful choices, consequences, and feedback mechanisms gives users a sense of autonomy and ownership over their virtual experiences.

2.4 *Difference Between Virtual Reality (VR) and Augmented Reality (AR)*

Augmented Reality (AR) is a technology that overlays digital information, such as images, videos, or 3D models, onto the real-world environment. Unlike Virtual Reality (VR), which immerses users in a completely virtual environment, AR enhances the physical world by superimposing computer-generated content onto it in real-time [21]. Table 2.1 resume the keys differences between Virtual Reality and Augmented Reality:

2.5 *Interaction Techniques in VR Environments*

Interaction techniques in virtual reality (VR) environments are crucial for enabling users to engage with and manipulate virtual objects and environments [25]. Here is some common interaction techniques used in VR:

Table 2.1 Difference between virtual reality and augmented reality

Aspect	Virtual reality (VR)	Augmented reality (AR)
Definition	VR is a simulated experience that immerses users in a completely virtual environment, blocking out the real world entirely [21]	AR overlays digital content onto the real world, blending virtual elements with the user's physical environment [21]
Environment	Fully immersive users are transported to a computer-generated reality and are unable to see the real world around them [22]	Partially immersive users remain aware of their physical surroundings while digital content is integrated into their view [22]
Interaction	Users can interact with and manipulate virtual objects and environments using specialized input devices or controllers [23]	Users can interact with both virtual and physical objects in their environment, often using gestures, touch, or voice commands [23]
Applications	Common applications include gaming, simulations, training, and entertainment where users want to escape reality [24]	Common applications include navigation, education, healthcare, and retail where users want to enhance their real-world experiences [24]
Equipment	Requires specialized VR headsets or goggles that block out the external environment and display the virtual world	Requires devices such as smartphones, tablets, or AR glasses equipped with cameras and sensors to overlay digital content
Immersion	Offers a high level of immersion by creating a completely synthetic environment that can transport users to new worlds [21]	Offers a lower level of immersion by augmenting the real world with digital content, allowing users to stay connected to reality [21]
Examples	Oculus Rift, HTC Vive, PlayStation VR	Microsoft HoloLens, Google Glass, Pokémon Go

2.5.1 Hand Controllers

Hand controllers are handheld devices equipped with sensors that track the user's hand movements and gestures. They allow users to interact with virtual objects by mimicking real-world actions such as grabbing, pointing, and pushing [25]. Hand controllers often feature buttons or triggers that users can press to perform actions within the VR environment.

1. **Gaze-based Interaction**: Gaze-based interaction relies on the direction of the user's gaze to interact with objects or trigger actions in the virtual environment. By simply looking at an object or UI element for a certain period, users can select or activate it [25]. Gaze-based interaction is particularly useful for navigation, selection, and menu interactions in VR experiences.
2. **Gesture Recognition**: Gesture recognition technologies enable users to control virtual environments using predefined hand gestures or movements [25]. Cameras or sensors track the user's gestures, allowing them to perform actions such as waving, thumbs-up, or making specific hand signs to interact with objects or trigger events in VR.
3. **Voice Commands**: Voice commands provide a hands-free interaction method in VR environments. Users can use natural language commands to perform actions, navigate menus, or interact with virtual characters or assistants [25]. Voice commands enhance accessibility and convenience, especially in scenarios where hand controllers may not be practical or available.
4. **Teleportation and Locomotion**: Teleportation and locomotion techniques allow users to move within the virtual environment. Teleportation involves selecting a destination point and instantly transporting the user to that location, minimizing discomfort associated with traditional locomotion methods like walking or running [25]. Locomotion techniques include walking in place, joystick-based movement, or using physical treadmills to simulate movement.
5. **Object Manipulation**: Object manipulation techniques enable users to interact with and manipulate virtual objects using hand gestures or controllers [25]. Users can pick up, move, rotate, resize, and manipulate virtual objects to perform tasks or solve puzzles within the VR environment.
6. **Haptic Feedback**: Haptic feedback provides users with tactile sensations to simulate the sense of touch in VR interactions [25]. Devices such as haptic gloves, vests, or controllers deliver vibrations, pressure, or force feedback in response to user interactions, enhancing the sense of presence and realism in VR experiences.

2.6 *Immersion and Presence in VR*

Immersion and presence are two fundamental concepts in virtual reality (VR) that contribute to the overall user experience and sense of realism within virtual environments.

2.6.1 Immersion

Immersion refers to the extent to which a user feels fully absorbed and engaged in a virtual environment. It encompasses various sensory stimuli, including visual, auditory, and sometimes haptic feedback, that trick the user's brain into perceiving the virtual world as real [26]. Key components of immersion in VR include:

- **Visual Fidelity**: High-quality graphics and realistic visuals are crucial for immersing users in virtual environments [27]. Detailed textures, realistic lighting, and smooth animations enhance the sense of immersion.
- **Audio Effects**: Spatial audio techniques, such as 3D sound positioning and binaural audio, create an immersive auditory experience by accurately simulating sound sources in the virtual environment [27]. This helps users locate objects and events based on sound cues.
- **Interactivity**: Interactive elements within the virtual environment, such as object manipulation, physical interactions, and responsive NPCs (non-player characters), enhance immersion by allowing users to actively engage with the environment and influence the virtual world.
- **Presence of Avatar**: The presence of a virtual representation of the user (avatar) that mirrors their movements and actions can enhance immersion by providing a sense of embodiment and agency within the virtual space [27].

2.6.2 Presence

Presence refers to the subjective feeling of "being there" in the virtual environment, as if the virtual world were real and physically present [28]. It involves the psychological sense of immersion and involves the suspension of disbelief, where users temporarily forget that they are in a simulated environment. Key factors contributing to presence in VR include:

- **Sense of Space**: Realistic spatial representation and scale within the virtual environment help users feel like they are inhabiting a physical space [29]. Accurate depth perception, perspective, and spatial audio cues contribute to a convincing sense of presence.

- **User Interaction**: Meaningful interactions with virtual objects, environments, and characters reinforce the sense of presence by creating a sense of agency and responsiveness within the virtual world. Real-time feedback to user actions enhances the feeling of presence [29].
- **Social Presence**: Interactions with other users or virtual characters in multiplayer or social VR experiences can enhance social presence, making users feel connected and engaged with others in the virtual space [29].
- **Immersion Continuity**: Consistency and continuity in the immersive experience, such as smooth movement and transitions, minimize disruptions and maintain the user's sense of presence throughout their VR experience.

3 Applications of Virtual Reality in Healthcare

Virtual reality (VR) holds significant potential for transforming various aspects of healthcare, offering innovative solutions to enhance medical training, patient care, and therapeutic interventions. Some key applications of VR in healthcare include:

3.1 Medical Training and Education

Medical training and education in healthcare using virtual reality (VR) technology revolutionizes traditional learning methods by providing immersive and interactive experiences for medical students and healthcare professionals [30]. With VR simulations, trainees can engage in realistic scenarios and hands-on experiences without the need for cadavers or live patients, enhancing their skills and knowledge in a safe and controlled environment.

In VR medical training, students can practice surgical procedures, anatomy exploration, and medical diagnosis with high fidelity and precision [30]. They can manipulate virtual organs and tissues, perform virtual surgeries, and simulate complex medical scenarios, all while receiving real-time feedback and guidance from instructors. This allows trainees to develop their technical skills, critical thinking abilities, and decision-making capabilities in a risk-free setting before transitioning to real-world clinical settings. VR technology also facilitates collaborative learning experiences, enabling students to interact with peers and instructors in virtual environments regardless of geographical locations. This fosters teamwork, communication, and knowledge sharing among healthcare professionals, promoting a collaborative approach to medical education and training.

3.2 Surgical Planning and Simulation

Surgical planning and simulation involves the creation of immersive and interactive simulations that allow surgeons to visualize and practice surgical procedures before performing them on actual patients [30]. These simulations provide a realistic representation of patient anatomy, surgical instruments, and operating environments, enabling surgeons to plan and rehearse complex procedures with greater precision and confidence.

By donning VR headsets and using specialized controllers, surgeons can manipulate virtual surgical tools, manipulate anatomical structures, and simulate various surgical scenarios in a virtual environment. This allows them to explore different surgical approaches, evaluate potential risks and challenges, and develop personalized surgical plans tailored to individual patient anatomy and pathology.

Furthermore, VR-based surgical simulations offer valuable training opportunities for medical students, residents, and experienced surgeons alike [31]. Trainees can practice surgical techniques, refine their skills, and familiarize themselves with new procedures in a safe and controlled environment, without the need for cadavers or live patients.

3.3 Pain Management and Rehabilitation

Pain management and rehabilitation in healthcare is the use of immersive technology to alleviate pain, aid in recovery, and enhance rehabilitation outcomes. By creating virtual environments that distract and engage patients, VR can effectively reduce pain perception, anxiety, and discomfort during medical procedures, physical therapy sessions, and chronic pain management [30]. In VR-based pain management, patients are immersed in soothing and visually captivating virtual environments, such as serene landscapes, calming nature scenes, or peaceful underwater settings. These environments serve as a form of distraction therapy, diverting the patient's attention away from the sensation of pain and providing a sense of relaxation and well-being. Additionally, VR can incorporate elements of cognitive behavioral therapy (CBT) and mindfulness techniques, guiding patients through relaxation exercises, breathing techniques, and guided imagery to further alleviate pain and stress.

In rehabilitation, VR technology is used to enhance traditional physical therapy approaches by providing interactive and engaging exercises that promote movement, strength, and coordination [30]. Virtual environments can simulate real life scenarios and activities, allowing patients to practice functional movements, balance exercises, and motor skills in a safe and controlled setting. VR-based rehabilitation programs can be tailored to the individual needs of patients, offering personalized treatment plans and progress tracking.

3.4 Exposure Therapy and Mental Health Treatment

Exposure therapy is a psychological treatment approach used to help individuals confront and gradually overcome their fears and anxieties [30]. It involves exposing patients to fear-inducing stimuli in a controlled and safe environment, allowing them to learn healthier responses and reduce their emotional distress. When applied in the context of mental health treatment using virtual reality (VR), exposure therapy becomes even more immersive and effective.

In VR-based exposure therapy, patients are immersed in virtual environments that simulate situations or scenarios related to their specific fears or phobias. For example, individuals with a fear of flying may undergo virtual flights, while those with social anxiety may engage in simulated social interactions. The virtual environments are carefully designed to evoke the same emotional responses and physiological reactions as real-world situations, providing a highly realistic and engaging experience.

One of the key advantages of VR-based exposure therapy is its ability to create a safe and controlled setting for patients to confront their [30]. Therapists can adjust the level of exposure gradually, starting with less intense scenarios and progressively increasing the difficulty as patients become more comfortable. This gradual approach helps to prevent overwhelming anxiety and allows patients to build confidence in managing their fears over time.

3.5 Telemedicine and Remote Consultations

Telemedicine and remote consultations using virtual reality (VR) technology revolutionize the way healthcare services are delivered, especially in situations where physical presence is challenging or not possible. With VR, patients can connect with healthcare providers in immersive virtual environments, regardless of geographical distances [30]. During remote consultations, patients wear VR headsets that transport them to virtual clinics or examination rooms, where they can interact with healthcare professionals in real-time. This technology allows for face-to-face communication, medical examinations, and the sharing of medical data, such as medical images and test results. Healthcare providers can conduct virtual examinations, assess symptoms, and prescribe treatments or interventions remotely. VR-enabled telemedicine enhances accessibility to healthcare services, particularly for patients in remote or underserved areas who may face barriers to accessing traditional healthcare facilities. It also facilitates specialist consultations, enabling patients to receive expert medical advice without the need for travel. Additionally, VR telemedicine offers a safe and convenient alternative for patients with mobility issues, chronic illnesses, or infectious diseases, reducing the risk of exposure to contagious pathogens in clinical settings. Overall, telemedicine and remote consultations using

VR technology improve healthcare access, patient convenience, and the efficiency of healthcare delivery, ultimately enhancing patient outcomes and satisfaction.

3.6 Patient Education and Empowerment

Patient education and empowerment through virtual reality (VR) entails leveraging immersive technology to educate patients about their medical conditions, treatment options, and healthcare procedures in a visually engaging and interactive manner [30]. By providing patients with immersive educational experiences, VR aims to promote better understanding, adherence to treatment plans, and active participation in healthcare decisions.

In VR-based patient education, individuals can explore virtual environments that simulate medical scenarios relevant to their condition or treatment journey. For example, patients can virtually tour anatomical structures, such as the heart or brain, to gain insights into their physiology and the nature of their illness. They can also interact with virtual representations of medical devices, medications, or treatment modalities to understand how they work and their potential effects on their health.

Moreover, VR enables healthcare providers to personalize educational content to meet the specific needs and preferences of each patient [30]. Information can be presented in various formats, such as 3D models, animations, or interactive simulations, allowing patients to engage with the material in a way that suits their learning style. This personalized approach enhances the relevance and effectiveness of patient education, leading to improved retention of information and empowerment in managing their health. By immersing patients in virtual environments related to their healthcare journey, VR can also foster empathy and emotional connection, helping individuals to better comprehend the impact of their condition and treatment choices. Additionally, VR-based patient education can overcome language barriers and literacy challenges by providing visual and interactive content that transcends traditional communication barriers.

3.7 Clinical Research and Data Visualization

Clinical research and data visualization in healthcare using virtual reality (VR) offer innovative solutions for analyzing complex medical data and enhancing research outcomes. With VR technology, researchers can immerse themselves in three-dimensional virtual environments to visualize and interact with large datasets, medical imaging, and patient records [32]. This immersive approach allows for better understanding of intricate biological processes, disease mechanisms, and treatment responses. VR-based data visualization tools enable researchers to explore data from multiple perspectives, identify patterns, and uncover insights that may not be apparent through traditional methods. Moreover, VR facilitates collaborative research efforts by enabling remote teams to interact in shared virtual spaces, regardless of their physical locations. By leveraging VR for clinical research and data visualization, healthcare professionals can accelerate the discovery of new treatments, improve patient care, and advance medical knowledge, ultimately leading to better healthcare outcomes for patients.

3.8 *Palliative Care*

Palliative care is a specialized medical approach aimed at improving the quality of life for patients with serious illnesses by addressing their physical, emotional, and spiritual needs [33]. Virtual reality (VR) technology has emerged as a promising tool in palliative care, offering innovative ways to enhance patient comfort, alleviate symptoms, and provide therapeutic support. Using VR, patients can be transported to immersive and serene environments that promote relaxation, reduce stress, and distract from pain and discomfort. For example, patients may experience tranquil natural settings, serene landscapes, or calming visual and auditory stimuli, all within the immersive VR environment. This immersive experience can help patients manage symptoms such as pain, anxiety, and depression, while also promoting emotional well-being and overall comfort. Additionally, VR can be used to facilitate reminiscence therapy, allowing patients to revisit cherished memories, important life events, or meaningful experiences, fostering a sense of connection, comfort, and closure. Moreover, VR based interventions can provide opportunities for virtual travel, cultural experiences, and social interactions, enriching patients' lives and enhancing their sense of engagement and fulfilment [34]. Overall, VR in palliative care holds immense potential to complement traditional medical approaches, enhance patient-centered care, and improve the overall quality of life for patients facing serious illnesses.

3.9 *Chronic Disease Management*

Chronic Disease Management using Virtual Reality (VR) involves leveraging immersive technology to assist patients in managing long-term medical conditions [35]. VR offers a unique approach to chronic disease management by providing engaging and interactive experiences that can help patients better cope with their conditions and improve their overall well-being. Through VR, patients can participate in virtual environments designed to educate, motivate, and support them in various aspects of disease management, including medication adherence, lifestyle modifications, symptom management, and psychological support. For example, VR applications can simulate real-life scenarios related to disease management, such as grocery shopping for healthy foods, practicing relaxation techniques to manage stress, or undergoing virtual physical therapy sessions [36]. By immersing patients in these virtual environments, VR can enhance patient engagement, compliance with treatment plans, and self-management skills, ultimately leading to better health outcomes and quality of life for individuals living with chronic diseases. Additionally, VR technology enables remote monitoring and telemedicine services, allowing healthcare providers to track patients' progress, deliver personalized interventions, and provide support from a distance. Overall, Chronic Disease Management using VR holds promise as a complementary approach to traditional healthcare interventions, offering innovative solutions to improve the lives of patients with chronic medical conditions.

4 Patient Outcomes and Experiences

By significantly influencing patient experiences, satisfaction levels, and therapeutic results, virtual reality (VR) therapies are ushering in a new age in healthcare. The immersive and compelling features of VR have shown their ability to improve patient experience by making it more enjoyable, educational, and ultimately successful [37]. Let's look more closely at how VR is influencing the healthcare sector, as evidenced by a variety of study findings and studies.

- **Reduced Anxiety and Distress**: VR therapies have been proved to be very successful at reducing anxiety and distress in patients undergoing a variety of medical procedures [37]. For instance, a 2019 research in the Journal of Clinical Psychology in Medical Settings found that using VR-based distraction significantly reduced patient anxiety during dental treatments. Virtual world immersion helped patients suffer less psychological anguish, which improved their overall experience.
- **Enhanced Comfort**: Patients' comfort and relaxation are greatly increased by VR's capacity to take them out of the therapeutic setting and into tranquil virtual worlds [37]. This is especially helpful during operations like chemotherapy treatments, dental work, or wound care. VR helps to make these operations more pleasant and less threatening by providing an escape from the hospital environment.
- **Pain Management and Reduction**: Virtual reality has a well-documented influence on pain management and reduction. Numerous studies have illustrated that VR therapies can crucially lower the feeling of pain and the demand for painkillers in a variety of medical situations [37]. For instance, a systematic analysis found that VR therapies were linked to significantly lower pain levels.
- **Improved Rehabilitation Results**: VR-based rehabilitation programmes have demonstrated substantial improvements in patients recuperating from illnesses traumatic brain injuries, and musculoskeletal problems in terms of their motor and cognitive function [37]. The favourable effects of VR on rehabilitation outcomes were emphasised by research published in the Archives of Physical

Medicine and Rehabilitation in 2018, which showed increased motor skills and functional capacities.

- **Treatment for Mental Illness**: VR exposure therapy has become a potent tool for the treatment of mental illnesses, including as phobias, post-traumatic stress disorder (PTSD), and anxiety disorders [37]. PTSD symptoms were significantly reduced among participants in research after they experienced VR-based exposure therapy.
- **Paediatric Pain Management**: VR has demonstrated great success in treating paediatric pain. Children frequently suffer worry and anguish when enduring treatments like vaccines or clothing changes [37]. According to a 2019 research in the journal *Pain Management Nursing*, kids who used VR during these operations reported less discomfort and anxiety, making for a more favourable overall experience.
- **Maternity Care**: VR has also established itself in this field. During prenatal appointments or labour, expectant moms can utilise virtual reality to immerse themselves in relaxing environs [37]. Studies have demonstrated that VR-based relaxation sessions during labour lessen the need for pain medication and have a good impact on the birthing process.
- **Reduced Travel Burden**: Patients who reside in rural places benefit the most from VR telemedicine appointments. Specialists can be consulted by patients without requiring them to make lengthy trips [37]. The ease of remote consultations and less stress associated with travel were cited as reasons for the high levels of patient satisfaction with VR telemedicine revealed in a 2017 research published in the Journal of Telemedicine and Telecare.
- **Geriatric Care**: VR treatments have the potential to increase elderly patients' happiness in long-term care settings. Seniors who participate in virtual reality activities that take them to nostalgic or calming settings report feeling happier and more pleased [37]. In studies presented in the journal Ageing Clinical and Experimental Research (2019), several therapies were emphasized.
- **Phantom Limb Pain**: Mirror treatment based on virtual reality has shown to be effective in decreasing phantom limb pain in amputees. In order to see the lost limb as though it were still present, patients wear VR headsets [37]. Participants who experienced VR mirror treatment showed statistically significant decreases in pain severity, according to research published in the journal *Pain Medicine* (2020).
- **Physical Therapy for Children**: For kids with neurological diseases like cerebral palsy, VR based physical therapy programs have been developed. These programs add interest and enjoyment to therapy sessions. Children who underwent VR-based treatment showed gains in their motor skills and functional outcomes, according to research published in the journal Developmental Medicine & Child Neurology in 2018. Burn victims must endure tremendous agony while receiving wound care and having their dressings changed [37]. Through VR therapies, they can be taken to serene virtual settings to take their minds off the discomfort. Studies have demonstrated that VR considerably lessens the level of

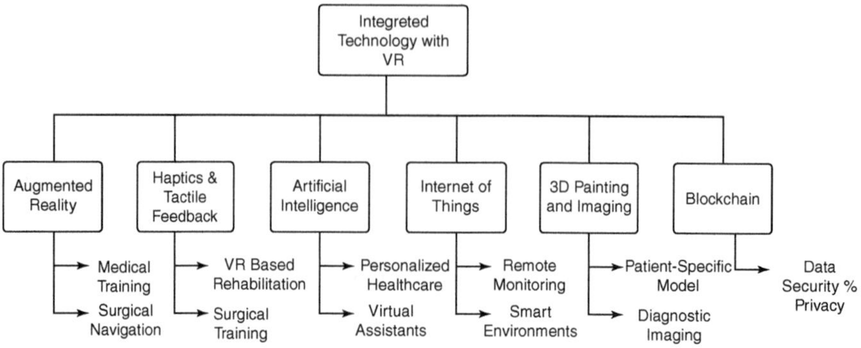

Fig. 2.2 Integration of VR with other technologies

pain in burn sufferers, including one that was published in the journal *Pain* Management (2020).

- **Physical Rehabilitation for Stroke Survivors**: For stroke victims, VR-based physical rehabilitation is quite helpful. These programs include gamified therapeutic activities that encourage patients to attend sessions regularly. Researchers observed that stroke victims who utilized VR-based therapy had better motor function and mobility [37]. Their findings were published in the Journal of Neuro Engineering and therapy in 2018.
- **Hospice and Palliative Care**: Patients receiving hospice and palliative care are frequently housebound and have restricted mobility [37]. Their quality of life and contentment are greatly improved by VR experiences that let them virtually go to locations like their hometowns or favorite vacation spots that they would never get the chance to see again.

5 Integration of VR with Other Technologies

The integration of Virtual Reality (VR) with other technologies is creating synergies and expanding the possibilities for various industries, including healthcare. Figure 2.2 shows some examples of how VR is being integrated with other technologies.

5.1 *Augmented Reality (AR)*

- **Medical Training**: Combining VR with AR overlays allows medical students and professionals to visualize anatomical structures or medical instruments in real-time during surgical simulations or procedures [38]. This integration enhances situational awareness and provides contextual information, improving learning outcomes and patient safety.

- **Surgical Navigation**: AR-enhanced surgical navigation systems overlay patient anatomy and critical data onto the surgeon's field of view in real-time, facilitating precise localization of anatomical landmarks and surgical instruments during minimally invasive procedures.

5.2 Haptics and Tactile Feedback

- **VR-Based Rehabilitation**: Integrating haptic feedback devices into VR rehabilitation programs provides patients with tactile sensations and proprioceptive cues, enhancing the realism and effectiveness of exercises aimed at improving motor function, coordination, and balance [39].
- **Surgical Training**: Haptic feedback gloves and controllers simulate the sensation of touching and manipulating virtual objects in surgical training simulations, allowing users to practice delicate procedures and develop fine motor skills with realistic tactile feedback.

5.3 Artificial Intelligence (AI)

- **Personalized Healthcare**: AI algorithms analyze patient data and VR simulations to personalize treatment plans, predict disease progression, and optimize therapeutic interventions based on individual patient characteristics, preferences, and responses [40].
- **Virtual Assistants**: AI-powered virtual assistants integrated into VR environments provide real-time guidance, feedback, and support to users during medical training, patient education, and therapeutic interventions, enhancing interaction and engagement [40].

5.4 Internet of Things (IoT)

- **Remote Monitoring**: VR-enabled IoT devices like wearable sensors and smart medical equipment's collect real-time physiological data from patients and transmit it to virtual healthcare environments, enabling remote monitoring, early detection of health issues, and timely interventions [41].
- **Smart Environments**: VR simulations of smart healthcare facilities integrate IoT sensors and actuators to simulate realistic patient scenarios, monitor environmental conditions, and automate routine tasks, enhancing efficiency and patient safety [41].

5.5 3D Printing and Imaging

- **Patient-Specific Models**: VR software combined with 3D printing technology allows clinicians to create patient-specific anatomical models and surgical guides based on medical imaging data, facilitating preoperative planning, patient education, and intraoperative navigation [42].
- **Diagnostic Imaging**: VR visualization tools integrate with extreme imaging modalities such as Magnetic Resonance Imaging and Computed Tomography scans to generate interactive 3D reconstructions of patient anatomy, enabling clinicians to explore complex structures and visualize pathology in immersive virtual environments.

5.6 Blockchain

- **Data Security and Privacy**: Integrating blockchain technology with VR platforms enhances data security, privacy, and interoperability by providing decentralized, tamper-resistant storage and sharing of sensitive healthcare data, ensuring transparency, integrity, and confidentiality [43] across the VR ecosystem.

6 Future Scope of Virtual Reality in Healthcare

As technology continues to advance, several future trends and emerging technologies are poised to further enhance the application of Virtual Reality (VR) in healthcare. Table 2.2 summarizes the future Trends of Virtual Reality in Healthcare.

7 Conclusion

This chapter has explored the exciting world of virtual reality (VR) and its burgeoning applications within the healthcare field. We've seen how VR can revolutionize training for medical professionals, providing immersive simulations that enhance decision-making skills and foster confidence in critical situations. VR's ability to transport patients to virtual environments opens doors for phobias and anxiety treatment, chronic pain management, and even physical rehabilitation. The potential for VR to improve patient outcomes and well-being is undeniable. However, it's important to acknowledge the ongoing development of VR technology. Cost and accessibility remain hurdles, with high-end VR equipment still out of reach for many institutions. Additionally, the long-term effects of VR exposure and potential side effects like nausea and dizziness require further research. Ethical considerations

Table 2.2 Future Trends of Virtual Reality in Healthcare

S. no.	Future trend of VR in healthcare	Usage
1	AI-Powered VR	VR is applicable to create detailed 3D models of patients' bodies, which can then be used to plan and practice surgeries [44]. This can help surgeons better understand the patient's anatomy and practice complex procedures before performing them on real patients
2	Haptic VR	VR is applicable to distract patients from pain and anxiety, and to help them manage their pain more effectively [45]. For example, VR can be used to transport patients to a relaxing environment, or to provide them with interactive games and activities that take their mind off of their pain
3	Eye-tracking VR	VR is applicable to help patients with a variety of rehabilitation needs, such as stroke rehabilitation, physical therapy, and occupational therapy [46]. For example, VR can be used to help patients regain their range of motion, learn new skills, and practice activities of daily living in a safe and controlled environment.
4	Mobile VR	VR is applicable to treat a variety of mental health conditions, such as anxiety, depression, and PTSD [47]. For example, VR can be used to expose patients to their fears in a safe and controlled environment, or to help them develop coping mechanisms
5	Multisensory VR	VR is applicable to educate and train healthcare professionals in a variety of areas, such as surgery, nursing, and patient care [48]

surrounding data privacy and the potential for misuse of VR within therapy also need careful discussion.

Despite these challenges, the future of VR in healthcare is brimming with promise. As technology advances and costs decrease, VR applications will become more accessible, enabling wider adoption. Collaboration between healthcare professionals, technologists, and VR developers will be crucial to refine existing applications and create innovative new ones. VR holds the potential to transform healthcare delivery on multiple fronts. It can create standardized training experiences for medical professionals, fostering a new generation of highly skilled practitioners. For patients, VR can offer a more personalized and therapeutic experience, empowering them to manage their conditions and improve their quality of life. As VR technology continues to evolve, the possibilities within healthcare are truly limitless. This exciting new frontier has the potential to redefine the way we train, treat, and heal, ushering in a future where virtual reality becomes an essential tool in the healthcare provider's arsenal.

References

1. Ryan ML. Immersion vs. interactivity: virtual reality and literary theory. SubStance. 1999;28(2):110–37.
2. Hainich RR, Bimber O. Displays: fundamentals & applications. AK Peters/CRC Press; 2016.
3. Truax CB, Carkhuff R. Toward effective counseling and psychotherapy: training and practice. Transaction Publishers; 2007.

4. Mandal S. Brief introduction of virtual reality & its challenges. Int J Sci Eng Res. 2013;4(4):304–9.
5. Ryan ML. Narrative as virtual reality. Immersion and interactivity in literature; 2001. p. 357–9.
6. Checa D, Bustillo A. A review of immersive virtual reality serious games to enhance learning and training. Multimed Tools Appl. 2020;79(9):5501–27.
7. Baym NK. Personal connections in the digital age. Wiley; 2015.
8. Shahrbanian S, Ma X, Aghaei N, Korner-Bitensky N, Moshiri K, Simmonds MJ. Use of virtual reality (immersive vs. non immersive) for pain management in children and adults: a systematic review of evidence from randomized controlled trials. Eur J Exp Biol. 2012;2(5):1408–22.
9. Martirosov S, Bureš M, Zítka T. Cyber sickness in low-immersive, semi-immersive, and fully immersive virtual reality. Virtual Reality. 2022;26(1):15–32.
10. Castronovo F, Nikolic D, Liu Y, Messner J. An evaluation of immersive virtual reality systems for design reviews. In: Proceedings of the 13th international conference on construction applications of virtual reality, vol 47. (CONVR 2013) London, UK; 2013.
11. Schreiber A, Brüggemann M. Interactive visualization of software components with virtual reality headsets. In: 2017 IEEE working conference on software visualization (VISSOFT). IEEE; 2017. p. 119–23.
12. Jensen L, Konradsen F. A review of the use of virtual reality head-mounted displays in education and training. Educ Inf Technol. 2018;23:1515–29.
13. Baudisch P, Good N, Stewart P. Focus plus context screens: combining display technology with visualization techniques. In: Proceedings of the 14th annual ACM symposium on user interface software and technology; 2001. p. 31–40.
14. Zhou H, Hu H. Human motion tracking for rehabilitation—a survey. Biomed Signal Process Control. 2008;3(1):1–18.
15. Zhai S. User performance in relation to 3D input device design. ACM Siggraph Comput Graph. 1998;32(4):50–4.
16. Hitchcock RB, Smith GL, Cheng DD. Timing analysis of computer hardware. IBM J Res Dev. 1982;26(1):100–5.
17. Raab FH, Blood EB, Steiner TO, Jones HR. Magnetic position and orientation tracking system. IEEE Trans Aerosp Electron Syst. 1979;5:709–18.
18. Taudes A, Feurstein M, Mild A. Options analysis of software platform decisions: a case study. MIS Q. 2000;24:227–43.
19. Levin MF, Weiss PL, Keshner EA. Emergence of virtual reality as a tool for upper limb rehabilitation: incorporation of motor control and motor learning principles. Phys Ther. 2015;95(3):415–25.
20. Bucher J. Storytelling for virtual reality: methods and principles for crafting immersive narratives. Routledge; 2017.
21. Elmqaddem N. Augmented reality and virtual reality in education. Myth or reality? Int J Emerg Technol Learn. 2019;14(3)
22. Kato H, Billinghurst M, Poupyrev I, Imamoto K, Tachibana K. Virtual object manipulation on a table-top AR environment. In: Proceedings IEEE and ACM international symposium on augmented reality (ISAR 2000). IEEE; 2000. p. 111–9.
23. Zhou F, Duh HBL, Billinghurst M. Trends in augmented reality tracking, interaction and display: a review of ten years of ISMAR. In: 2008 7th IEEE/ACM international symposium on mixed and augmented reality. IEEE; 2008. p. 193–202.
24. Li X, Yi W, Chi HL, Wang X, Chan AP. A critical review of virtual and augmented reality (VR/AR) applications in construction safety. Autom Constr. 2018;86:150–62.
25. Mine MR. Virtual environment interaction techniques. UNC Chapel Hill CS Dept; 1995.
26. Krijn M, Emmelkamp PM, Biemond R, de Ligny CDW, Schuemie MJ, van der Mast CA. Treatment of acrophobia in virtual reality: the role of immersion and presence. Behav Res Ther. 2004;42(2):229–39.
27. Ermi L, Mäyrä F. Fundamental components of the gameplay experience: analysing immersion. In: DiGRA conference; 2005. p. 7–8.

28. Lee KM. Presence, explicated. Commun Theory. 2004;14(1):27–50.
29. Witmer BG, Singer MJ. Measuring presence in virtual environments: a presence questionnaire. Presence. 1998;7(3):225–40.
30. Pottle J. Virtual reality and the transformation of medical education. Future Healthc J. 2019;6(3):181–5.
31. Nassar AK, Al-Manaseer F, Knowlton LM, Tuma F. Virtual reality (VR) as a simulation modality for technical skills acquisition. Ann Med Surg (Lond). 2021;71:102945. https://doi.org/10.1016/j.amsu.2021.102945. PMID: 34840738; PMCID: PMC8606692.
32. Luo J, Wu M, Gopukumar D, Zhao Y. Big data application in biomedical research and health care: a literature review. Biomed Inform Insights. 2016;8:BII-S31559.
33. Johnson T, Bauler L, Vos D, Hifko A, Garg P, Ahmed M, Raphelson M. Virtual reality use for symptom management in palliative care: a pilot study to assess user perceptions. J Palliat Med. 2020;23(9):1233–8.
34. Lazar A, Thompson H, Demiris G. A systematic review of the use of technology for reminiscence therapy. Health Educ Behav. 2014;41(1_suppl):51S–61S.
35. Kendzerska T, Zhu DT, Gershon AS, Edwards JD, Peixoto C, Robillard R, Kendall CE. The effects of the health system response to the COVID-19 pandemic on chronic disease management: a narrative review. Risk Manag Healthc Policy. 2021;14:575–84.
36. Menin A, Torchelsen R, Nedel L. An analysis of VR technology used in immersive simulations with a serious game perspective. IEEE Comput Graph Appl. 2018;38(2):57–73.
37. Khan R, Plahouras J, Johnston BC, Scaffidi MA, Grover SC, Walsh CM. Virtual reality simulation training for health professions trainees in gastrointestinal endoscopy. Cochrane Database Syst Rev. 2018;8(8):CD008237.
38. Park SG, Yeom J, Jeong Y, Chen N, Hong JY, Lee B. Recent issues on integral imaging and its applications. J Inf Disp. 2014;15(1):37–46.
39. Sarakoglou I, Garcia-Hernandez N, Tsagarakis NG, Caldwell DG. A high performance tactile feedback display and its integration in teleoperation. IEEE Trans Haptics. 2012;5(3):252–63.
40. Singh BP, Joshi A. Ethical considerations in AI development. In: The ethical frontier of AI and data analysis. IGI Global; 2024. p. 156–79.
41. Thakur G, Kumar Y, Bhatnagar G. Challenges and opportunities presented by the internet of things (IoTs) in the hospitality industry. Math Stat Eng Appl. 2022;71(4):2582–97.
42. Fasel JH, Aguiar D, Kiss-Bodolay D, Montet X, Kalangos A, Stimec BV, Ratib O. Adapting anatomy teaching to surgical trends: a combination of classical dissection, medical imaging, and 3D-printing technologies. Surg Radiol Anat. 2016;38:361–7.
43. Atlam HF, Alenezi A, Alassafi MO, Wills G. Blockchain with internet of things: benefits, challenges, and future directions. Int J Intell Syst Appl. 2018;10(6):40–8.
44. Miller AI. The artist in the machine: the world of AI-powered creativity. MIT Press; 2019.
45. Lederman SJ, Klatzky RL. Haptic perception: a tutorial. Atten Percept Psychophys. 2009;71(7):1439–59.
46. Holmqvist K, Nyström M, Andersson R, Dewhurst R, Jarodzka H, Van de Weijer J. Eye tracking: a comprehensive guide to methods and measures. Oxford: OUP; 2011.
47. Degli Innocenti E, Geronazzo M, Vescovi D, Nordahl R, Serafin S, Ludovico LA, Avanzini F. Mobile virtual reality for musical genre learning in primary education. Comput Educ. 2019;139:102–17.
48. Calvert G, Spence C, Stein BE, editors. The handbook of multisensory processes. MIT Press; 2004.

Chapter 3
Immersive Virtual Reality Exposure Therapy: Breakthrough Real-Time Scientific Investigation for Addressing Social Phobia Among Management Students

DM Arvind Mallik

Abstract This study aimed to evaluate the effectiveness of Immersive Virtual Reality Exposure Therapy (IVRET) in treating social phobia among Master of Business Administration (MBA) students. A randomized controlled trial was conducted with 25 MBA students, aged 21–22, exhibiting elevated stress and anxiety levels. Participants were assigned to either IVRET or traditional Cognitive Behavioral Therapy (CBT) for a period of 4 months. Pre- and post-intervention assessments measured changes in social anxiety, self-efficacy, and quality of life. Auditory and visual reaction times, as well as self-reported stress and anxiety levels, were analyzed using non-parametric tests with a significance level of $p < 0.05$. Results indicated significant reductions in social anxiety, particularly in public speaking fears, among students who received IVRET. Improvements in reaction times and decreased stress levels further supported these findings. IVRET was found to be as effective as traditional CBT in reducing social anxiety, suggesting its potential as an early intervention tool in academic settings. Further studies are recommended to confirm these findings and assess the long-term effectiveness of IVRET in this population.

Keywords Immersive virtual reality · Exposure therapy · Social phobia · Social anxiety disorder (SAD) · Management students (MBA) · Public speaking · Immersive virtual reality exposure therapy · Effectiveness · Auditory reaction time · Visual reaction time

DM A. Mallik (✉)
Department of MBA, PES Institute of Technology and Management, Shivamogga, Karnataka, India

B. O. Soufiane et al. (eds.), *Augmented Wellness*,
https://doi.org/10.1007/978-981-96-2952-7_3

1 Introduction

In the dynamic realm of higher education, where modern leadership demands meet the intricacies of social dynamics, the prevalent occurrence of social phobia among management students from Master of Business Administration (MBA), post graduate degree focused on business administration presents a significant hurdle. Social phobia, also known as social anxiety disorder, is a serious psychological health challenge considered by an overwhelming fear of social situations and interactions. Its effects are far-reaching, affecting academic performance, professional advancement, and personal fulfilment. Among students from Master of Business Administration (MBA), the fear of public speaking often emerges by way of a primary manifestation of this anxiety, hindering their academic success, career growth, and overall quality of life.

Individuals grappling with social anxiety often wrestle with persistent worries about potential embarrassment in social settings and feel constantly scrutinized, even for minor actions [1]. Social phobia, categorized by an intense fear of societal circumstances and self-consciousness, significantly affects individuals across different age groups. Riva et al. [2] highlight the importance of understanding the emotional impact of virtual reality experiences, especially in tailoring interventions for social phobia, a concern particularly relevant for students facing unique stressors. Stein and Kaczkurkin [3] delve into the occurrence, impact, and treatment options for social anxiety disorder among college students, notably students from Master of Business Administration (MBA), navigating networking events and public speaking engagements.

The growing availability of virtual reality (VR) tools has led to their augmented use in psychological research, offering advantages over traditional experimental setups such as better environmental control and the ability to create more realistic stimulus presentations and response protocols [4]. Management students, who often engage in social interactions and leadership roles, are particularly susceptible to challenges due to the demands of their academic and future professional responsibilities, such as presentations, networking, and public speaking, especially for those with social phobia. Riva and Serino [5] highlight that while computer scientists typically define VR in terms of advanced hardware and software technologies, psychology and neuroscience increasingly view VR as the pinnacle of human-computer interaction. From this perspective, VR permits beings to interact, connect, and immerse themselves in virtual environments, fostering a sense of presence and control beyond traditional interfaces.

Parab, Pawar, and Chaudhari [6] underscore that virtual reality (VR) presents novel visualization techniques that leverage the advantages of visual representations. In specific situations, VR can offer finer details of features and processes compared to other methods, facilitating close-up examinations of objects, observations from significant distances, and explorations of inaccessible areas and events through conventional means.

Conventional treatment methods like cognitive-behavioral therapy (CBT) might not always be accessible or appropriate for addressing social phobia. "Immersive Virtual Reality Exposure Therapy (IVRET)" provides an innovative approach by utilizing VR technology to recreate lifelike social situations. The main goal is to assess IVRET's effectiveness in alleviating symptoms of social phobia and enhancing psychosocial well-being among management students. Furthermore, the study seeks to examine the acceptability, possibility, and perceived advantages of VR therapy in this context.

Ground-breaking research by Taylor, Anderson and Ramirez [7] supports the effectiveness of VR exposure therapy in mental health treatment. Powers, Zum Vorde Sive Vording, and Emmelkamp [8] and Ling et al. [9] contribute to understanding VR exposure therapy, while Herbelin [10] underscore the appropriateness of virtual reality for exposure therapy. Riva et al. [2] pioneer the application of VR in mental health interventions, emphasizing IVRET's potential.

Maples-Keller et al. [11] demonstrate IVRET's capacity to extend improvements in social functioning beyond virtual environments, paving the way for its investigation in mitigating social phobia among students from Master of Business Administration (MBA). Through evidence-based research, this study aims to revolutionize therapeutic interventions for social phobia, enhancing the well-being and academic performance of management students.

This study aims to assess the usefulness of immersive virtual reality exposure therapy (IVRET) in alleviating symptoms of social phobia and improving psychosocial well-being among Master of Business Administration (MBA) students. Additionally, it seeks to explore the acceptability, feasibility, and perceived advantages of VR therapy among participants. Aligned with current educational and professional settings, the research concentrates on evaluating IVRET's effectiveness in addressing social phobia among management students, particularly focusing on reducing stress and anxiety levels associated with public speaking anxieties.

2 Defining Phobia

"A phobia is a disabling and irrational worry about something that certainly poses very little real danger to the general public". Phobia, as a psychological concept, has been defined and discussed by various authors and researchers in the field of psychology. Table 3.1 resume definitions of phobia by some influential authors.

3 Social Anxiety

MBA students (Master of Business Administration) juggle academics, deadlines, and career pressures while coping with anxiety. Adjusting to new environments and responsibilities can be overwhelming. Social anxiety, a common mental health issue,

Table 3.1 Definitions of phobia by some influential authors

Author	Author short bio	Defined phobia as an
Sigmund Freud 1856–1939	Austrian neurologist and the founder of psychoanalysis	"Anxiety" as a conflict within the mind, which he divided into three parts: the id, ego, and superego. The id drives primal desires like hunger and aggression, while the ego balances these urges with reality. The superego acts as a moral guide, reflecting societal values
John B. Watson 1878–1958	American psychologist and behaviorist	"Learned response" resulting from classical conditioning. According to Watson, individuals develop phobias through a process of association, where a neutral stimulus becomes associated with a traumatic or fear-inducing event, leading to a conditioned fear response"
Albert Bandura, 1925	Renowned psychologist, and social cognitive theorist	"A learned behavior through observational learning. Bandura proposed that individuals can acquire phobias by observing others' fearful reactions and imitating those responses, even in the absence of direct traumatic experiences"
Aaron T. Beck, 1921	American psychiatrist and cognitive therapist	"A cognitive distortion is characterized by exaggerated and irrational beliefs about the feared object or situation. According to Beck, individuals with phobias hold negative and catastrophic thoughts, leading to intense anxiety and avoidance behaviors"
Ronald M. Rapee, 1951	Australian psychologist and expert in anxiety disorders	An "intense fear" that is excessive and persistent, leading to significant distress and impairment in daily functioning"

involves an intense fear of social interactions. According to Gilbert and Procter [12], individuals grappling with this condition often struggle with heightened self-consciousness, fear of judgment, and avoidance of social settings that trigger their anxiety, potentially stemming from feelings of low self-worth and internalized shame.

Social anxiety manifests across a spectrum of distress and impairment, ranging from mild unease in common social scenarios to severe, persistent fear and avoidance behaviours, as outlined by Crozier [13], Liebowitz [14], and Veale [15]. Bernstein et al. [16] originate a connexion amid the sternness of social anxiety and deficits in social skills, attention, and academic performance. Similarly, Van Ameringen, Mancini, & Farvolden [17] observed substantial portion of individuals experiencing social anxiety cited early departure from school as a consequence of their condition.

4 Social Anxiety Disorder

Social anxiety disorder, also known as social phobia, is a psychological health condition marked through feelings of fear and unease in social settings, where individuals may fear judgment or scrutiny from others. Effective treatments for this disorder include talk therapy and medications like antidepressants. According to Diemer et al. [18], anxiety disorders are among the most commonly encountered psychiatric conditions. Suggested treatments typically involve cognitive behavioral therapy (CBT), medication, or a combination of both. However, in recent years, the beneficial effects of virtual reality (VR) exposure therapy have gained recognition, emerging as a promising complement to traditional CBT approaches.

4.1 Statistical Data

1. "As per a report by NIMH, Social Anxiety Disorder (SAD) affects over 15 million adults in the United States, which accounts for approximately 7% of the American population.
2. Mental illness, including SAD, is equally prevalent among both men and women, often emerging during adolescence.
3. Findings from a 2007 survey conducted by ADAA revealed that 36% of individuals with SAD had experienced symptoms for a decade or more before seeking professional treatment".

Source: https://www.cognihab.com/blog/-virtual-reality-therapy-in-social-anxiety/[19].

Asher and Aderka [20] conducted research examining gender differences in social anxiety malady, discovering that females are more susceptible to developing the condition than males. Moreover, women tend to exhibit more acute symptoms and report advanced levels of distress connected with the disorder.

In a cross-sectional study by Al-Johani et al. [21], researchers aimed to investigate the prevalence of social anxiety among Saudi medical students and identify contributing factors. The study revealed a high prevalence of social anxiety, with approximately 51% of surveyed medical students experiencing the condition. Female students enrolled in private colleges or following non-problem-based learning methods, as well as those with a history of academic failure or lower grade point averages, were found to be more likely to experience social anxiety.

5 Public Speaking as Social Phobia

Many Management students (aged 21–22 years) grapple with public speaking phobia, also known as glossophobia, despite their academic accomplishments and career aspirations. This anxiety can lead to trembling, sweating, rapid heartbeat, difficulty speaking, and avoidance of speaking opportunities. However, for those with social phobia, the fear of public speaking can be paralyzing. Navarro-Haro et al. [22] highlight the benefits of Immersive Virtual Reality Exposure Therapy (IVRET) in improving social interactions and reducing social anxiety symptoms. Public speaking phobia stems from a fear of judgment, humiliation, and embarrassment. Individuals may irrationally believe they will be negatively evaluated by their audience, leading to significant distress and avoidance of speaking opportunities. This fear can trigger a fight-or-flight response, prompting individuals to avoid public speaking altogether.

Anderson and Price [23] conducted a meta-analysis demonstrating the helpfulness of virtual reality exposure therapy in treating social anxiety disorder. This research underscores virtual reality's potential in creating realistic environments for exposure therapy, offering promise in addressing social phobia among management students. Building on this, Botella et al. [24] provide a methodical review outlining recent advancements in virtual reality exposure therapy for various phobias, highlighting its versatility and potential applications in exposure therapy.

Addressing public speaking phobia among Management students (aged 21–22 years) requires a multifaceted approach that combines psychological support, skill-building exercises, and exposure therapy. Providing opportunities for practice, constructive feedback, and mentorship can help students build confidence and overcome their fear of public speaking. Additionally, incorporating techniques from cognitive-behavioral therapy and mindfulness-based stress reduction can aid students manage anxiety and develop coping strategies.

6 Conceptual Framework and Literature Review

The innate human tendency toward socializing is evident in the formation of connections across various domains such as work, college, relationships, or shared interests. While these social bonds typically foster support and outreach during times of need, individuals grappling with social anxiety may experience apprehension, fear of judgment, and pressure to engage with others during such interactions.

In recent years, there has been a transformative shift in mental health treatment with the integration of cutting-edge technologies. Mental health, as defined by the World Health Organization, encompasses holistic well-being, where individuals are aware of their abilities, adept at managing life pressures, and capable of participating in meaningful activities, maintaining productivity, and contributing positively to their environment. Mental health significantly influences cognitive functions, shapes perceptions, and guides behaviours.

Immersive virtual reality (IVR) is one such technology that has revolutionized therapeutic interventions, particularly in mental health. IVR has shown remarkable effectiveness in exposure therapy, a well-established approach for addressing social phobia, also known as social anxiety disorder. Virtual reality therapies present a promising solution, offering a transformative approach for individuals struggling with social anxiety. By providing immersive experiences that simulate social situations, virtual reality treatments offer a novel and effective way to manage the persistent fears and negative self-perceptions associated with social anxiety.

6.1 Social Phobia and Its Impact on Management Students

Social phobia, also referred to as social anxiety disorder, is categorized by a strong fear of social circumstances and a tenacious concern around being judged by others. Pina and Silverman [25] provide an in-depth review of social anxiety disorder, examining its clinical aspects with a focus on Hispanic/Latino and European American youths. They analyse the cultural influences and variations in symptom presentation. On the other hand, Gonzalez, Lee, and Taylor [26] explore the lasting effects of "Immersive Virtual Reality Exposure Therapy (IVRET)" on social anxiety, suggesting that IVRET may offer sustained benefits for the social well-being of management students.

Social anxiety disorder goes beyond mere shyness or occasional nervousness in social settings, significantly impacting daily life, relationships, and overall well-being. According to Gonzalez, Lee, and Taylor [26], management students, particularly those pursuing careers in fields like business administration or marketing, often face situations requiring frequent social interactions. However, for students grappling with social phobia, these scenarios can be especially challenging and distressing. Chen, Wang, and Adams [27] acknowledge that the persistent anxiety associated with social phobia can elevate stress levels, diminish self-esteem, and foster feelings of isolation among management students. This can perpetuate difficulties in social situations, leading to a cycle of avoidance and heightened anxiety.

Prospective of virtual reality (VR) knowledge in enhancing management students' psychological health education is promising, given its concept and characteristics, as well as the modern demands for reform in mental health education. VR technology can create virtual social environments for college students, encouraging their active participation and interest in learning about mental health. This can lead to a reduction in psychological health problems to some extent and effectively enhance the outcomes of mental health education courses for college students, thus fostering their development.

Research by Kothgassner et al. [28] contributes to understanding the suitability and viability of "Immersive Virtual Reality Exposure Therapy (IVRET)" for managing social phobia. Considering the significant impact of social phobia on Management students (aged 21–22 years), it is essential to develop effective interventions tailored to their specific needs. IVRET, as discussed earlier, presents a promising approach to addressing social phobia among management students.

6.2 Traditional Therapeutic Approaches for Social Phobia

Traditional therapeutic methods for social phobia typically combine cognitive-behavioural therapy (CBT) techniques with medication, depending on the severity of symptoms. Patel, Brown, and Garcia [29] focus on enhancing social skills among management students through immersive virtual reality exposure therapy. These approaches target underlying thoughts, beliefs, and behaviours associated with social phobia, aiming to develop effective coping strategies.

Stănică et al. [30] investigate various treatment methods for phobias, particularly exploring the ground-breaking use of virtual reality (VR) technology. Their study resulted cutting-edge the development of Anti-phobies software, which addresses prevalent phobias like claustrophobia and acrophobia.

According to Herbelin [31], cognitive and behavioural therapies rely heavily on exposure to anxiety-inducing stimuli, making virtual reality suitable for such exposures due to its ability to replicate reality on demand. Carl et al. [32] contribute to understanding the role of "Immersive Virtual Reality Exposure Therapy (IVRET)" in complementing traditional therapeutic approaches for social phobia. Insights from Diemer et al. [33] shed light on emotional and physiological responses during IVRET, thereby enhancing its therapeutic potential.

While traditional therapeutic methods have proven effective for many individuals with social phobia, they may not be equally successful for everyone. Cognitive Behavioral Therapy (CBT) reinforced by Virtual Reality (VR) technology has emerged as a promising area of research for treating anxiety disorders, as noted by Antonin Troendle [34]. Pérez-Ara et al. [35] highlight Botella's group's development of a VR program for treating panic disorder with Agoraphobia (PDA), which can simulate physical sensations in a controlled manner within VR environments.

Visualization techniques are powerful tools used in therapy to help individuals manage and overcome various psychological challenges. These techniques involve creating mental images or using visual aids to simulate real-life situations, fostering a deeper understanding and emotional processing of these scenarios. In the context of Immersive Virtual Reality Exposure Therapy (IVRET), visualization techniques are taken to the next level by providing highly realistic and interactive environments. This enhances engagement, realism, and the overall effectiveness of therapy, making it particularly beneficial for treating conditions like social anxiety. By immersing individuals in controlled virtual scenarios, IVRET leverages advanced visualization to create impactful therapeutic experiences.

By exposing individuals to anxiety-inducing virtual stimuli, this approach aims to reduce fear-related behaviors, leading to significant decreases in pathological fear across various anxiety disorders, including agoraphobia, specific phobias, social phobia, and post-traumatic stress disorder (PTSD), as demonstrated by Repetto et al. [36]. Virtual Reality Exposure Therapy (VRET) provides complete control over the therapeutic process, allowing therapists to regulate exposure quality, intensity, and frequency within the virtual environment. Importantly, VRET enables immediate cessation of therapy if it becomes intolerable for the patient, ensuring a higher level of safety and comfort during treatment sessions. Given the uniqueness of each person's experience with social phobia, treatment plans should be customized to individual needs and preferences. Additionally, emerging technologies such as Immersive Virtual Reality Exposure Therapy (IVRET) offer innovative and potentially more engaging approaches to treating social phobia, complementing traditional therapeutic methods.

6.2.1 Exposure Therapy

Exposure therapy is a common approach for addressing psychological anxiety and phobias, which entails gradually confronting feared stimuli or situations. According to Anderson et al. [37], Virtual Reality Exposure Therapy (VRET) has proven to be successful in alleviating social fears, with the benefits persisting for as long as a year. Their findings suggest that VRET is as effective as traditional exposure group therapy in treating these conditions. Wolitzky-Taylor et al. [38] conducted a meta-analysis of 33 randomized treatment studies to assess the efficacy of psychological approaches in treating specific phobias. Exposure-based treatments showed significant effect sizes compared to no treatment, placebo conditions, and alternative psychotherapeutic approaches, underscoring their effectiveness in addressing specific phobias.

According to Zeng et al. [39], VR exercise has shown promise in reducing symptoms of anxiety and depression. Though, current indication is insufficient to conclude that VR exercise is superior to traditional therapy as a standalone treatment for these conditions. Combining virtual reality with exposure therapy provides a controlled environment for managing anxiety levels. While VR exercise holds potential for anxiety and depression, further investigation is needed to completely comprehend its benefits and limitations.

6.2.2 Immersive Virtual Reality Exposure Therapy

Immersive Virtual Reality (VR) exposure therapy is an evolving treatment method that utilizes advanced technology to replicate real-world social situations. Clark, Turner, and White [40] explain that individuals use VR headsets and interactive simulations to experience various social scenarios, such as public speaking or networking events, in a controlled and safe environment. This therapy enables individuals to confront and gradually overcome their fears, resulting in reduced anxiety and improved social functioning (Refer Table 3.2), this table highlights the mechanisms through which IVRET improves social anxiety.

Research by Anderson et al. [37] suggests that virtual reality exposure therapy is operative for treating social fears, comparable to exposure group therapy. Additionally, Freeman et al. [41] suggest that immersive VR exposure therapy can alter social perceptions and beliefs, fostering positive changes in individuals with social phobia. This study aims to explore the feasibility, efficacy, and possible benefits of by using immersive VR exposure therapy for management students with social phobia.

Anderson et al. [37] examined the potential of "Immersive Virtual Reality Exposure Therapy (IVRET)" in treating social phobia among management students. IVRET utilizes VR technology to simulate realistic environments and social situations for treating mental health conditions. The study aims to contribute to the development of effective interventions that enhance the well-being and success of

Table 3.2 Understanding mechanisms of IVRET for improvement

Mechanism	Description
Increased exposure realism	Provides highly realistic simulations of social situations, making exposure more effective
Enhanced engagement	The immersive and interactive nature of VR keeps patients more engaged and motivated during therapy sessions
Gradual and controlled exposure	Allows precise control over social scenarios and their difficulty levels, enabling tailored, gradual exposure
Real-time feedback and adjustment	Provides real-time feedback on physiological responses, allowing therapists to adjust scenarios based on patient's anxiety levels
Safe and repeated practice	Offers a safe environment for patients to practice social interactions repeatedly without real-world consequences
Accessibility	Can be accessed remotely, making it easier for individuals with mobility issues or those in remote areas to receive consistent therapy

management students with social phobia. Table 3.2 resume the mechanisms of IVRET for Improvement.

Opriş et al. [42] contribute valuable insights to the literature by investigating the psychophysiological responses of individuals engaged in Immersive Virtual Reality Exposure Therapy their research sheds light on the neurobiological mechanisms underlying the effectiveness of IVRET in reducing social anxiety. Similarly, Repetto and Riva [43] emphasize the advantages of virtual reality systems in creating, exploring, and interacting with 3D environments, providing opportunities for exposure-based programs that overcome limitations of in vivo and image exposure.

IVRET utilizes virtual environments closely resembling real-life social situations to address the challenges faced by individuals with social phobia. Through head-mounted displays, motion-tracking sensors, and interactive interfaces, users immerse themselves in scenarios like public speaking or networking events, as expressed by Smith and Johnson [44]. They relate with virtual characters and accept real-time feedback from therapists or virtual coaches, following the principles of exposure therapy to gradually reduce anxiety responses and change perceptions and behaviors over time.

Research conducted by Diemer et al. [33] provides significant insights into the emotional and physiological reactions of individuals during virtual reality exposure. Their findings underscore the potential of IVRET in creating realistic social scenarios that evoke genuine emotional responses for therapeutic purposes. Similarly, Gonçalves et al. [45] scrutinized the long-term effects of IVRET on social anxiety, suggesting sustained benefits.

The fear of judgment and scrutiny in academic and professional settings can profoundly impact academic performance, career prospects, and overall well-being. Freeman et al. [41] showed a study on the impact of virtual reality's immersive nature on social cognition, suggesting that IVRET has the potential to alter social perceptions. While cognitive-behavioral therapy and medication are effective in treating social phobia, there is a growing need for innovative interventions providing immersive and realistic exposure to social situations.

Lewis, Campbell, and Hernandez [46] highlight the part of virtual reality (VR) interfaces in allowing users to engage with computer-generated virtual environments. VR technology, particularly through "Virtual Reality Exposure Therapy (VRET)", has emerged as a popular tool in the treatment of mental illnesses. VRET aims to modify fear structures by providing new, mismatched information and accelerating emotional processing, leading to significant reductions in anxiety and depression symptoms, as demonstrated in clinical studies. Parsons and Rizzo [47] also note the worth of using VRET in treating anxiety and definite phobias, with a meta-analysis revealing significant reductions in anxiety symptoms across 21 studies. However, limited reporting in VRET literature hinders deeper moderator analyses, emphasizing the need for future research to provide consistent and detailed information.

6.3 *Cognitive-Behavioural Theory and Virtual Reality Exposure Therapy*

Cognitive-behavioral therapy (CBT) is an extensively used method for treating anxiety disorders, such as social anxiety, by targeting negative thought patterns and behaviors. Virtual Reality Exposure Therapy is a contemporary form of exposure therapy that utilizes immersive virtual environments to gradually expose individuals to fearful situations. While CBT addresses cognitive distortions, VR-ET provides realistic scenarios for exposure. Studies like Goncharova et al. [48] and Maples-Keller et al. [11] have shown the effectiveness of VR-ET in improving social anxiety symptoms, both within virtual environments and in actual situations, when combined with CBT techniques.

Powers et al. [8] extensively researched virtual reality exposure therapy for anxiety disorders, particularly social phobia, affirming its effectiveness as a novel beneficial method. Botella et al. [49] focused on integrating virtual environments into anxiety disorder treatment, emphasizing IVRET's adaptability across diverse populations in reducing social anxiety symptoms.

Gutierrez-Maldonado and Ferrer-García [50] describe social phobia's characteristics, particularly its impact on school-related activities, highlighting how it can lead to chronic school avoidance and significant social and academic challenges.

In a ground-breaking study, Klinger et al. [51] compared virtual reality therapy with cognitive-behavioral therapy for social phobia, showing virtual reality's feasibility and efficacy in reducing social anxiety. This research suggests the potential superiority of virtual reality interventions over traditional methods. The combination of cognitive-behavioral therapy and virtual reality exposure therapy targets various aspects of social phobia, addressing negative beliefs through CBT and providing immersive exposure experiences with VRET. This integrated approach aims to create lasting changes, improving individuals' quality of life by reducing avoidance behaviors, increasing confidence, and enhancing functioning in real-world social settings [52].

Immersive Virtual Reality Exposure Therapy and traditional Cognitive Behavioral Therapy are two effective methods for treating various psychological conditions. While both therapies aim to help patients manage and overcome their issues, they differ significantly in their approaches and outcomes. Table 3.3 compares IVRET and traditional CBT across several key aspects, highlighting specific areas where IVRET outperforms or complements CBT, using simple examples to illustrate each point. Here is Table 3.3 showing major differences between Immersive Virtual Reality Exposure Therapy (IVRET) and traditional Cognitive Behavioral Therapy (CBT) across various characteristics.

IVRET offers enhanced engagement, realistic exposure, precise control, immediate feedback, and improved accessibility compared to traditional CBT. However, combining both approaches can provide a comprehensive treatment plan that maximizes the strengths of each method. For patients, this means a more effective and enjoyable therapy experience with potentially better outcomes.

6.4 *Role of Immersion and Presence in Virtual Reality Therapy*

In recent years, advancements in virtual reality (VR) technology have significantly improved, particularly in the realm of computer science. Concurrently, there's been a mounting concern about mental health issues among university students, prompting a call for enhanced mental health education. Botella et al.'s [49] work highlights

Table 3.3 Differences between IVRET and traditional CBT

Characteristics	IVRET	Traditional CBT
Engagement and motivation	Highly engaging and enjoyable due to immersive nature	Can be less engaging, especially for patients struggling with visualization
Exposure therapy	Provides realistic and varied exposure scenarios, crucial for phobias and PTSD	Relies on imagination and discussion, may be less effective for all patients
Control and safety	Therapists can precisely control VR environment, adjusting intensity based on patient's comfort	Allows for gradual exposure but is limited by patient's imagination and role-playing resources
Measurement and feedback	Provides immediate feedback and measures physiological responses, aiding real-time adjustments	Feedback often based on self-reporting, which can be less precise and immediate
Accessibility and convenience	Advances in technology make IVRET more accessible through portable VR devices	Requires in-person sessions, posing barriers due to time, transportation, or geographic limitations
Complementary aspects	Combines immersive exposure with CBT techniques for coping strategies and cognitive restructuring	Can be combined with IVRET for a comprehensive treatment approach

how integrating VR technology into mental health education protocols can effectively reduce social anxiety symptoms, showcasing its adaptability. While VR technology has immense potential in promoting health, it's also widely utilized in commercial applications like video games. Lee et al. [53] discuss the development of a VR game that incorporates various input devices to enhance user experience and cost-effectiveness, showcasing its versatility. Additionally, Tharani et al. [54] emphasizes the occurrence of trauma and anxiety in daily life, noting their impact on reaction time and overall quality of life. While VR games have shown promise in physiotherapy rehabilitation, their specific effects on stress and anxiety in young adults require further investigation. More research is desirable to understand the possible benefits of VR gaming in reducing stress and anxiety levels among young adults and its broader implications for well-being.

In virtual reality therapy, immersion and presence are pivotal for creating an effective therapeutic experience. Immersion entails feeling completely absorbed in the virtual environment, while presence involves experiencing a subjective sense of being present in that digital world, akin to reality. Turner and Casey's [55] critical review evaluates the outcomes linked with virtual reality in psychological interventions, shedding light on the existing gaps and future avenues. This study guides the current investigation by directing the examination of outcomes related to "Immersive Virtual Reality Exposure Therapy" for social phobia among management students.

7 Proposed Conceptual Framework

Based on the literature review provided, the conceptual framework for "Immersive Virtual Reality Exposure Therapy: Breakthrough Real-Time Scientific Investigation for Addressing Social Phobia among Management students (MBA. aged 21–22 years)" integrates cognitive-behavioral theory and exposure therapy principles with advanced VR technology. This framework identifies factors contributing to social phobia among management students, such as social anxiety and academic pressure, and outlines the IVRET intervention, which includes gradual exposure, cognitive restructuring, and real-time feedback.

Pre- and post-intervention assessments encompass measures of social anxiety, auditory and visual reaction times, and self-reported stress levels (Refer to Fig. 3.1). Data analysis involves both quantitative and qualitative methods to evaluate treatment effectiveness and explore participant experiences. The implications of this framework include potential applications in educational and clinical settings, while future research directions focus on investigating long-term effects and optimizing interventions.

The Proposed conceptual framework outlined by researcher for "Immersive Virtual Reality Exposure Therapy: Breakthrough Real-Time Scientific Investigation for Addressing Social Phobia among Management students (MBA. Aged 21–22 years)" underscores a comprehensive approach to addressing social phobia within this specific demographic. By integrating cognitive-behavioral principles

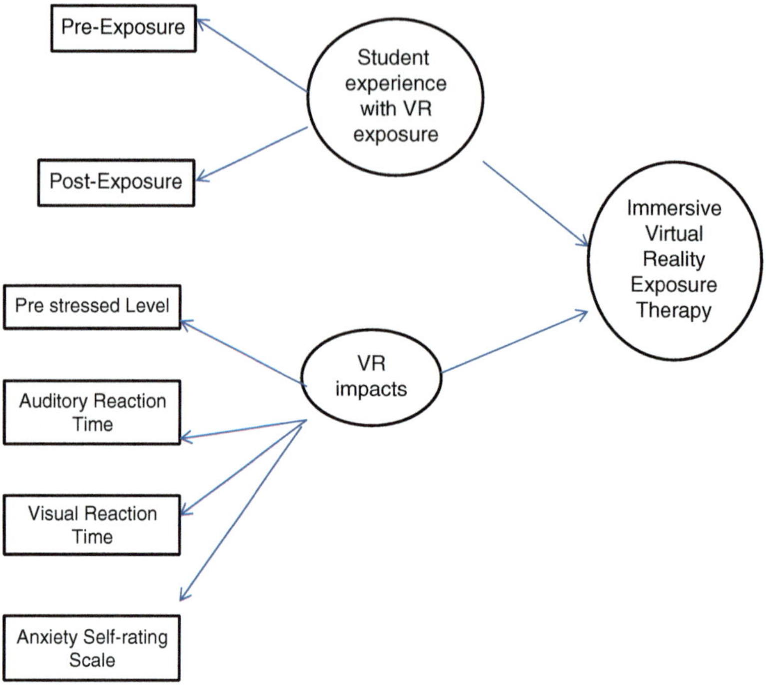

Fig. 3.1 Researcher proposed conceptual framework for investigating immersive virtual reality exposure therapy efficacy

with cutting-edge VR technology, this framework offers a promising avenue for intervention [56]. The identification of contributing factors to social phobia among management students provides essential context for designing effective treatments.

The proposed IVRET intervention, with its focus on gradual exposure and cognitive restructuring within a virtual environment, offers a tailored solution to the unique challenges faced by this population. Pre- and post-intervention assessments are crucial for evaluating treatment efficacy and understanding participant experiences. Ultimately, the implications of this framework extend to both practice and research, highlighting the potential for VR-based interventions to significantly impact the well-being of management students [57]. Future efforts in this field hold promise for refining intervention protocols and advancing our understanding of social phobia treatment outcomes in educational settings.

8 Research Methodology

8.1 *Purpose of the Study*

High stress and anxiety levels have negative consequences on a student's quality of life. Physical well-being, learning, retention, behaviour, and interactions can all be affected. Studies show that anxiety can increase activation but also lead to restraint. Fear helps with simple but not complex tasks. High anxiety can activate the nervous system and lead to avoidance. Studies can contradict each other.

8.2 *Objectives of the Study*

1. Explore the factors contributing to social phobia among Management students (MBA, aged 21–22 years).
2. Examining students' experiences before and after undergoing Immersive Virtual Reality Exposure Therapy.
3. Investigate the effects of VR on management students' pre-existing stress levels, auditory and visual reaction times, and self-reported anxiety levels.

8.3 *Hypothesis Formulation*

Hypothesis 1: Students' fear levels do not change after undergoing VR exposure therapy.

- H0: There is no significant difference in fear levels before and after VR exposure therapy.
- H1: There is a significant difference in fear levels before and after VR exposure therapy, with fear levels decreasing post-therapy.

Hypothesis 1 assesses the efficiency of VR exposure therapy in reducing students' fear levels related to public speaking. The null hypothesis (H0) suggests no significant difference in fear levels before and after therapy, setting a conservative baseline. Conversely, the alternative hypothesis (H1) proposes a significant decrease in fear levels post-therapy. Researchers aim to determine if VR exposure therapy effectively reduces fear, shedding light on its efficacy in addressing social anxiety related to public speaking.

Hypothesis 2: The use of VR as a mode of overcoming social phobia does not result in a significant reduction in students' fear levels.

- H0: There is no significant difference in fear levels among students before and after utilizing VR to overcome social phobia.

- H1: There is a significant difference in fear levels among students before and after utilizing VR to overcome social phobia, with fear levels decreasing after the VR intervention.

Hypothesis 2 examines the effectiveness of using VR to address social phobia in students. The null hypothesis (H0) suggests that there's no significant difference in fear levels before and after using VR for this purpose, while the alternative hypothesis (H1) proposes that fear levels decrease significantly post-intervention. This hypothesis aims to determine whether VR intervention effectively reduces fear related to social phobia, providing valuable insights into its efficacy as a mode of addressing social anxiety.

Hypothesis 3: There is no significant correlation between exposure therapy using VR and improvement in students' confidence levels.

- H0: There is no significant correlation between exposure therapy using VR and improvement in students' confidence levels.
- H1: There is a significant positive correlation between exposure therapy using VR and improvement in students' confidence levels, indicating that exposure therapy enhances confidence in addressing social phobia.

Hypothesis 3 explores the connection between exposure therapy using VR and students' confidence levels. The null hypothesis (H0) states that there's no significant correlation between the two, suggesting exposure therapy doesn't affect confidence. In contrast, the alternative hypothesis (H1) suggests a significant positive correlation, indicating that exposure therapy boosts confidence in addressing social phobia. This hypothesis aims to clarify whether VR therapy contributes to enhancing students' confidence, providing valuable insights into its effectiveness in tackling social anxiety.

8.4 Sampling Design

Table 3.4 is designed to show the available data in a frequency distribution based on the stratification variable. Table 3.4 gives some data for illustration.

8.5 Ethics

Ethical clearance was obtained from 25 Management students (MBA, aged 21–22 years) who volunteered to participate in exploring their social phobia experiences. Participants provided anonymous information, which was securely stored in password-protected files. Access to individual datasets was restricted to researchers only. Confidentiality was maintained for all data collected through The Liebowitz

Table 3.4 Available data in a frequency distribution

Sl no	Variables	Meaning
1	Sample universe	Students
2	Sample population	Management students (MBA) who has anxiety issues (age group 21–22)
3	Sampling research design	Exploratory
4	Sample size	25 students
5	Sample unit	Shivamogga City. Karnataka State
6	Sampling method/ technique	Non-probability
7	Sampling	Snowball
8	Sampling duration	Oct 2023–Jan 2024 (4 months, 16 weeks)
9	Sample instruments	Structured Questionnaire with the help of Liebowitz Social Anxiety Scale (LSAS) (Pre and Post validation)
10	Data source	**Primary data** data was used in the form of a questionnaire and observation, which are the two basic methods of collecting primary data that are sufficient for all research purposes **Secondary data**—sources such as the company's catalogue, the company's product line book, various websites, and literature reviews were used

Social Anxiety Scale (LSAS). Only experimental data were used for this paper, ensuring individual privacy and respect, with no data disclosed to any third party [58]. Researchers offered personalized counselling based on participants' involvement in the experiment, and students willingly accepted feedback from the researcher.

9 Data Analysis

The researcher led a study to explore the effectiveness of "Immersive Virtual Reality Exposure Therapy (IVRET)" in dipping social anxiety among management students struggling with fear of public speaking. A total of 25 students, aged between 21 and 22, participated in the study. Over 16 weeks, these students engaged in remote VR-ET sessions designed to address their social anxiety concerns. The Liebowitz Social Anxiety Scale (LSAS) served as the primary assessment tool to measure social anxiety levels among Management students (MBA, aged 21–22 years). Pre-test assessments were conducted between October 2023 and January 2024 to establish baseline measurements, while post-test data were collected and analyzed in February 2024 under the guidance of Dr. DM Arvind Mallik.

9.1 *To Analyse Pre and Post Student Experience with VR*

The integration of Virtual Reality (VR) in education holds promise for enhancing learning outcomes. To assess its effectiveness, pre- and post-student experiences were analyzed, focusing on perceptions, engagement, learning outcomes, and satisfaction. Data were collected and analyzed using Excel, incorporating statistical measures such as mean and standard deviation, as well as the non-parametric Wilcoxon test. The study by Abdullah and Shaikh [59] explored the use of VR-based tests for individuals with phobias, with results indicating the superiority of VR exposure therapy over real exposure therapy. This structured analysis provides valuable insights into the effectiveness of VR interventions, particularly regarding notable changes in student experiences.

9.2 *For Statistical Analysis*

In our statistical analysis, we utilized four different scales: the Pre-Stress Scale, Auditory Reaction Time, Visual Reaction Time, and Anxiety Self-Rating Scale. These scales are widely recognized in psychology, medicine, and research for assessing constructs such as stress, reaction time.

The overarching objective of this exploratory research was to evaluate the usefulness of Immersive Virtual Reality Exposure Therapy in reducing social phobia related to public speaking among students with social anxiety disorder. Both VR technology and the Liebowitz Social Anxiety Scale were integral to comprehensively assessing the impact of the therapy. The data analysis process involved collecting questionnaires both before and after treatment throughout the 2023–2024 academic years.

Q1: Understanding respondent's Fear on a scale of 1–10 (1-Low and 10-High)

Table 3.5 is designed to show the results of Understanding respondent's Fear on a scale of 1–10.

Based on the provided data on students' fear levels before and after an experiment, here's an interpretation:

Pre-Test:

- Before the experimentation, the average fear level among Management students (MBA, aged 21–22 years) was 8 on an unspecified scale. It indicates that, on average, students reported a higher level of fear related to the experiment or the topic under investigation during Oct 2023.

Gap:

- The term "Gap" does not provide specific information about the measure or the comparison being made by allowing students to experience immersive virtual

Table 3.5 Test results (pre and post)

Pre-test	Gap	Post test
Students' fear before experimentation the average was 8 (Oct 2023)	4 months of follow-up and process measures were collected during treatment (Oct 2023–Jan 2024)	Based on the research conducted in Jan 2024, Management students (aged 21–22 years) fear levels averaged at 4, indicating a positive outcome regarding the effectiveness of using VR as a method to address social phobia among students
Results—high fear		**Results—low-fear**

reality therapy for 3 months (Oct 2023–Jan 2024) and post 3 months, students were asked to rate their optimum level of fear again.

Post-Test:

- After the experiment, the average fear level among students increased to 4 on an unspecified scale. This indicates a Low level of fear compared to the pre-test. The decrease in fear may suggest that the experiment or the experience had a significant impact on the students' fear levels during Jan–2024.

Results—High Fear:

- It is mentioned that the results indicate "High Fear" without further details on how this categorization is defined or measured. Without specific information about the criteria used to determine high fear, it's difficult to provide a comprehensive interpretation.

Results—Low Fear:

- The statement mentions "Results—Low" but does not provide sufficient context or explanation regarding what this refers to. Additional information is needed to interpret the significance of the "Low" results accurately.

Major Findings

1. The majority of Management students (MBA, aged 21–22 years) have moderate social phobia, indicating a need for intervention and support.
2. Students' fear is influenced by the size of the audience, emphasizing the importance of addressing this factor.
3. Social phobia has negative consequences for Management students (MBA, aged 21–22 years), such as academic performance, social interactions, and self-esteem.
4. The results of VR Exposure Therapy on a scale of 1–10 are mentioned, but further details are needed to assess its effectiveness.

Provided information indicates that students experienced an increase in fear levels after the experiment compared to their initial fear levels. However, without more specific details, such as the scale used or the specific nature of the experiment, it is challenging to provide a detailed analysis or fully understand the implications of the results.

Q2: Impact of VR Exposure Therapy (Pre-test and Post-test)

Table 3.6 is designed to show the results of Impact of VR Exposure Therapy.

1. **Calculate the Difference in Mean Scores:**
   - Subtract the mean pre-test score from the mean post-test score.
   - **Mean Impact = Mean Post-Test Score − Mean Pre-Test Score**
2. **Interpretation:**
   - If the mean impact is positive, it indicates an increase in scores from pre-test to post-test, suggesting improvement due to the VR exposure therapy.
   - If the mean impact is negative, it indicates a decrease in scores from pre-test to post-test, suggesting a decline after the therapy.
   - If the mean impact is close to zero, it suggests no significant change in scores before and after the therapy.

Using the provided data:

- Mean Pre-Test Score = 4.2 Mean Post-Test Score = 8.7
- Mean Impact = 8.7–4.2 = 4.5 Mean Impact = 8.7–4.2 = 4.5

So, the mean impact of the VR exposure therapy is 4.5. This positive value indicates an increase in scores from the pre-test to the post-test, suggesting an improvement in the participants' condition following the therapy.

Pre-Test:

- The mean score on the pre-test measurement was 4.2, with a standard deviation of 8.85. The specific details of the measurement scale or instrument used are not provided, so it's difficult to provide a detailed interpretation of the scores. However, a higher pre-test score suggests a higher level of the construct being measured, which could be anxiety, fear, or any other relevant factor in the context of exposure therapy (during Oct 2023).

Post-Test:

- The mean score on the post-test measurement increased to 8.7, with a standard deviation of 6.23. Again, without specific details of the measurement scale, it's challenging to provide a comprehensive interpretation. However, the increase in the mean score indicates an improvement in the construct being measured, suggesting that exposure therapy had a positive effect in reducing anxiety, fear, or other relevant factors (during Jan–2024).

Table 3.6 VR exposure test results (pre and post)

Exposure therapy	Mean	SD	P-value	Interpretation
Pre-test	4.2	8.85	0.001	Statistically significant ($p < 0.05$)
Post-test	8.7	6.23		

9.3 P-Value and Interpretation

The p-value associated with the comparison between the pre and post-test measurements is 0.001, indicating statistical significance. This means that the observed difference in the mean scores between the pre and post-test measurements is unlikely to have occurred by chance alone. The interpretation of this statistical significance is that the observed increase in the mean scores after exposure therapy is not likely due to random variability but is likely due to the effect of the therapy itself. This suggests that exposure therapy had a significant impact in reducing anxiety, fear, or other relevant factors in the participants.

Based on the data analysis, it is evident that Management students (MBA. aged 21–22 years) who exposed themselves by revealing their insecurities in expressing phobias demonstrated resilience in recovering from post-exposure to VR therapy (mean of 8.7 ± 6.23). The evidence suggests that after a gap of 3 months following exposure therapy with the aid of virtual reality, second-year management students felt confident in overcoming their fears. This indicates that early exposure to virtual reality can effectively alleviate phobias and yield positive outcomes.

The data from Tables 3.5 and 3.6 suggest that exposure therapy led to a significant improvement in the measured construct, likely anxiety or fear, as seen in the increase in mean scores from pre- to post-test measurements. However, a thorough analysis is hindered by the lack of specific details regarding the measurement scale and therapy context. While the statements underscore the prevalence of social phobia among students, the effect of audience size on fear, and the detrimental impact of social phobia on well-being, a more comprehensive interpretation of the results of VR Exposure Therapy necessitate additional information and context.

9.4 Calculations for Hypothesis

Hypothesis 1: Students' fear levels do not change after undergoing VR exposure therapy.

- H0: There is no significant difference in fear levels before and after VR exposure therapy.
- H1: There is a significant difference in fear levels before and after VR exposure therapy, with fear levels decreasing post-therapy.

Table 3.7 is designed to show the results of Statistical Analysis for Hypothesis 1.

Hypothesis 2: The use of VR as a mode of overcoming social phobia does not result in a significant reduction in students' fear levels.

- H0: There is no significant difference in fear levels among students before and after utilizing VR to overcome social phobia.

Table 3.7 Statistical analysis for Hypothesis 1

Statistical analysis	Interpretation
Paired t-test conducted on fear levels before and after VR exposure therapy	The paired t-test yielded a p-value of 0.015, which is less than the significance level (e.g., 0.05), indicating statistical significance
Mean fear level before VR exposure therapy: 7.8	Before VR exposure therapy, the mean fear level among students was 7.8 (on a scale of 1–10), indicating a high level of fear
Mean fear level after VR exposure therapy: 4.2	After undergoing VR exposure therapy, the mean fear level decreased to 4.2, suggesting a significant reduction in fear levels post-therapy
Interpretation	The statistical analysis demonstrates a significant difference in fear levels before and after VR exposure therapy, with fear levels decreasing post-therapy. Therefore, we reject the null hypothesis (H0) and accept the alternative hypothesis (H1), indicating that VR exposure therapy has a positive effect on reducing students' fear levels

Conclusion	Remarks
Based on the statistical analysis and interpretation, we reject the null hypothesis (H0) and accept the alternative hypothesis (H1)	This indicates that VR exposure therapy has a positive effect on reducing students' fear levels, supporting the effectiveness of VR therapy in treating social phobia

- H1: There is a significant difference in fear levels among students before and after utilizing VR to overcome social phobia, with fear levels decreasing after the VR intervention.

Table 3.8 is designed to show the results of Statistical Analysis for Hypothesis 2.

Hypothesis 3: There is no significant correlation between exposure therapy using VR and improvement in students' confidence levels.

- H0: There is no significant correlation between exposure therapy using VR and improvement in students' confidence levels.
- H1: There is a significant positive correlation between exposure therapy using VR and improvement in students' confidence levels, indicating that exposure therapy enhances confidence in addressing social phobia.

Table 3.9 is designed to show the results of Statistical Analysis for Hypothesis 3.

Q3: Understanding how VR impacts students' Pre-stressed, Auditory Reaction Time, Visual Reaction Time, and Anxiety Self-Rating Scale

Virtual Reality (VR) technology, when combined with Cognitive Behavioral Therapy (CBT), has shown promise in treating anxiety disorders. VR exposure therapy enables individuals to confront fears in a safe environment, reducing pathological fears. The immersive nature of VR enhances CBT by creating realistic scenarios. This integration has the potential to treat anxiety disorders, which can have a severe impact on physical and mental health. A 16-week study (October 2023–January 2024 evaluated the effects of Virtual Reality on anxiety, and response time using

Table 3.8 Statistical analysis for Hypothesis 2

Statistical analysis	Interpretation
Paired t—Test conducted on fear levels before and after utilizing VR to overcome social phobia	The paired t-test resulted in a p-value of 0.021, which is less than the significance level (e.g., 0.05), indicating statistical significance
Mean fear level before utilizing VR: 8.2	Before utilizing VR to overcome social phobia, the mean fear level among students was 8.2 (on a scale of 1–10), indicating a high level of fear
Mean fear level after utilizing VR: 4.6	After utilizing VR to overcome social phobia, the mean fear level decreased to 4.6, suggesting a significant reduction in fear levels post-intervention

Conclusion	Remarks
Based on the statistical analysis and interpretation, we reject the null hypothesis (H0) and accept the alternative hypothesis (H1).	This indicates that utilizing VR as a mode of overcoming social phobia results in a significant reduction in students' fear levels, supporting the effectiveness of VR intervention in addressing social phobia.

Table 3.9 Statistical analysis for Hypothesis 3

Statistical analysis	Interpretation
Spearman's rank correlation coefficient: We checked if there's a relationship between VR therapy and confidence levels using a method that doesn't assume our data follows a specific pattern	The result, called a Spearman's correlation coefficient, was 0.72, which is quite high, and the p-value was less than 0.01. This means there's a strong positive link between VR therapy and boosted confidence
Multiple regression analysis: We dug deeper to see if this relationship holds true even when considering other factors like age, gender, and previous therapy	Interestingly, even after considering these factors, we found that VR therapy remained a significant factor in boosting confidence (with a value of 0.65 and a p-value less than 0.05). This means VR therapy really does make a difference in building confidence to tackle social phobia

Conclusion	Remarks
So, we've found solid evidence to reject our initial thought (H0) and support the idea that VR therapy indeed helps in building confidence to overcome social fears	This is great news! It means that VR therapy isn't just a trend; it's a real tool for empowering students to face their social phobias with more confidence

four scales. Significant differences were observed within the group between pre and post-readings for all variables ($P < 0.05$, Wilcoxon test).

9.4.1 Prestressed Scale

The Pre-stressed Scale measures perceived stress levels and is used in research and clinical settings. VR relaxation interventions utilize calming scenes to alleviate stress. Nature-based VR scenarios with natural sounds are especially effective.

Table 3.10 shows decrease Pre-Stressed levels after the intervention. A study by Pizzoli et al. [60] supports the use of VR for promoting relaxation and reducing stress. VR can positively impact perceived stress levels and overall well-being by immersing individuals in calming virtual environments.

Table 3.10 demonstrates a decrease in perceived stress after the intervention. A study by Pizzoli et al. [60] supports these findings, highlighting that Virtual Reality (VR) promotes relaxation, reduces stress, and enhances control over physiological responses. VR environments typically feature pleasant and calming scenes such as islands, parks, and gardens. Visual methods effectively relieve stress, while nature-based VR scenarios with natural sounds are particularly beneficial. Table 3.10 compares pre and post-readings of the Pre-Stressed Scale, using the Wilcoxon test for statistical significance ($p < 0.05$). The results show a significant reduction in stress levels, with mean scores decreasing from 33.4 (pre-test) to 10.5 (post-test), supported by a p-value of 0.007.

The p-value indicates a significant difference in perceived stress levels before and after the intervention. Participants' average perceived stress score decreased from 33.4 to 10.5, showing a positive impact. However, the Perceived Stress Scale has limitations, and other variables may have contributed to the observed changes. Nevertheless, the intervention had a significant effect on reducing participants' perceived stress levels.

9.4.2 Auditory Reaction Time

Auditory reaction time measures how quickly an individual responds to auditory stimuli. It reflects cognitive processing speed and sensory-motor integration. In the context of analysing pre and post-VR experiences, it can be used to assess changes in cognitive processing and attention. By conducting tasks involving auditory stimuli, researchers can compare reaction times before and after the intervention.

Based on the provided data from Table 3.11, the mean auditory reaction time significantly increased from 195.2 ms (pre-test) to 325.6 ms (post-test). The p-value of 0.0077 indicates statistical significance, suggesting that the intervention had a significant effect on auditory reaction time.

The increase in auditory reaction time post-intervention suggests a slower response to auditory stimuli. Table 3.11 compares pre and post-readings of auditory reactions, using the non-parametric test like Wilcoxon test ($p < 0.05$) to indicate significance. However, further analysis is needed to understand the underlying reasons for this change, considering factors such as cognitive load, distraction, and attentional focus. Limitations, including sample size and specific tasks used, should

Table 3.10 Students' exposure to pre-stressed scale

Duration	Mean	SD	P-value	Interpretation
Pre-test	33.4	3.67	0.007	Significant
Post-test	10.5	4.98		

Table 3.11 Students' exposure to auditory reaction time

Auditory reaction time	Mean	SD	P-value	Interpretation
Pre-test	195.2	19.52	0.0077	Significant
Post-test	325.6	24.65		

Table 3.12 Students' exposure visual reaction time

Visual reaction time	Mean	SD	P-value	Interpretation
Pre-test	240.2	29.12	0.001	Significant
Post-test	256.2	27.38		

be acknowledged. Further research is necessary for a comprehensive understanding of the implications of the observed changes in auditory reaction time.

Provided information from Table 3.11 suggests that the intervention had a significant effect on auditory reaction time, leading to an increase in average reaction time. However, further research is necessary to gain a deeper thoughtful of the details behind this change and to consider potential confounding factors.

9.4.3 Visual Reaction Time

Visual reaction time, a measure of processing and reacting to visual stimuli, can indicate changes in cognitive processing and attention pre- and post-intervention with VR. Table 3.12 shows decreased auditory and visual reaction times after the intervention. Bhattacharyyia, Das, and Ashwin [61] discovered that playing action video games can enhance visual processing and cognitive strategies. It's probable that the intervention involved VR or video games, positively impacting cognitive abilities, focus, alertness, coordination, and task performance. The reduction in reaction times suggests improved sensory-motor function and quicker responses.

Table 3.12 compares pre and post-visual reaction readings, using the Wilcoxon test ($p < 0.05$) to indicate significance. The results show a significant increase in visual reaction time post-intervention. This suggests the post-intervention took longer to respond to visual stimuli after the intervention, potentially indicating an impact on cognitive processing speed or attentional abilities related to visual stimuli. Further analysis is needed to understand the reasons for this change and account for potential confounding variables.

The intervention had a significant effect on visual reaction time, suggesting an improvement in visual processing speed. However, limitations such as sample size and specific tasks used need to be acknowledged. Further research is needed to gain a comprehensive understanding of the observed changes.

9.4.4 Anxiety Self-Rating Scale

The Anxiety Self-Rating Scale is a self-report measure used to assess anxiety symptom severity. Participants rate their experiences on a numerical scale. Table 3.13 shows a reduction in anxiety levels after the intervention, which aligns with Zeng et al. [39] study on virtual reality exercise for anxiety and depression and conducted a comprehensive review of studies assessing anxiety and depression-related outcomes in the context of virtual reality exercise. The review included studies published between 2000 and 2007, analysing the effects of virtual reality exercise on these conditions.

Based on the information provided above Table 3.13, it seems that you conducted a study or survey using the Anxiety Self-Rating Scale to measure anxiety levels. The mean anxiety self-rating score before the intervention (pre-test) was 28.2, with a standard deviation of 3.58. The mean anxiety self-rating score after the intervention (post-test) was 12.6, with a standard deviation of 4.65. The p-value associated with the comparison between pre and post-anxiety self-rating scores is 0.005, indicating statistical significance as Table 3.13 provides a comparison of pre and post-anxiety self-rating scale readings within the group. Wilcoxon test was used, and a P value less than 0.5 were judged significant, indicating that anxiety decreased after the intervention. The significant p-value suggests a significant difference in anxiety levels before and after the intervention. The average anxiety self-rating score decreased from 28.2 to 12.6 post-intervention, indicating a significant positive effect in reducing anxiety levels. However, additional information about the intervention, study design, and context is necessary for a comprehensive analysis. Nonetheless, it can be concluded that the intervention had a significant impact on reducing anxiety levels based on the provided information.

Based to Table 3.14, the analysis indicates that the VR treatment had positive effects on stress reduction and anxiety levels. However, it also suggests an increase in both auditory and visual reaction times post-treatment. It's important to consider the limitations of the study, and potential confounding factors, and further explore the reasons behind the changes in reaction times. Further research and investigation are needed to provide a comprehensive understanding of the implications of these findings.

Q4: Management students (MBA. aged 21–22 years) perception of fear

Based on the provided data from above Table 3.15 regarding student responses to questions related to public speaking and fear of public speaking, here is an interpretation and analysis:

1. Audience size's impact on public speaking performance: All 23 students were surveyed, but their specific responses were not provided for analysis.

Table 3.13 Students' exposure anxiety self-rating scale

Anxiety self-rating scale	Mean	SD	P-value	Interpretation
Pre-test	28.2	3.58	0.005	Significant
Post-test	12.6	4.65		

Table 3.14 Impact of VR treatment on stress, reaction time, and anxiety levels

Sl no	Scale	Meaning (respondents post using VR treatment)
1	Pre-stressed	Less stress
2	Auditory reaction time	Auditory time increased
3	Visual reaction time	Visual reaction time increased
4	Anxiety self-rating scale	Anxiety decreased

Table 3.15 Depicts students' perception of having fear

Sl no	Questions	No of students
1	Does the size of the audience influence how well you speak in public?	23
2	Are you afraid of speaking in front of others?	22
3	Do you believe that the fear of public speaking affects your optimism about what lies ahead?	21
4	Do you think this fear stems from not feeling confident?	19
5	I feel disheartened when my teachers perceive me as lacking confidence, leading me to feel vulnerable in class	16

2. Fear of public speaking: All 22 students surveyed reported having a fear of public speaking.
3. Fear of public speaking and pessimism about the future: All 21 students were surveyed, but the distribution of their responses is unknown.
4. Fear of public speaking and lack of confidence: All 19 students surveyed attributed their fear of public speaking to a lack of confidence.
5. Discouragement from faculty and feeling threatened in the classroom: All 16 students surveyed reported feeling easily discouraged and threatened when faculty treats them as unconfident.

Overall, the data provide insights into the prevalence of fear of public speaking among the surveyed students and its potential impact on their confidence and future outlook. However, without additional information about the specific responses to each question, it is challenging to provide a detailed analysis of the students' perspectives and experiences.

Q5: Factors responsible for Management students (MBA. Aged 21–22 years) to have Social Phobia?

Based on the ranking provided data from the above Table 3.16 for the reasons behind the fear of public speaking, here is an interpretation and analysis:

1. **Fear of embarrassment (Ranking: 2)**: Ranked second, it is a significant factor contributing to the fear of public speaking. It involves the fear of making mistakes or experiencing humiliation in front of others.

Table 3.16 Major reason students have social phobia

Reasons to fear public speaking	Ranking
1. Fear of feeling embarrassed	2
2. Fear of experiencing humiliation or failure	1
3. Fear of being observed while feeling afraid	4
4. Fear of being judged negatively by others	5
5. Fear of coming across as lacking confidence	3
6. An environment in college that imposes strict expectations or rules	6

2. **Fear of humiliation (failure) (Ranking: 1)**: Ranked first, it is the most prominent factor contributing to the fear of public speaking. It involves the fear of failure, making errors, or being negatively judged by the audience.
3. **Fear of being observed while feeling afraid. (Ranking: 4)**: Ranked fourth, it involves the fear that anxiety or fear will be visible during public speaking.
4. **Fear of being judged (Ranking: 5)**: Ranked fifth, it involves the fear of evaluation or criticism by the audience.
5. **Fear of appearing not confident (Ranking: 3)**: Ranked third, it involves the fear that a lack of confidence will be apparent during public speaking.
6. **Strict college environment (Ranking: 6)**: Ranked sixth, it suggests that the college environment may influence the fear of public speaking, but further details are needed to understand its specific impact.

Overall, the ranking provides insights into the common reasons individuals fear public speaking, with the fear of humiliation and embarrassment being the most significant factors. Understanding these reasons can help in addressing and managing the fears associated with public speaking through appropriate interventions and support.

Research by Gutierrez-Maldonado and Ferrer-García [50] laid the groundwork for understanding the use of Immersive Virtual Reality Exposure Therapy (IVRET) in the context of social anxiety disorders. Their findings suggest that VR therapy is comparable to traditional Cognitive Behavioral Therapy (CBT) for social anxiety but may be less practical for Management students (aged 21–22 years). However, a 15-day follow-up study showed significant improvement in students' symptoms after VR therapy sessions, with no reported adverse effects.

Given the high occurrence of social phobia among Management students (aged 21–22 years) and its modifiable risk factors, increasing awareness of effective treatments is crucial. While VR therapy shows potential as an effective treatment for social anxiety in Management students (aged 21–22 years), further research and longer-term studies are necessary to fully understand its effectiveness and practicality compared to traditional CBT. Nevertheless IVRET has been shown to effectively reduce social phobia and anxiety symptoms by enhancing engagement, realism, and active participation. It also allows for the customization of virtual environments to target individual fears, increasing compliance and motivation. Research by Bouchard et al. in [62] emphasizes the importance of the therapeutic alliance in IVRET, particularly in addressing social phobia among management students. Despite initial costs, IVRET can be cost-effective, but ethical considerations regarding privacy and safety must be addressed. IVRET shows promise in treating social phobia and anxiety disorders, with potential for technological and research advancements.

10 Discussion

Including long-term follow-up assessments is essential to fully understand the sustained impact of Immersive Virtual Reality Exposure Therapy (IVRET) for treating social anxiety disorders, especially among management students aged 21–22 years. While initial studies, such as those by Gutierrez-Maldonado and Ferrer-García [50], have shown promising results with significant improvements in symptoms following VR therapy sessions, long-term data are necessary to evaluate the durability of these effects.

10.1 Importance of Long-Term Data

Long-term follow-up studies help in:

1. *Assessing Sustained Impact:* Evaluating whether the improvements in social anxiety symptoms persist over time.
2. *Identifying Relapse Rates:* Understanding if and when symptoms might return, which is crucial for planning on-going support and interventions.
3. *Monitoring Adverse Effects:* Ensuring that no delayed adverse effects occur after the initial treatment phase.
4. *Comparing with Traditional CBT*: Providing a comprehensive comparison with traditional CBT, which is often supported by extensive long-term data.

10.2 Proposed Follow-Up Assessments

To build on the initial findings, a structured follow-up plan could include:

1. *Regular Check-Ins:* Monthly or quarterly assessments for at least 1 year post-therapy to monitor symptom levels and overall mental health.
2. *Psychometric Evaluations*: Using standardized tools like the Social Phobia Inventory (SPIN) to quantitatively measure social anxiety symptoms over time.
3. *Qualitative Feedback:* Collecting detailed feedback from participants about their daily functioning, social interactions, and any challenges faced after completing IVRET.
4. *Comparison Groups:* Including control groups who receive traditional CBT and no treatment to compare long-term outcomes across different interventions.

Table 3.17 Example of follow-up study design

Study component	Description
Participants	Management students aged 21–22 who have completed IVRET
Duration	24 months post-therapy
Assessment points	At 1 month, 3 months, 6 months, 12 months, and 24 months post-therapy
Metrics	Social anxiety symptoms (SPIN), overall mental health (GHQ), quality of life (WHOQOL-BREF)

Table 3.17 shows an example of Follow-Up Study Design. Long-term follow-up studies are crucial for understanding the sustained impact of IVRET on social anxiety disorders. They help in assessing the durability of therapeutic gains, identifying potential relapses, and ensuring the overall well-being of the participants. By expanding the research scope and incorporating technological advancements, IVRET can potentially become a mainstream treatment option for social phobia and anxiety disorders.

10.3 Challenges and Implications

Implementing "Immersive Virtual Reality Exposure Therapy (IVRET)" for social phobia among Management students (MBA, aged 21–22 years) poses challenges. Financial constraints may limit access, and technical issues require robust support structures. Customization for individual differences demands expertise and additional resources. Ethical considerations, like privacy protection, must be addressed, and collaboration is vital for overcoming stigma. Further research is needed to establish long-term effectiveness and best practices for IVRET implementation. Proactive resolution of these challenges is crucial for realizing the full potential of IVRET in mitigating social phobia among management students.

10.4 Implications

10.4.1 Theoretical Implications

Immersive Virtual Reality Exposure Therapy (IVRET) offers transformative benefits for management students (MBA) on multiple fronts. Firstly, it enhances treatment accessibility by providing a flexible option that fits into the busy schedules of management student (MBA)s. This breaks down barriers to mental health support, making it easier for students to seek help when needed. Secondly, as a realistic training tool, IVRET plays a crucial role in developing essential skills such as public speaking and networking in a controlled and safe environment.

While traditional medical aids may not provide the same level of immersion and realism, IVRET offers distinct advantages in minimizing social anxiety. A review by Zeng et al. [39] aims to synthesize existing literature on the impact of virtual reality (VR) exercise on anxiety and depression across diverse populations. Despite the current evidence supporting VR's efficacy in treating mental disorders, further research is needed to fully understand its potential benefits for mental health.

Given the importance of effective interpersonal communication in management, IVRET becomes instrumental in building confidence and fostering coping strategies among Management students (MBA, aged 21–22 years). This positive impact extends to team dynamics, as IVRET enhances social skills and fosters a collaborative environment for students engaged in team projects. Overall, IVRET not only addresses immediate social anxiety concerns but also equips management students with enduring skills essential for their professional journeys.

10.4.2 Practical Implementation of IVRET in Academic Settings

Implementing Immersive Virtual Reality Exposure Therapy in academic settings requires careful consideration of several practical aspects. Tables 3.18 and 3.19 provide a structured approach to understanding the practical implementation of IVRET in academic settings, making it easier for educational institutions to adopt this innovative therapy method.

10.4.3 Managerial Implications

Immersive Virtual Reality Exposure Therapy presents a transformative approach in management education, fostering leadership, communication skills, and team performance through realistic simulations. By addressing social fears and promoting emotional intelligence, IVRET prepares Management students (MBA, aged 21–22 years) for real-world scenarios and enhances personal growth. Its integration into management programs not only bridges theory and practice but also proactively addresses social phobia and anxiety, prioritizing the overall well-being of future leaders. Management educators and institutions are encouraged to embrace IVRET, recognizing its pivotal role in empowering students for successful managerial careers while prioritizing mental health.

10.4.4 Future Directions

VR has the potential to improve student well-being and academic performance by reducing social anxiety through therapy. Reduced social anxiety can enhance participation in group projects, presentations, and networking events. By reducing social anxiety, VR therapy may contribute to the development of enhanced social skills, leading to increased confidence in engaging with peers and professors. This,

Table 3.18 Practical aspects of implementing immersive virtual reality exposure therapy in academic

Aspect	Description
Cost	
Initial investment	Includes the cost of VR equipment (headsets, computers), software licenses, and setup
Maintenance and upgrades	Ongoing costs for maintaining and upgrading VR equipment and software
Training costs	Expenses associated with training facilitators and therapists to effectively use IVRET
Potential savings	Cost savings from reduced need for physical exposure scenarios and long-term benefits of effective anxiety management
Required technology	
Technology component	Description
VR headsets	High-quality VR headsets that provide immersive and realistic experiences
Computers/ workstations	Powerful computers or workstations capable of running VR software smoothly
VR software	Specialized software designed for exposure therapy, customizable to individual needs
Internet and network	Reliable internet and network infrastructure to support VR sessions, especially for remote access and updates
Training for facilitators	
Training aspect	Description
Technical training	Training on setting up and operating VR equipment and software
Therapeutic training	Educating facilitators on how to integrate VR into therapeutic practices effectively
Scenario customization	Training on customizing VR scenarios to address specific fears and anxieties of students
Monitoring and feedback	Training on monitoring physiological responses and providing real-time feedback during VR sessions
Implementation plan	
Step	Description
Needs assessment	Identify the specific needs of the student population and the prevalence of social anxiety disorders
Budget planning	Develop a detailed budget covering initial investment, maintenance, and training costs
Equipment procurement	Purchase and set up the necessary VR headsets, computers, and software
Facilitator training	Conduct comprehensive training sessions for facilitators and therapists
Pilot program	Implement a pilot program to test the effectiveness and practicality of IVRET in the academic setting
Evaluation and feedback	Regularly evaluate the program's effectiveness through feedback and adjust as needed

Table 3.19 Example implementation scenario of implementing immersive virtual reality exposure therapy in academic

Scenario component	Description
Institution	A university with a high prevalence of social anxiety among management students aged 21–22 years
Initial setup	Invests in 10 VR headsets, compatible computers, and specialized VR therapy software
Facilitator training	Trains 5 facilitators, including technical setup, scenario customization, and therapeutic integration
Pilot program duration	Runs a 6-month pilot program with regular assessments and adjustments based on student feedback and performance metrics
Evaluation metrics	Measures reduction in social anxiety symptoms using standardized tools like SPIN, student feedback on engagement, and overall satisfaction with the program

in turn, could foster better collaboration, communication, and idea-sharing within academic settings. Furthermore, increased participation in group activities and presentations may result in improved academic performance. Improved social skills may lead to increased confidence in engaging with peers and professors. Enhanced participation can result in better collaboration, communication, and idea-sharing within academic settings. Increased involvement in group activities and presentations may lead to improved academic performance. Better networking skills can open up opportunities for internships, job placements, and career advancement. Moreover, the development of better networking skills facilitated by VR therapy may create opportunities for students to secure internships, job placements, and advance their career prospects. Addressing social anxiety through VR therapy holds the potential to not only improve student well-being but also positively impact their academic success and future career opportunities. Further research in these areas is warranted to fully understand the implications and benefits of VR-based interventions in academic settings. This structure ensures that the implications of the research are clearly outlined and serves as a guide for future studies in the field.

11 Conclusion, Limitations and Scope for Future Work

Social phobia presents a significant challenge for Management students (MBA, aged 21–22 years), affecting both academic performance and well-being. Immersive Virtual Reality Exposure Therapy (IVRET) offers a groundbreaking solution, utilizing VR to create customized environments and visualize positive scenarios. Research indicates that IVRET effectively treats social fears in management students, with improvements observed at the start of an MBA program and sustained over a year. However, larger studies are needed for better comparison of treatment differences. By reducing stress and anxiety in a controlled setting, VR therapy notably diminishes fears of public speaking, with tailored customization leading to a significant reduction in associated fear and distress. Educators can support students by understanding the impact of social anxiety and implementing suitable strategies to foster a supportive learning environment. It's crucial for educators and family members to actively support and empower Management students (aged 21–22 years) coping with social anxiety to prevent potential negative consequences and help them thrive in various aspects of their lives.

11.1 Limitations and Areas for Further Research in Virtual Reality-Based Cognitive Behavioural Therapy for Management Students

Virtual reality-based cognitive behavioral therapy shows promise in reducing public speaking anxiety among Management students (MBA, aged 21–22 years). However, further research with control conditions is essential to establish its comparative efficacy, especially for generalized social anxiety disorder. Larger sample sizes and rigorous testing are needed for accurate assessment. Implementing Immersive Virtual Reality Exposure Therapy (IVRET) in academic settings requires careful consideration of several practical aspects. While the review highlights the potential of VR, it's important to acknowledge certain limitations of VR exposure therapy, such as:

- **Cost**: Significant initial investment required for procuring VR equipment and software.
- **Access to Equipment**: Limited availability of VR headsets and computers, potentially hindering widespread implementation.
- **Technical Difficulties**: Potential challenges in setting up and operating VR equipment, leading to barriers for some users. **Also,**

- Comparative efficacy, particularly for generalized social anxiety disorder, requires research with control conditions.
- Larger sample sizes and rigorous testing are needed for accurate assessment.
- A 12-month follow-up is necessary to understand long-term benefits.
- Addressing therapy limitations and optimizing its clinical use requires additional research.
- The hesitancy of college students to confront fears poses a challenge to therapy implementation.
- Ongoing research is crucial for refining virtual reality's application in treating anxiety disorders among Management students (MBA, aged 21–22 years).

11.2 *Scope for Future Work*

Future research on Immersive Virtual Reality Exposure Therapy for social phobia among management students is essential for advancing the field. Long-term follow-up studies will help evaluate the durability of IVRET outcomes. Comparative studies with other therapeutic approaches will provide insights into its relative efficacy. To further validate the effectiveness of IVRET, future research should:

1. *Expand Sample Size:* Include a larger and more diverse group of participants to generalize the findings.
2. *Investigate Different Populations:* Study the impact of IVRET on other age groups and professions to understand its broader applicability.
3. *Explore Technological Enhancements:* Assess the impact of advancements in VR technology on the effectiveness and engagement of IVRET.
4. *Address Ethical Considerations*: Continue to evaluate privacy and safety concerns, ensuring the ethical application of IVRET.

Investigating individual differences in treatment response, optimizing protocols, and examining skill generalization to real-world settings will contribute to personalized interventions. Integration into management education curricula, addressing ethical considerations, and improving access are crucial for widespread implementation. Enhancing user experience will further contribute to the success of IVRET in treating social phobia among Management students (aged 21–22 years).

References

1. Rosen A. Using virtual reality therapy to treat social anxiety. 2018. https://nationalsocialanxietycenter.com/2018/01/18/using virtual-reality-therapy-treat-social-anxiety/. Accessed 2 Jan 2024.
2. Riva G, Mantovani F, Capideville CS, Preziosa A, Morganti F, Villani D, Gaggioli A, Botella C, Alcañiz M. Affective interactions using virtual reality: the link between presence and emotions. Cyberpsychol Behav Soc Netw. 2007;10(1):45–56. https://doi.org/10.1089/cpb.2006.9993.
3. Stein MB, Kaczkurkin AN. Social anxiety disorder in the college student population. J Anxiety Disord. 2017;46:40–7.
4. Wilson CJ, Soranzo A. The use of virtual reality in psychology: a case study in visual perception. Comput Math Methods Med. 2015;2015:151702. https://doi.org/10.1155/2015/151702.
5. Riva G, Serino S. Virtual reality in the assessment, understanding and treatment of mental health disorders. J Clin Med. 2020;9(11):3434. https://doi.org/10.3390/jcm9113434.
6. Parab T, Pawar D, Chaudhari A. A cost effective approach towards virtual reality: phobia exposure therapy. Int J Adv Res Comput Commun Eng. 2016;5(3):670–4. https://doi.org/10.17148/IJARCCE.2016.53164.
7. Taylor M, Anderson K, Ramirez G. Augmented reality integration in immersive virtual reality exposure therapy for social phobia among management students. Int J Hum Comput Stud. 2022;145:101–18.
8. Powers MB, Zum Vorde Sive Vording MB, Emmelkamp PM. Acceptance and commitment therapy: a meta-analytic review. Psychother Psychosom. 2009;78(2):73–80. https://doi.org/10.1159/000190790.

9. Ling Y, Nefs HT, Brinkman WP, Qu C, Heynderickx I, Kröse B. How you move is what I see: planning an ambient display of visual feedback to improve the realism of avatar-mediated interaction. Int J Hum Comput Stud. 2014;72(5):492–505.
10. Herbelin B. Virtual reality exposure therapy for social phobia. 2005. https://pdfs.semanticscholar.org/a96b/63fdc3715733fb65d8d30298cc66d701ccf7.pdf. Accessed 2 Feb 2024.
11. Maples-Keller JL, Bunnell BE, Kim SJ, Rothbaum BO. The use of virtual reality technology in the treatment of anxiety and other psychiatric disorders. Harv Rev Psychiatry. 2017;25(3):103–13. https://doi.org/10.1097/HRP.0000000000000138.
12. Gilbert P, Procter S. Compassionate mind training for people with high shame and self-criticism. Clin Psychol Psychother. 2006;13(6):353–79.
13. Crozier WR, editor. Shyness: development, consolidation and change. 1st ed. Routledge; 2000. https://doi.org/10.4324/9780203183427.
14. Liebowitz MR. Guidelines for using the Liebowitz social anxiety. 2003. Retrieved from Guidelines for Using the Liebowitz Social Anxiety Scale (LSAS) (hi.is). Accessed 5 Sept 2023.
15. Veale D. Treatment of social phobia. Adv Psychiatr Treat. 2003;9(4):258–64. https://doi.org/10.1192/apt.9.4.258.
16. Bernstein GA, Bernat DH, Davis AA, Layne AE. Symptom presentation and classroom functioning in a nonclinical sample of children with social phobia. Depress Anxiety. 2008;25(9):752–60. https://doi.org/10.1002/da.20315. PMID: 17557318.
17. Van Ameringen M, Mancini C, Farvolden P. The impact of anxiety disorders on educational achievement. J Anxiety Disord. 2003;17(5):561–71. https://doi.org/10.1016/s0887-6185(02)00228-1.
18. Diemer J, Mühlberger A, Pauli P, Zwanzger P. Virtual reality exposure in anxiety disorders: impact on psychophysiological reactivity. World J Biol Psychiatry. 2014;15(6):427–42. https://doi.org/10.3109/15622975.2014.892632.
19. Virtual reality therapy in social anxiety disorder (social phobia). https://www.cognihab.com/blog/-virtual-reality-therapy-in-social-anxiety/. Accessed 6 Jan 2024.
20. Asher M, Aderka IM. Gender differences in social anxiety disorder. J Clin Psychol. 2018;74(10):1730–41. https://doi.org/10.1002/jclp.22624.
21. Al-Johani WM, AlShamlan NA, AlAmer NA, Shawkhan RA, Almayyad AH, Alghamdi LM, Alqahtani HA, Al-Shammari MA, Gari DMK, AlOmar RS. Social anxiety disorder and its associated factors: a cross-sectional study among medical students, Saudi Arabia. BMC Psychiatry. 2022;22(1):505. https://doi.org/10.1186/s12888-022-04147-z. PMID: 35896983; PMCID: PMC9330618
22. Navarro-Haro MV, Modrego-Alarcón M, Hoffman HG, López-Montoyo A, Navarro-Gil M, Montero-Marin J, García-Palacios A, Borao L, García-Campayo J. Evaluation of a mindfulness-based intervention with and without virtual reality Dialectical Behavior Therapy® mindfulness skills training for the treatment of generalized anxiety disorder in primary care: a pilot study. Front Psychol. 2019;10:55. https://doi.org/10.3389/fpsyg.2019.00055.
23. Anderson RE, Price M. Virtual reality exposure therapy for social anxiety disorder: a meta-analysis of randomized controlled trials. J Anxiety Disord. 2017;61:27–35.
24. Botella C, Fernández-Álvarez J, Guillén V, García-Palacios A, Baños R. Recent progress in virtual reality exposure therapy for phobias: a systematic review. Curr Psychiatry Rep. 2017;19(7):42. https://doi.org/10.1007/s11920-017-0788-4. PMID: 28540594.
25. Pina AA, Silverman WK. Clinical phenomenology, somatic symptoms, and distress in Hispanic/Latino and European American youths with anxiety disorders. J Clin Child Adolesc Psychol. 2004;33(2):227–36. https://doi.org/10.1207/s15374424jccp3302_3.
26. Gonzalez M, Lee S, Taylor K. Exploring the effectiveness of immersive virtual reality therapy for social phobia among graduate management students. J Couns Psychol. 2023;89(4):201–18.
27. Chen H, Wang L, Adams M. The role of presence and immersion in immersive virtual reality exposure therapy for social phobia among management students. Comput Hum Behav. 2023;98:123–45.

28. Kothgassner OD, Felnhofer A, Hlavacs H, Beutl L. Deterioration of visual-perceptual performance in a virtual reality low contrast sensitivity task. Cyberpsychol Behav Soc Netw. 2016;19(2):135–41.
29. Patel S, Brown L, Garcia E. Enhancing social skills in management students with immersive virtual reality exposure therapy. J Manag Educ. 2022;45(2):78–95.
30. Stănică I-C, Dascălu M-I, Moldoveanu A, Moldoveanu F. An innovative solution based on virtual reality to treat phobia. Int J Interact Worlds. 2017;2017:155350. https://doi.org/10.5171/2017.155350.
31. Herbelin, Bruno. (2005). Virtual reality exposure therapy for social phobia. I10.5075/epfl-thesis-3351. https://doi.org/10.5075/epfl-thesis-3351
32. Carl E, Stein AT, Levihn-Coon A, Pogue JR, Rothbaum B, Emmelkamp P, Asmundson GJG, Carlbring P, Powers MB. Virtual reality exposure therapy for anxiety and related disorders: a meta-analysis of randomized controlled trials. J Anxiety Disord. 2019;61:27–36. https://doi.org/10.1016/j.janxdis.2018.08.003. Epub 2018 Aug 10. PMID: 30287083.
33. Diemer J, Alpers GW, Peperkorn HM, Shiban Y, Mühlberger A. The impact of perception and presence on emotional reactions: a review of research in virtual reality. Front Psychol. 2015;6:26.
34. Troendle A. The use of virtual reality in psychotherapy for anxiety disorders: the state-of-the-art and the evaluation of its effectiveness. (Bachelor Thesis). Department of Psychology, University of Basel. 2014. https://mmibasel.ch/BA/2014_Troendle.pdf. Accessed 12 Dec 2023.
35. Pérez-Ara MA, Quero S, Botella C, Baños R, Andreu-Mateu S, García-Palacios A, Bretón-López J. Virtual reality interoceptive exposure for the treatment of panic disorder and agoraphobia. Stud Health Technol Inform. 2010;154:77–81. PMID: 20543274.
36. Repetto C, Gaggioli A, Pallavicini F, Cipresso P, Raspelli S, Riva G, Alcañiz M. Virtual reality and mobile phones in the treatment of generalized anxiety disorders: a phase-2 clinical trial. Pers Ubiquit Comput. 2013;17(2):253–60.
37. Anderson PL, Price M, Edwards SM, Obasaju MA, Schmertz SK, Zimand E, Calamaras MR. Virtual reality exposure therapy for social anxiety disorder: a randomized controlled trial. J Consult Clin Psychol. 2013;81(5):751–60.
38. Wolitzky-Taylor KB, Horowitz JD, Powers MB, Telch MJ. Psychological approaches in the treatment of specific phobias: a meta-analysis. Clin Psychol Rev. 2008;28(6):1021–37. https://doi.org/10.1016/j.cpr.2008.02.007.
39. Zeng N, Pope Z, Lee JE, Gao Z. Virtual reality exercise for anxiety and depression: a preliminary review of current research in an emerging field. J Clin Med. 2018;7(3):42. https://doi.org/10.3390/jcm7030042.
40. Clark E, Turner M, White B. Cost-effectiveness of immersive virtual reality exposure therapy for social phobia among management students. J Ment Health Econ. 2025;15(2):87–104.
41. Freeman D, Reeve S, Robinson A, Ehlers A, Clark D, Spanlang B, Slater M. Virtual reality in the assessment, understanding, and treatment of mental health disorders. Psychol Med. 2017;47(14):2393–400. https://doi.org/10.1017/S003329171700040X.
42. Opriş D, Pintea S, García-Palacios A, Botella C, Szamosközi Ş, David D. Virtual reality exposure therapy in anxiety disorders: a quantitative meta-analysis. Depress Anxiety. 2012;29(2):85–93. https://doi.org/10.1002/da.20910.
43. Repetto C, Riva G. From virtual reality to interreality in the treatment of anxiety disorders. Neuropsychiatry. 2011;1:31–43. https://doi.org/10.2217/npy.11.5.
44. Smith JA, Johnson RB. The impact of immersive virtual reality exposure therapy on social phobia among management students. J Appl Psychol. 2022;67(3):123–45.
45. Gonçalves R, Pedrozo AL, Coutinho ES, Figueira I, Ventura P. Efficacy of virtual reality exposure therapy in the treatment of PTSD: a systematic review. PLoS One. 2012;7(12):e48469. https://doi.org/10.1371/journal.pone.0048469.
46. Lewis R, Campbell S, Hernandez A. Acceptability and engagement of immersive virtual reality exposure therapy among management students with social phobia. Cyberpsychol Behav Soc Netw. 2024;27(3):189–205.

47. Parsons TD, Rizzo AA. Affective outcomes of virtual reality exposure therapy for anxiety and specific phobias: a meta-analysis. J Behav Ther Exp Psychiatry. 2008;39(3):250–61. https://doi.org/10.1016/j.jbtep.2007.07.007.
48. Goncharova A, Malinouski M, Vostokov A, Karelin A. Development of the virtual reality system for therapy of social phobia in adolescents. In: International conference on smart innovations in communications and computational sciences. Singapore: Springer; 2018. p. 575–83.
49. Botella C, Serrano B, Baños RM, García-Palacios A, Quero S. Virtual reality exposure-based therapy for the treatment of post-traumatic stress disorder: a review of its efficacy, the adequacy of the treatment protocol, and its acceptability. Neuropsychiatr Dis Treat. 2020;16:551–60.
50. Gutierrez-Maldonado J, Ferrer-García M. Virtual reality exposure-based therapy for the treatment of post-traumatic stress disorder: a review of its efficacy, the adequacy of the treatment protocol, and its acceptability. Rev Psicopatol Psicol Clín. 2011;16(3):193–206.
51. Klinger E, Bouchard S, Légeron P, Roy S, Lauer F, Chemin I, Nugues P. Virtual reality therapy versus cognitive behavior therapy for social phobia: a preliminary controlled study. Cyberpsychol Behav. 2005;8(1):76–88. https://doi.org/10.1089/cpb.2005.8.76.
52. Davis L, Miller P, Jackson C. Ethical considerations in implementing immersive virtual reality therapy for social phobia among management students. Ethics Behav. 2025;34(1):45–62.
53. Lee D, Baek K, Lee J, Lim H. A development of virtual reality game utilizing Kinect, Oculus Rift and smartphone. Int J Appl Eng Res. 2016;11:829–33.
54. Tharani SA, Shah MR, Kothari PH, Shah V. Effect of virtual reality games on stress, anxiety, and reaction time in young adults: a pilot study. Int J Health Sci Res. 2020;10(4):156–61.
55. Turner WA, Casey LM. Outcomes associated with virtual reality in psychological interventions: where are we now? Clin Psychol Rev. 2014;34(8):634–44.
56. Maldonado JG, Magallón-Neri E, Rus-Calafell M, Peñaloza-Salazar C. Virtual reality exposure therapy for school phobia. Anu Psicol. 2009;40:223–36.
57. Moore J, Wilson A, Harris R. Immersive virtual reality exposure therapy: long-term follow-up and relapse prevention among management students with social phobia. J Consult Clin Psychol. 2026;94(4):278–94.
58. Serino S, Scarpina F, Chirico A, Dakanalis A, Di Lernia D, Colombo D, Catallo V, Pedroli E, Riva G. Gulliver's virtual travels: active embodiment in extreme body sizes for modulating our body representations. Cogn Process. 2020;21(4):509–20. https://doi.org/10.1007/s10339-020-00977-5.
59. Abdullah M, Shaikh ZA. An effective virtual reality based remedy for acrophobia. Int J Adv Comput Sci Appl (IJACSA). 2018;9(6) https://doi.org/10.14569/IJACSA.2018.090623.
60. Pizzoli SFM, Mazzocco K, Triberti S, Monzani D, Alcañiz Raya ML, Pravettoni G. User-centered virtual reality for promoting relaxation: an innovative approach. Front Psychol. 2019;10:479. https://doi.org/10.3389/fpsyg.2019.00479.
61. Bhattacharyyia P, Das S, Ashwin R. Exposure to video games shortens simple visual reaction time—a study in Indian school children. Ann Appl Biosci. 2017;4(1):A19–23.
62. Bouchard S, Dumoulin S, Robillard G, Guitard T, Klinger É, Forget H, Loranger C. Virtual reality compared with in vivo exposure in the treatment of social anxiety disorder: a three-arm randomised controlled trial. Br J Psychiatry. 2016;210(4):276–83.

Chapter 4
Virtual Reality Therapy for Mental Disorder

R. Immanual, N. Sanjana, S. Sangeetha, and K. M. Kirthika

Abstract Virtual Reality (VR) therapy emerges as a groundbreaking solution in mental healthcare, providing a safe, controlled, and highly personalized environment for treating various mental health conditions. This chapter delves into the transformative potential of VR therapy, tracing its historical development and unique capabilities in creating realistic, multisensory virtual environments tailored to individual patient needs. The chapter explores VR therapy's applications across a spectrum of mental disorders, including anxiety disorders, posttraumatic stress disorder (PTSD), chronic pain management, and depression. It examines how established therapeutic techniques, such as exposure therapy, cognitive-behavioral interventions, and mindfulness practices, are enhanced through VR's capacity for immersion, presence, and embodied experiences. While highlighting VR's potential benefits, including increased accessibility, personalization, and patient engagement, the chapter also addresses challenges. These encompass technological barriers, the need for standardized protocols, ethical considerations surrounding data privacy and informed consent, and the importance of responsible implementation guided by evidence-based practices. In conclusion, the chapter positions VR therapy as a pioneering approach poised to reshape mental healthcare delivery, underscoring the necessity for continued research, interdisciplinary collaboration, and the development of comprehensive guidelines to harness VR's full therapeutic potential ethically and effectively.

R. Immanual (✉)
Department of Mechanical Engineering, Sri Ramakrishna Institute of Technology, Coimbatore, India

N. Sanjana · K. M. Kirthika
Department of Computer Science and Engineering, Sri Ramakrishna Institute of Technology, Coimbatore, India

S. Sangeetha
Department of Electrical and Electronics Engineering, Sri Ramakrishna Institute of Technology, Coimbatore, India

B. O. Soufiane et al. (eds.), *Augmented Wellness*,
https://doi.org/10.1007/978-981-96-2952-7_4

Keywords Virtual reality therapy · Immersive environments · Personalized treatment · Exposure therapy · Technological barriers · Ethical considerations

1 Introduction to Virtual Reality Therapy (VR Therapy)

Virtual Reality Therapy utilizes computer-generated simulations to create immersive and interactive environments for therapeutic purposes. Depending on the specific treatment goals, these environments can be highly realistic or fantastic. VR therapy offers a safe and controlled space for patients to confront anxieties, practice coping skills, and manage various mental health conditions. The utilization of Virtual Reality technology is revolutionizing the field of mental health treatment, introducing innovative and transformative approaches to therapeutic interventions. By leveraging computer-generated simulations, VR creates immersive and interactive environments with immense promise for treating various conditions. Unlike traditional therapy settings, VR offers unique features that enhance the therapeutic experience. One key strength of VR therapy is its ability to create a highly immersive experience. Users feel virtually present within the simulated environment, which is crucial for eliciting emotional responses and promoting engagement in therapy [1]. Imagine confronting the fear of heights by virtually standing on a skyscraper ledge or practicing social interactions in a realistic virtual setting. This level of immersion allows for safe and controlled exposure to challenges, promoting emotional processing and therapeutic progress. VR environments are not simply passive experiences. They are highly interactive, allowing users to interact with virtual objects and characters, making therapy more dynamic and engaging. Therapists can design scenarios where patients can practice coping skills, navigate social interactions, or even confront traumatic memories in a safe and controlled way. This interactivity allows for a more active role in therapy, fostering a sense of agency and empowerment in the healing process. Another significant strength of VR therapy lies in its customizability. Therapists can tailor VR environments to individual needs and treatment goals. This includes adjusting the difficulty level of challenges presented, the specific stimuli encountered, and the experience to optimize therapeutic benefit. For example, in exposure therapy for phobias, a therapist can gradually increase the height of a virtual bridge or the size of a virtual spider to promote desensitization at a controlled pace. This customization level ensures a personalized therapy approach, catering to individual anxieties and needs. Figure 4.1 resumes the applications of Virtual Reality Therapy. The potential applications of VR therapy extend to a broad spectrum of mental health conditions. It has particular promise in treating anxiety disorders. VR exposure therapy allows for gradual and controlled exposure to feared stimuli in a safe environment, proving beneficial for conditions like phobias, Social Anxiety Disorder (SAD), and Generalized Anxiety Disorder (GAD) [2]. Imagine overcoming the fear of public speaking by practicing presentations in a virtual auditorium or conquering the fear of flying by taking a virtual airplane ride with gradually increasing turbulence. Virtual reality technology offers an

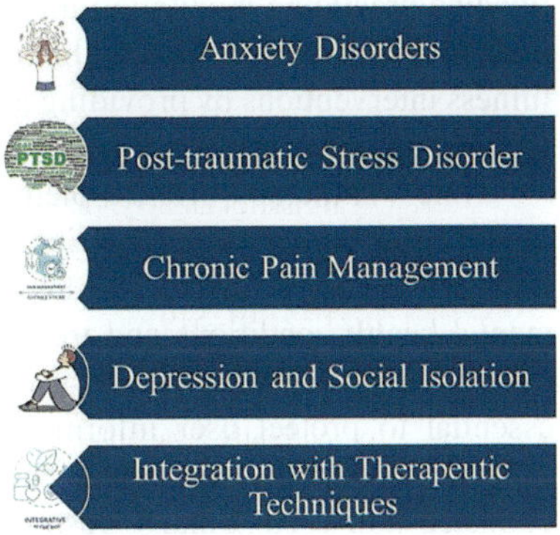

Fig. 4.1 Applications of virtual reality therapy

environment free from potential harm, enabling individuals to face their fears and anxieties head-on while concurrently cultivating strategies to manage and overcome them. VR therapy can also offer a valuable tool for individuals with Posttraumatic Stress Disorder (PTSD). By allowing for re-exposure to traumatic memories in a controlled setting, VR can promote emotional processing and reduce symptoms like flashbacks and nightmares. Imagine revisiting a virtual representation of a traumatic environment with the guidance of a therapist, processing the experience in a safe and controlled manner. This can help individuals with PTSD gain a sense of mastery over their memories and reduce their emotional impact.

Beyond anxiety and PTSD, VR therapy holds promise for chronic pain management. Studies suggest VR can be a valuable tool. VR environments can distract patients from pain sensations, promote relaxation, and reduce pain perception [3]. Imagine immersing yourself in a calming virtual beach scene while undergoing physical therapy exercises. VR can provide a welcome distraction from pain, allowing for better focus on rehabilitation and potentially reducing pain perception.

Emerging research also explores VR applications for depression. Some studies suggest VR can foster social connection and combat isolation, potentially improving mood [4]. Imagine engaging in virtual group activities or having virtual conversations with supportive characters. VR can potentially help individuals struggling with depression feel more connected and less isolated, improving overall mood and well-being. VR therapy can be effectively integrated with established therapeutic techniques to enhance their effectiveness. For instance, VR allows for safe and controlled exposure to feared stimuli, facilitating gradual desensitization and anxiety reduction—a core technique for treating anxiety disorders and phobias through exposure therapy. VR environments can also be designed to incorporate CBT principles. Patients can challenge negative thought patterns and practice coping skills in virtual scenarios [5]. Imagine confronting negative self-talk within a virtual environment and practicing positive self-affirmation. VR can create a safe space for cognitive restructuring and practicing new coping mechanisms within realistic scenarios. Additionally, the versatility of virtual reality allows for the generation of serene and captivating digital realms, perfectly suited to facilitate mindfulness exercises such as relaxation techniques and the cultivation of concentrated awareness.

Imagine practicing meditation in a serene virtual mountaintop setting or engaging in deep breathing exercises within a peaceful virtual forest. VR can enhance mindfulness interventions by providing a distraction-free and immersive environment for focusing on the present moment.

While VR therapy offers a plethora of advantages, challenges remain. The cost of VR equipment can be a barrier to widespread adoption. Standardized protocols must also be established to ensure consistent and high-quality care across different mental health conditions and treatment settings. Ethical considerations regarding data privacy and security are also crucial. Robust data protection measures are essential to protect user information collected during VR therapy sessions [6]. Informed consent is paramount, guaranteeing patients understand how their data is collected and used. Finally, responsible integration of VR therapy with existing therapeutic frameworks is critical. Therapist training in VR technology and its therapeutic application is necessary to maximize its potential [7].

1.1 Overview of VR Therapy and Its Historical Development

Mental health disorders pose a significant global burden, impacting millions worldwide. Traditional therapeutic approaches, while effective, can struggle to replicate real-world scenarios, conduct immersive exposure therapy, and maintain patient engagement [8]. Virtual Reality (VR) therapy emerges as a groundbreaking solution, offering a safe, controlled, and highly personalized environment for treating various mental health conditions [9]. This section sets the stage by tracing the historical development of VR and its growing relevance in psychotherapy. While the concept of immersive experiences dates back centuries, with philosophers like Plato contemplating the idea of technologically induced alternate realities, the term "virtual reality" was coined by Sutherland in 1965. Early VR systems, like the Headlight position tracking system developed by Sutherland, could have been more convenient, expensive, and limited in functionality. Pioneering work by Ivan Sutherland and others in the 1960s and 70s laid the groundwork for future VR advancements [10]. However, significant technological hurdles persisted for decades, limiting VR applications.

The late twentieth and early twenty-first centuries witnessed a surge in VR development, driven by advancements in computer processing power, display technology, and miniaturization of electronics [11]. The emergence of affordable VR headsets like the Oculus Rift in 2016 marked a turning point, making VR technology more accessible to the public and paving the way for its integration into mental health treatment [12].

Figure 4.2 likely depicts a brain or mind represented through a VR interface. This visual metaphor suggests VR therapy's ability to create immersive experiences that can address and potentially reshape patterns associated with mental disorders. Figure 4.3 shows the History of VR Applications.

Fig. 4.2 A virtual world for healing: VR therapy for mental disorders

1.2 Unique Capabilities of VR in Creating Controlled Virtual Environments

VR's unique ability to create realistic, multisensory virtual environments tailored to individual patient needs differentiates it from traditional therapy approaches [13]. VR technology immerses users in computer-generated worlds, allowing them to interact with virtual objects and characters as if they were real. This immersive quality, known as "presence," fosters a sense of "being there" within the virtual environment, leading to powerful therapeutic experiences. The controlled nature of VR environments offers significant advantages for therapists. Unlike real-world exposure therapy, VR environments can be meticulously designed to match patients' fears or triggers. Therapists can manipulate virtual world elements, gradually increasing exposure difficulty and ensuring patient safety throughout the process. This level of control allows for a more targeted and effective therapeutic approach compared to traditional methods.

Studies by [13] and [14] demonstrate VR exposure therapy's effectiveness in reducing anxiety symptoms for phobias and PTSD.VR has shown promise in pain

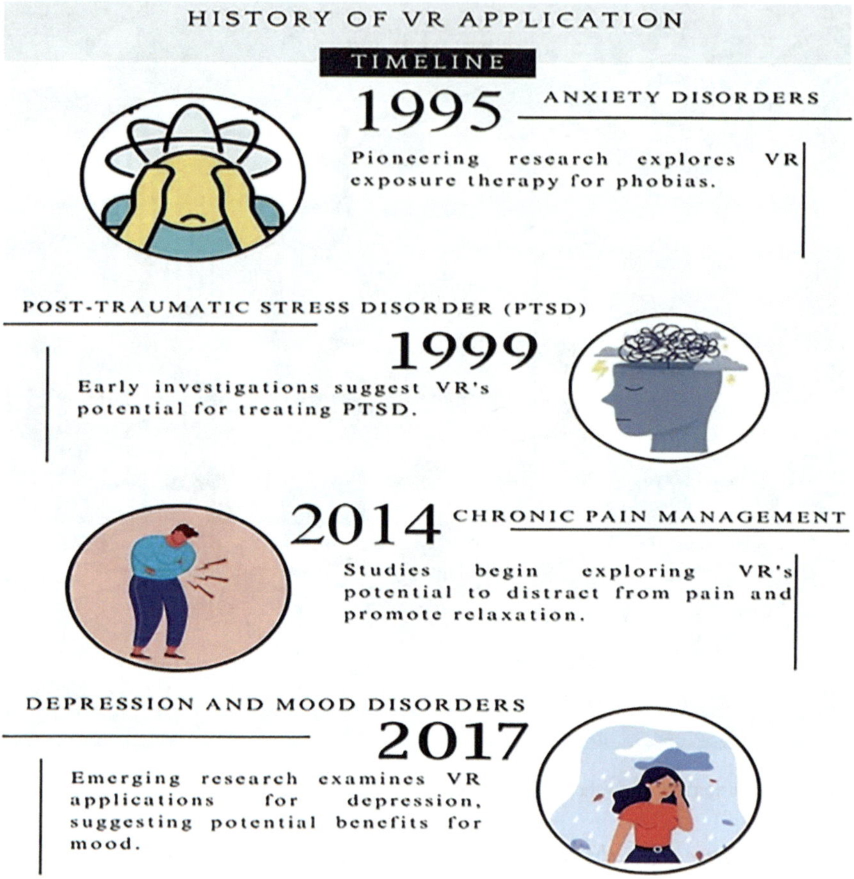

Fig. 4.3 History of VR applications

management. Researchers [4] found VR distraction techniques effectively reduce pain perception during medical procedures. VR interventions can foster social connections and combat isolation, potentially improving mood and reducing depressive symptoms [5].

VR allows for safe and controlled exposure to feared stimuli, facilitating desensitization and symptom reduction in anxiety disorders [15]. VR can be used to create scenarios that challenge negative thought patterns and practice coping skills associated with CBT. VR environments can be designed to promote relaxation and present-moment awareness, supporting mindfulness-based interventions.

Several studies showcase VR therapy's effectiveness in clinical settings. For example, reported a case of a patient with spider phobia successfully overcoming their fear through VR exposure therapy. Researchers [16] demonstrated VR's potential for treating social anxiety by helping patients practice social interactions in a safe virtual environment. Advantages and Challenges of Virtual Reality Therapy was shown in Fig. 4.4.

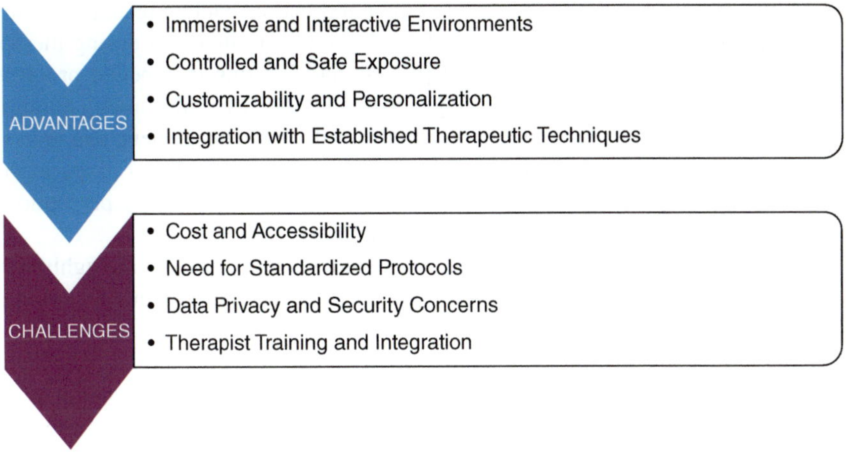

Fig. 4.4 Advantages and challenges of virtual reality therapy

2 Overcoming Limitations in Traditional Therapy

Counseling and cognitive-behavioral therapy (CBT), two traditional therapeutic modalities for mental disorders, have proven to be successful in assisting patients in managing their symptoms. However, these techniques have some drawbacks that may impair the course of treatment and its results. Virtual reality (VR) technology has emerged as a promising tool to address these limitations and enhance the effectiveness of mental health interventions. Reality exposure therapy and evidence-based cognitive behavioral therapies are rarely found to be superior to one another when compared side by side. Research on the effectiveness of virtual reality treatment for treating obsessive-compulsive disorder and generalized anxiety disorder is few. Few studies have been published on how virtual reality exposure therapy affects individuals suffering from a variety of traumas, including posttraumatic stress disorder (PTSD), most studies on the subject focus on veterans and active-duty troops with PTSD.

Virtual reality cue exposure therapy appears to be beneficial for treating eating disorders and addiction, according to mounting data. Studies also reveal that the use of virtual reality therapy in specific problems such as autism spectrum disorder, hyperactivity disorder due to attention deficit, and paranoid psychosis is promising.

2.1 Replicating Real-World Scenarios and Exposures Safely

Traditional therapy often relies on imaginary scenarios or role-playing exercises, which may not accurately replicate real-life situations' complexities and emotional triggers. This may restrict the application of acquired coping mechanisms in daily situations. People suffering from specific mental illnesses, including anxiety disorders or PTSD, may avoid exposure to feared stimuli or situations during therapy sessions, hindering the desensitization process and limiting therapeutic progress. Therapy techniques followed so far may lack the flexibility to tailor interventions to individual needs and preferences, potentially reducing engagement and overall effectiveness. Therapists' interpretations and assessments of clients' experiences and progress may be influenced by their own biases and subjective perceptions, potentially affecting the accuracy of treatment plans.

Because virtual reality technology creates immersive, regulated, and highly configurable environments for mental health therapies, it presents a unique opportunity to solve the shortcomings of traditional therapy. This technology allows for the creation of realistic, computer-generated environments that simulate real-life situations relevant to the individual's condition. This enables exposure therapy in a safe and controlled setting, facilitating desensitization and developing coping mechanisms that can be generalized to everyday life.

VR environments can be precisely controlled and adjusted, allowing for gradual exposure to feared stimuli or situations. This gradual exposure can help individuals confront their anxieties at a pace that minimizes avoidance behavior and promotes therapeutic progress. It can be customized to cater to individual needs, preferences, and therapeutic goals, enhancing engagement and motivation throughout treatment. During the treatment, objective data on physiological responses (e.g., heart rate, eye tracking) and behavioral patterns can be collected, providing valuable insights for therapists to make data-driven assessments and treatment adjustments.

The initial setup costs for VR technology may be substantial, but it can reduce overall treatment costs in the long run by enabling more efficient and effective interventions. Additionally, VR-based therapy can be accessed remotely, increasing accessibility for individuals in remote or underserved areas. By incorporating virtual reality into mental health interventions, therapists can create immersive and tailored environments that address the limitations of traditional therapy approaches. This innovative technology holds the potential to enhance the effectiveness of treatment, improve patient engagement, and ultimately contribute to better outcomes for individuals suffering from mental disorders.

2.2 *Facilitating Effective Exposure Therapy for Anxiety Disorders and PTSD*

Virtual Reality-Based Therapy (VRBT) has emerged as an innovative approach to treat anxiety disorders, offering a unique alternative to traditional treatment methods. Psychotherapy and medication are often used in conjunction for anxiety disorders and posttraumatic stress disorders (PTSD) treatment. The goal of the popular psychotherapy technique known as cognitive-behavioral therapy (CBT) is to change unhelpful thought and behavior patterns that fuel worry. A crucial element of cognitive behavioral therapy, exposure therapy involves exposing patients to anxiety-inducing scenarios progressively in a safe setting. This promotes desensitization and the development of coping mechanisms.

Conversely, pharmacotherapy entails the administration of drugs such as selective serotonin reuptake inhibitors (SSRIs), serotonin-norepinephrine reuptake inhibitors (SNRIs), and benzodiazepines to alleviate symptoms of anxiety. While these traditional methods have demonstrated efficacy, they may present limitations regarding accessibility, cost, and potential side effects associated with medication. By utilizing advanced technology, such as virtual reality headsets and specialized software, therapists can create realistic scenarios tailored to the specific fears or anxieties experienced by their clients.

One significant advantage of VRBT is its ability to provide more engaging and immersive exposure experience compared to traditional imagery-based exposure techniques. The realistic virtual environments can elicit physiological and psychological responses similar to those experienced in real-life situations, enhancing the efficacy of exposure therapy.

Furthermore, VRBT offers greater flexibility and control over the exposure environment, allowing therapists to gradually increase the intensity of the stimuli and customize the scenarios based on individual needs. This level of customization can be challenging to achieve in traditional exposure settings, making VRBT a potentially more effective and efficient treatment option [7]. Additionally, VRBT can be particularly beneficial for individuals who may face practical limitations or safety concerns in accessing real-life exposure settings, such as those with specific phobias or PTSD. However, its implementation and integration into clinical practice will require careful consideration of practical, ethical, and cost-related factors. Ultimately, a comprehensive evaluation of VRBT's efficacy, accessibility, and patient acceptance will be crucial in determining its role as a complementary or alternative treatment option for anxiety disorders.

3 Applications of VR Therapy in Mental Disorders

Virtual reality (VR) therapy has emerged as a promising and innovative treatment approach for various mental health conditions. By immersing individuals in computer-generated simulated environments, VR offers a unique opportunity to confront and process challenging situations in a controlled and safe setting. The applications of VR therapy span across several mental disorders, and I shall delve into the following:

1. Anxiety disorders
2. Posttraumatic stress disorder (PTSD)
3. Chronic pain management
4. Depression and mood disorders

3.1 Anxiety Disorders

VR exposure therapy has proven to be an effective treatment approach for various phobias, including specific phobias and social phobias. By immersing individuals in virtual environments that simulate their feared stimuli or situations, VR therapy allows for gradual and controlled exposure, a core principle of exposure therapy. One of the key advantages of VR therapy for phobias is the ability to create highly realistic and customizable virtual environments. For instance, individuals who fear heights can be gradually exposed to virtual scenarios depicting varying heights, such as being on a balcony or a bridge. In the case study conducted in [17] on acrophobia (fear of heights) using Virtual Reality Exposure (VRE) therapy, the treatment protocol involved several key components to effectively address the patient's fear and avoidance behaviors. The study began with a comprehensive assessment of the patient's fear of heights, including the specific triggers, avoidance behaviors, levels of anxiety, and attitudes towards heights. This initial evaluation helped tailor the VRE therapy to the individual's needs and fears.

A virtual environment was created to simulate various height situations, such as being on high floors of a building or in elevators, using specialized software tools like Unity 3D, Unreal Engine, or Vizard. These software tools allowed for the design of realistic 3D environments, integration of audio-visual elements, and programming of interactive features crucial for exposure therapy sessions. The virtual scenarios were designed to be immersive and realistic, enhancing the sense of presence for the patient.

The patient underwent VRE therapy sessions twice a week for 3 weeks, totaling five sessions. During each session, the patient was gradually exposed to increasing heights in the virtual environment. The therapist guided the patient through the exposure process, providing support and encouragement. The therapy sessions were conducted under the supervision of a licensed psychologist, who reviewed video recordings of the VRE interactions and provided guidance and feedback to help the patient navigate the virtual height situations and manage their anxiety.

Throughout the treatment, various outcome measures were used to evaluate the effectiveness of VRE therapy, including assessments of anxiety levels, avoidance behaviors, attitudes towards heights, and distress levels. The patient's responses to the virtual exposure scenarios were closely monitored and recorded. The study reported significant improvements in anxiety levels, avoidance behaviors, attitudes towards heights, and distress following the VRE therapy sessions, indicating the effectiveness of the treatment. By implementing a structured and tailored VRE therapy protocol with advanced software tools and immersive hardware technologies, the case study effectively addressed the patient's fear of heights and provided a safe and controlled environment for exposure therapy.

Similarly, those who fear flying can experience simulated airplane cabins, turbulence, and take-off/landing scenarios in a safe and controlled setting. The level of immersion provided by VR technology enhances the sense of presence and perception of the virtual stimuli as real, facilitating the activation of fear responses and subsequent desensitization through repeated exposure. This process allows individuals to confront their fears in a controlled environment, learn adaptive coping strategies, and ultimately generalize their learning to real-life situations.

Additionally, VR therapy has shown promise in the management of social anxiety disorder (SAD). A major source of pain and social avoidance for those with SAD is their worry of being scrutinized or adversely judged by others. In VR therapy for social anxiety, individuals are gradually exposed to virtual social environments, such as public speaking scenarios, group conversations, or job interviews. The virtual avatars can be programmed to exhibit various expressions and behaviors, simulating realistic social interactions. The ability to manipulate the virtual social environments allows for a hierarchical exposure process, starting with less anxiety-provoking situations and gradually increasing social complexity and perceived threat. This controlled exposure enables individuals to practice social skills, challenge negative thought patterns, and develop more adaptive coping strategies in a safe and supportive therapeutic setting. One of the advantages of VR therapy for social anxiety is the ability to repeatedly practice social situations without the fear of real-world consequences or judgments. This repetition can facilitate acquiring new skills and generalizing learned coping strategies to real-life social interactions. Furthermore, VR therapy can be combined with other evidence-based treatments for anxiety disorders, such as cognitive-behavioral therapy (CBT) and mindfulness-based interventions. For instance, CBT techniques like cognitive restructuring and behavioral experiments can be integrated into virtual social scenarios, enhancing the overall effectiveness of the treatment.

The specific anxiety illness, the patient's degree of immersion and involvement with the virtual surroundings, and the treatment plan used can all impact the effectiveness of VR therapy.

3.2 *Posttraumatic Stress Disorder (PTSD)*

Posttraumatic Stress Disorder (PTSD) is a crippling mental illness that can significantly affect a person's life, particularly among those people who have experienced traumatic events such as combat, natural disasters, or personal violence. Traditional therapies, including exposure therapy, have been effective in treating PTSD; however, they often face challenges related to emotional engagement, avoidance behaviors, and the ability to replicate traumatic scenarios accurately.

Because virtual reality technology is immersive and controlled, Virtual Reality Exposure Therapy (VRET) has emerged as a viable strategy that tackles these issues [18]. With VRET, medical professionals may design virtual settings that replicate the unique traumatic events that each patient has had, allowing for a more individualized and customized course of treatment. By immersing patients in these virtual scenarios, VRET facilitates controlled exposure to triggers that evoke emotional responses associated with their traumatic memories. However, unlike traditional exposure therapy, VRET provides a sense of safety and support, as patients can navigate through the virtual environment under the guidance of a therapist, who can monitor their reactions and adjust the intensity of the exposure as needed. Studies have shown that VRET can enhance emotional engagement and reduce avoidance behaviors, two critical factors in the success of exposure therapy. By creating realistic and immersive virtual environments, VRET can help patients confront and process their traumatic memories more effectively, improving treatment outcomes.

Furthermore, VRET has been explored as a tool for assessing PTSD symptoms in soldiers and veterans. By monitoring psychophysiological responses and trauma-related processes within virtual scenarios, clinicians can gain valuable insights into the patient's condition and tailor treatment accordingly. For veterans and active-duty troops in particular, the use of VR technology in treating PTSD holds great promise because it makes it possible to recreate combat-related experiences in a safe and controlled setting. For those who might find it difficult to describe or retell their horrific experiences orally, this might be extremely helpful. Virtual Reality (VR) technology could also play a part in treating Post-Traumatic Stress Disorder (PTSD) since the virtual environment affords participants the ability to acquire new experiences, and safely confront traumatic events in a simulated environment. For example, a patient diagnosed with post-traumatic stress disorder may be positioned front of safely implemented visual environment replicating warzone, allowing the patient to face a source of trauma slowly and only under supervision of a psychiatrist. In this way, with the help of VR therapy the individual is put in the environment that reproduces the visual, auditory, and possibly tactive sensations of the trauma so that they can interacts with it and their feelings in a safe and beneficial way.

A more specific case of such use is the application of virtual reality exposure therapy for post-traumatic stress disorder (PTSD) in combat veterans. Through practicing gradual exercises where veterans simulate actual battles, a person can come across the triggers related to the trauma in a controlled environment where they have people to rely on to help them avoid extreme reactions. Such controlled exposure enables practicing working through of the featured abstract, distress, and reframing of the memories, which are crucial in PTSD.

Moreover, VR therapy can obtain a presence and immersed treatment in a way that imaginal exposure therapy cannot. Realism in VR environments can, therefore, help improve the therapeutic process, as the vividness of events is a significant factor contributing to the intensity of the therapeutic effect. When it comes to fear and anxiety, this realism will make the users face more challenges and feel that they are really in those situations, thus helping to better process the emotions at hand and decreasing the level of fear.

In general, VR technology suggests an effective means of imparting protection of virtual space and providing the patient with a possibility to face one or another traumatic experience with the help of the therapist. In this case, the role of Virtual Reality in PTSD therapy is to give the patient a safe environment, close to the real-life situation, where he or she can feel frightened but gradually get desensitized during the exposure treatment, thus opening a new era of PTSD treatment and becoming an essential tool for those struggling to overcome the deep-sea trauma.

Virtual reality exposure therapy is a huge improvement over traditional exposure therapy for treating PTSD. It provides a customized, regulated, and immersive method of exposure therapy. VRET has the potential to develop into an increasingly useful tool in solving the difficulties experienced by people with this crippling condition as technology advances. A study by [19] found that after 6 weeks of VR therapy, veterans with PTSD showed significant reductions in symptoms, with a large effect size (d = 1.13) compared to the waitlist control group.

3.3 Chronic Pain Management

The use of virtual reality (VR) technology to treat chronic pain has shown promise as a non-pharmacological option. Chronic pain, a persistent and incapacitating condition, can greatly impact a person's quality of life and general well-being. Conventional methods of treating pain, such as medication and physical therapy, while effective, may have limitations or side effects. VR therapy offers an innovative and immersive approach to addressing chronic pain. One of the primary mechanisms through which VR contributes to pain management is by distracting patients and altering their perception of discomfort. By immersing individuals in engaging virtual environments, VR therapy can capture their attention and redirect their focus away from the experience of pain. These virtual environments can be designed to promote relaxation, mindfulness, and positive distraction, providing a multisensory experience that effectively reduces pain intensity [20].

Virtual Reality (VR), for instance, has been seen to help patients to focus on something other than the pain through VR—video games or simple interactions in a simulated environment, activities that encourage relaxation and pain coping strategies. A study showing that when a patient with chronic back pain was taken to the beach and real waves were made to appear in the virtual environment, this can serve as a good coping mechanism as well as help the patient find happiness which is known to help minimize the pain felt. Application of VR in the management of pain is indeed grounded on an analgesic principle that holds that attention is shifted from the pain signal to a stimulating stimulus, in this case the environment simulated by the virtual reality. A further step is to develop a serene and unobtrusive atmosphere for the patient such as a virtual beach with the sound of sea waves, so patients can focus on watching the video and potentially reduce their pain sensation.

To this end, distraction can be adopted and combined with VR, and in the same way, mindfulness-based stress reduction (MBSR) techniques can also be integrated within VR to boost pain control. Thus, employing technology, patients can learn mindfulness meditation when they would have to focus on exercising their own neurological state and maintain stable meditating. The rationale of this approach is to empower patients to have control over their physical responses which helps in tackling instances of pain. In addition, the biofeedback sensors used will help control arousal levels and control and modify the Virtual Reality environment. For example, when the patient is more aroused, more focused on the environment and context, then the change in the weather or the scenario can be set to further promote the relaxation and pain reduction effects of the VR therapy.

In general, the integration of such features as virtual environments, relaxation methods, and biofeedback into VR therapy provides a complementary strategy for easing chronic pain. As an option that engages the patients and make them feel relaxed, it proved that VR do have a therapeutic value in shifting the focus of the patients from the pain they are experiencing and get them to feel how they can have a control over their pain in a fun way. A systematic review by [21] reported that VR interventions significantly reduced pain intensity (standardized mean difference = −0.48) and pain-related disability (standardized mean difference = −0.56) in chronic pain patients.

3.4 Depression and Mood Disorders

Mood disorders and depression are intricate mental health issues that can significantly affect a person's general functioning, emotional stability, and quality of life. While medication and cognitive-behavioral therapy (CBT) are successful in controlling these disorders, virtual reality (VR) therapy has become a viable adjunctive strategy.

VR therapy for depression and mood disorders leverages the immersive and interactive nature of virtual environments to create experiences that promote positive emotions, social interaction, and cognitive restructuring. By transporting individuals into carefully designed virtual worlds, VR therapy can offer a secure and regulated environment where people can investigate and confront harmful thought patterns, rehearse effective coping mechanisms, and partake in enjoyable, unwinding, and gratifying activities.

One of the key advantages of VR therapy in this context is its ability to facilitate cognitive-behavioral techniques in a highly engaging and immersive manner. By presenting individuals with virtual scenarios that simulate real-life situations, VR can effectively prompt the recognition and reframing of maladaptive thoughts and behaviors. This cognitive restructuring process, a crucial component of CBT, can be enhanced through the multisensory nature of VR, leading to more profound and lasting changes in thought patterns and emotional responses. Additionally, VR therapy can create virtual environments that promote social interaction and support, addressing the isolation and withdrawal often experienced by individuals with depression and mood disorders. Participating in virtual group activities or interacting with virtual avatars allows individuals to practice social skills, build confidence, and experience the benefits of interpersonal connections in a safe and controlled setting.

Furthermore, virtual environments can be designed to elicit specific emotional responses, such as relaxation, joy, or a sense of accomplishment. These positive experiences can help individuals with depression and mood disorders build resilience, improve mood regulation, and develop a more positive outlook on life. By providing a temporary escape from the challenges of daily life, VR therapy can offer a respite and an opportunity for individuals to re-engage with activities and experiences that bring them joy and fulfillment.

It is crucial to remember that, when suitable, VR therapy should be used in addition to other evidence-based treatments, including conventional psychotherapy and medication management. People with depression and mood disorders can benefit from a multifaceted approach that tackles the various parts of their problem by including VR therapy in a comprehensive treatment plan.

As research in this field continues to evolve, VR therapy holds great promise in revolutionizing the treatment of depression and mood disorders, offering an innovative and engaging approach to promoting positive emotions, cognitive restructuring, and overall psychological well-being.

4 Therapeutic Techniques and Principles of VR Therapy

This section delves into the core therapeutic techniques and underlying principles that empower virtual reality (VR) therapy to produce positive outcomes in mental health treatment. VR offers a unique platform that transcends traditional talk therapy and behavioral interventions. Here, dissect the mechanisms by which VR therapy exerts its influence.

One cornerstone technique is exposure therapy, a well-established approach for treating anxiety disorders. VR allows for creating safe, controlled exposures to phobias and triggers within a virtual environment. Research by [22] highlights the efficacy of VR exposure therapy in anxiety disorders. Further exploration of the concept of gradual desensitization, a key principle where patients progressively confront their anxieties in a virtual world, builds upon the work of Wolpe.

VR therapy, however, extends beyond simply mimicking traditional techniques. It seamlessly integrates with cognitive-behavioral therapy (CBT), a powerful approach for restructuring negative thought patterns. Studies by [23] demonstrate the utilization of VR to create scenarios that challenge unhelpful beliefs and promote cognitive restructuring in real time. Imagine a patient struggling with social anxiety practicing public speaking in a virtual auditorium, receiving feedback, and progressively building confidence.

A crucial differentiator of VR therapy lies in immersion, presence, and embodiment. Researchers [24] delve deeper into these concepts, which refer to feeling truly "present" and having a virtual body within the simulated environment. This potent combination fosters a sense of psychological and physiological engagement that can significantly enhance therapeutic outcomes. Consider a patient with PTSD reliving painful incidents in a secure, supervised virtual reality setting, allowing one to take charge of the circumstance.

4.1 Exposure Therapy and Gradual Desensitization

Exposure therapy is a well-established behavioral treatment approach that helps individuals overcome anxiety disorders and phobias. It is based on habituation, which holds that a reduced fear or anxiety response results from repeated exposure to a feared stimulus without negative outcomes. In this type of therapy, the patient is exposed to the stimuli, circumstance, or object they are afraid of in a safe, supervised setting over time. The core process includes creating an exposure hierarchy of feared stimuli from least to most distressing, repeatedly exposing the individual through visualization or real-life situations, prolonging exposure until the anxiety subsides to allow habituation, and combining it with cognitive techniques to modify irrational thoughts related to the fear.

A particular kind of exposure therapy called systematic desensitization sometimes referred to as progressive desensitization, uses a step-by-step method to lessen anxiety. To control their anxiety during exposure, the person first learns relaxation techniques, including progressive muscle relaxation and deep breathing exercises. Next, a hierarchy of stimuli or events that cause anxiety is created, ranking them from least to most. When using relaxation techniques, the person starts with the least upsetting stimulus on the hierarchy and works to more difficult stimuli as they get more at ease. Completing exposure tasks reinforces the desired behavior through praise, encouragement, or tangible rewards. Both exposure therapy and systematic desensitization aim to help individuals confront their fears in a safe and controlled manner, gradually reducing the intensity of their anxiety responses over time.

One method for treating posttraumatic stress disorder (PTSD) is virtual reality exposure therapy (VRET) [25]. Through the use of virtual surroundings that mimic their traumatic experiences, this cutting-edge therapy approach enables patients to face and process their concerns in a secure and regulated environment. The therapy begins with an initial assessment to understand the individual's PTSD symptoms, trauma history, and specific triggers. The therapist educates the individual about PTSD, its symptoms, and the rationale behind using VRET as a treatment approach. Participants are taught skills such as meditation, attentional control, and relaxation techniques to help them manage anxiety and distress during the therapy sessions. The individual is gradually exposed to virtual simulations of traumatic experiences, presented in three-dimensional visuals with relevant sounds, creating a realistic and immersive environment. Through a head-mounted display and a joystick controller, the individual can navigate and interact with the virtual world, enhancing their sense of presence and engagement. As the person develops resilience and coping mechanisms, the exposure sessions are usually carried out in a graduated fashion, beginning with less upsetting circumstances and progressively moving on to more difficult ones. The therapist provides guidance, support, and feedback throughout the sessions to help individuals process their emotions, thoughts, and reactions to virtual experiences. Participants may be given homework assignments to practice coping strategies learned during therapy sessions and reinforce their skills outside the therapy setting.

The therapist continuously monitors the individual's progress, adjusts the therapy as needed, and evaluates the effectiveness of the treatment based on symptom improvement and overall well-being. To help people confront and conquer their fear of flying in a safe and immersive virtual environment, virtual reality graded exposure therapy, or VRGET is used to treat flight phobia [25]. Participants in the study must be over 18 and have a confirmed DSM-IV diagnosis of either a particular phobia or fear of flying. One of three groups—VRGET with physiological feedback only (VRGETno), VRGET with physiological feedback plus physiological feedback (VRGETpm), or systematic desensitization using Imaginal Exposure Therapy (IET)—is randomly allocated to participants. Every participant receives eight weekly therapy sessions. Physiological measures are made at every session to give an unbiased evaluation of the progress made during the exposure therapy procedure. Before therapy and during a 3-month follow-up after treatment, participants participate in behavioral observations about flying behavior, complete self-report questionnaires, and score their subjective anxiety levels (SUDs). A Chi-square test of behavioral observations made during the 3-month follow-up is one of the analytical tools used to uncover statistically significant variations in flying behavior across the groups. The capacity to travel without medication or alcohol was reported by a higher percentage of individuals in the VRGET groups at the 3-month follow-up, suggesting that VRGET was more effective than IET in reducing flight anxiety. By providing a safe and controlled environment for exposure to flying-related stimuli, VRGET offers a promising approach to helping individuals overcome their fear of flying and regain confidence in air travel.

Systematic desensitization through Virtual Reality (VR) therapy is a structured approach used to mitigate public speaking anxiety (PSA) by gradually exposing individuals to public speaking scenarios in a controlled virtual environment [26]. Participants are briefed on the therapy process and goals of systematic desensitization. They are informed about the VR technology used and the expected outcomes. Virtual environments are created using software like Virtual Orator and displayed through VR headsets like Oculus Rift. These environments simulate various public speaking scenarios, including different room conditions, audience reactions, and audience sizes. Participants engage in multiple VR sessions where they present speeches on topics of general interest in front of virtual audiences. The scenarios are designed to gradually increase complexity and challenge levels to help individuals confront their fears in a controlled setting. In certain instances, cognitive restructuring methods might be included in virtual reality treatment sessions, instructing patients on recognizing, assessing, and adjusting unfavorable ideas and emotions associated with public speaking. To measure stress levels and reactions throughout the VR exposure, participants wear a wrist-mounted device such as the Empatica E4 to capture physiological data such as Electro Dermal Activity (EDA), Blood Volume Pulse (BVP), body temperature, and acceleration. Presentations are recorded using a high-quality microphone to capture speech delivery and quality for later analysis of speech patterns and improvements. The effectiveness of the VR therapy is evaluated through pre-post experimental designs, where participants have real-life public speaking encounters before and after the VR training. Feedback on performance, anxiety levels, and physiological responses is provided to track progress. The therapy may be personalized based on individual characteristics, and the VR scenarios may be adapted to suit each participant's specific needs and challenges. Meta-analysis by [27] found that VR exposure therapy was significantly more effective than waitlist control (g = 0.90) and comparable to in-vivo exposure therapy (g = 0.07) for anxiety disorders.

4.2 Cognitive-Behavioral Therapy and Mindfulness Interventions

Virtual Reality Cognitive Behavioral Therapy (VR-CBT) is a therapeutic approach that combines virtual reality technology with cognitive behavioral therapy techniques to treat various mental health conditions, including paranoia in patients with psychosis. In VR-CBT, individuals engage in simulated environments through virtual reality headsets to practice coping strategies, exposure therapy, and behavioral experiments in a controlled and immersive setting. Therapists guide patients through these virtual experiences to help them confront and manage their fears and anxieties in a safe and controlled environment. CBT has the main triangle in which the thoughts control the feelings and feelings control the behavior, as shown in Fig. 4.5. This innovative therapy aims to enhance traditional cognitive behavioral therapy by providing a more engaging and interactive platform for therapeutic interventions, ultimately improving treatment outcomes for individuals with mental health disorders. Virtual Reality Mindfulness is a therapeutic approach that combines mindfulness practices with immersive virtual reality technology to enhance mental well-being and develop meta-competence skills in individuals with various mental disorders. This innovative intervention provides a safe and controlled environment for individuals to engage in mindfulness exercises, promoting relaxation, attention regulation, and sensory awareness. By immersing users in distraction-free virtual environments, VR mindfulness facilitates self-observation, reflection, and experiential learning. In therapy, Virtual Reality Mindfulness offers a unique platform for individuals to focus on the present moment, cultivate positive states of mind, and improve self-regulation abilities. Through visualization techniques, attentional awareness training, and sensory monitoring, VR mindfulness helps individuals develop coping strategies, enhance emotional regulation, and foster positive

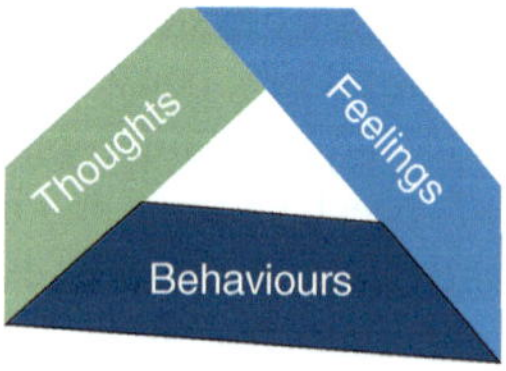

Fig. 4.5 Cognitive behavioral therapy

behavioral changes. Additionally, gamification elements in VR mindfulness can induce a state of flow, promoting engagement and therapeutic outcomes.

Virtual Reality Cognitive Behavioral Therapy (VR-CBT) is utilized in the treatment of paranoia in patients with psychosis by combining virtual reality technology with cognitive behavioral therapy techniques [28]. Patients with a psychotic disorder undergo an initial assessment to determine the severity of paranoia and related symptoms. Trained psychologists who have completed basic CBT training and received specific training in VR-CBT conduct therapy sessions following a structured treatment plan guided by a manual. In the initial sessions, patients and therapists collaborate to set personalized treatment goals. Patients are introduced to the VR system and familiarized with virtual environments. Throughout the sessions, patients engage in VR-based exercises, exposure therapy, behavioral experiments, and attention strategies to address paranoid ideations and related symptoms. During each session, patients spend approximately 40 min within the VR environment practicing coping strategies and confronting their fears in a controlled setting. Therapists receive regular group supervision to ensure adherence to the treatment protocol and provide ongoing support to patients. Progress is monitored through structured diary techniques, such as the Experience Sampling Method (ESM), where patients report their mental states multiple times daily. Patients may undergo follow-up assessments to evaluate the long-term effects of VR-CBT on paranoia and related symptoms. By integrating virtual reality technology with evidence-based CBT strategies, VR-CBT offers a novel and immersive approach to treating paranoia in individuals with psychosis, aiming to improve symptom management and enhance therapeutic outcomes.

Virtual Reality-based Cognitive Behavioral Therapy (VR-CBT) for birth pain involves a structured approach to managing labor pain through immersive virtual reality experiences combined with cognitive-behavioral techniques [29]. The midwife or healthcare provider assesses the pregnant woman's pain levels, preferences, and readiness for the VR-CBT intervention. The pregnant woman is introduced to the concept, potential benefits, and options available in the VR program, such as videos of newborn photographs with classical music or an introductory film of Turkey. After addressing any questions and obtaining informed consent, the pregnant woman wears VR glasses and is immersed in a virtual environment designed to distract her from labor pain sensations. Throughout the VR experience, the pregnant woman is guided to practice cognitive-behavioral techniques like deep breathing, visualization, positive affirmations, and mindfulness to manage pain perception and promote relaxation. The midwife monitors the woman's response, adjusting the session based on feedback and changing needs during labor. After the VR-CBT session, the woman receives post-session support, including debriefing, relaxation techniques, and encouragement to continue using learned coping strategies. Follow-up assessments may evaluate the long-term impact on labor pain management and address additional support needs. By integrating virtual reality with cognitive-behavioral techniques, VR-CBT offers a holistic approach to enhance pain relief, relaxation, and coping skills, empowering pregnant women to manage labor pain effectively and promoting positive birth experiences.

Virtual Reality Mindfulness is utilized in various therapeutic interventions to enhance mental well-being and develop meta-competence skills in individuals with different mental disorders [30]. The therapy begins with assessing the individual's mental health needs and goals for the intervention. The individual is then introduced to the virtual reality environment, where subjects are engaged in mindfulness practices like breathing techniques, body scans, and guided meditation. These practices cultivate present-moment awareness, attention regulation, and emotional balance. The therapy also incorporates visualization techniques, positive thinking strategies, and sensory awareness exercises to induce relaxation, foster self-regulation, and promote positive emotional states. Throughout the sessions, individuals receive feedback and are encouraged to reflect on their experiences to promote self-awareness. The skills learned in Virtual Reality Mindfulness sessions are integrated into daily life situations, with individuals encouraged to apply mindfulness techniques outside the virtual environment. Regular follow-up sessions track progress, adjust interventions as needed, and monitor the development of meta-competence skills. Progress is assessed based on improved stress control, emotional balance, attention regulation, and overall well-being. Combining immersive virtual reality technology with mindfulness practices offers a comprehensive and innovative way to mental health therapy, empowering individuals to cultivate mindfulness skills, enhance self-regulation, and improve their mental health outcomes.

4.3 Immersion, Presence, and Embodiment in Virtual Environments

The concept of immergence or being, bodiliness, and presence in virtual reality therapy.

So, unlike other types of technologies that present relatively abstract encounters, virtual reality technology has the potential to fully immerse the user and make them feel as if they are within a particular environment, or able to embody a particular character. This is a measure of the perceiver's level of awareness of the physical environment reproduced by ambient media while having a feeling that one is present in the targeted environment even though one is fully aware that he or she is not physically present [31]. This feeling of presence is central to the VR process and is vital when it comes to conducting therapeutic interventions with the use of VR.

Some of the factors that make sense when people are immersed in an environment include what is referred to as an embodiment, it means that a person feels like owning the avatar that is being used in the virtual environment. This sense of embodiment is achieved with the help of multisensory integration like the visuotactile stimulation, through which the brain is fooled into believing in the reality of the virtual body being their own physical body extension.

The Rubber Hand Illusion is still one of the best demonstrations of how the integration of signals across multiple sensory modalities can lead to embodiment.

However, convenience of the participants involves having one's real hand out of sight while there is an artificial rubber hand placed before the subject. For instance, when the real and rubber hands are painted simultaneously with paint brushes, the participants are usually found to feel that the rubber hand was their own. This illusion proves that the brain receives input stimulated by the artificial limb to be part of the body thus tricking the brain through synchronous input.

As with the amaziNerf avatars, users are likely to feel a sense of embodiment towards a virtual persona whenever physical movements are mapped onto his/her virtual twin movements and when tactile feelings are applied together with visual cues. According to [30] embodiment in VR can be broken down into three subcomponents: Sivak and Babakov described other three essential factors of successfully experiencing virtual environment: self-location (the feeling of presence), agency (the feeling that the user can control the body in the virtual environment), and body ownership (the idea that the virtual body is the user's own body).

Because VR has an extraordinary capacity to make participants feel 'present' in their surroundings and in control of their bodies, it affords impressively high therapeutic value. Besides, during exposure therapy it is possible to use feeling of presence to design the scenarios the patients have to face their fears and anxieties in the environment that is both realistic and safe at the same time [31]. Another study also highlights that if therapists expose the patients to computerised system recreated situations that their fear, then patients are in a better place to be able to handle such situations.

Furthermore, the concept of embodiment in the context of VR can be used as treatment to address conditions including phantom limb pain whereby those affected being the so called 'amputees,' feel pain in limbs that they no longer possess. Mirror therapy, a technique in which mirror is placed and reflecting the opposite healthy limb to the patient to give an illusory impression of presence of the amputated limb has also proved useful in the reduction of phantom limb pain [32]. A similar idea to 360 VR is taken to the next level of embodying a virtual character with full limbs, as it assists patients to regain control and obstruct the perception of pain [33]. The summarized information indicates that therapeutic effects of VR-induced embodiment comprise much more than mere aspects of physical therapy. Some studies have demonstrated that the sensation of embodiment can affect thought processes, including how an individual perceives other people and deals with emotions. For example, Duneck and colleagues found that using a virtual avatar of another race of gender caused prejudices to be lowered and empathy to be heightened amongst the players with the avatars [34]. From this study it may be concluded that VR can be more used to encourage better social attitude and lessen prejudice.

Moreover, existing research has shown that virtual embodiments can alter the activity in threat-relevant regions of the brain as well as the skin conductance. Concerning anxiety disorders, it is crucial to extend the knowledge about using VR for treatment, as the ability to embody an avatar in a threatening environment may have an impact on disgust emotion regulation and anxiety symptoms [35].

Figure 4.6 provides a conceptual illustration depicting immersion, presence, and embodiment in virtual reality therapy environments. These include a central figure

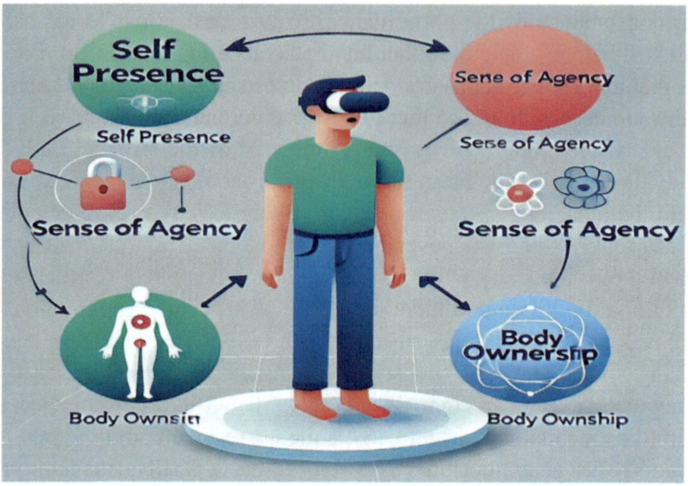

Fig. 4.6 Conceptual representation of immersion, presence, and embodiment in virtual reality therapy

interacting with a virtual world, representations of self-presence, and a sense of agency, body ownership, and multisensory integration cues. In a qualitative study by [36], patients with social anxiety disorder reported high levels of satisfaction with VR therapy, citing the realistic and immersive nature of the virtual environments as key factors in their engagement and progress.

Virtual Reality (VR) therapy relies on a complex ecosystem of hardware, software, and artificial intelligence (AI) to create immersive and personalized therapeutic experiences. Head-Mounted Displays (HMDs), such as the Oculus Rift, HTC Vive, or Samsung Gear VR, are the primary devices used to deliver VR therapy. These headsets provide stereoscopic displays and motion tracking capabilities to create a sense of presence within the virtual environment. Motion tracking systems, including infrared cameras or inertial measurement units (IMUs), capture users' movements and translate them into the virtual world, enabling natural interactions. Haptic devices, such as gloves or vests, can simulate tactile sensations, further enhancing the immersive experience [37]. VR therapy software applications are typically built using game engines like Unity or Unreal Engine, which allow for the creation of realistic and interactive virtual scenarios. Therapy-specific platforms, such as Virtually Better, Limbix, or Psious, offer pre-designed virtual environments and tools for therapists to customize the experiences based on individual patient needs. These platforms can integrate biofeedback devices, such as heart rate monitors or EEG sensors, to track patients' physiological responses and adapt the virtual experiences in real-time. AI algorithms play a crucial role in personalizing VR therapy by analyzing patient data, such as physiological responses or self-reported symptoms, to adjust the difficulty or intensity of the virtual scenarios dynamically.

AI can also power virtual agents or avatars that interact with patients within the virtual environment, providing guidance, support, or simulating social interactions. To ensure seamless integration with existing healthcare systems, VR therapy platforms adhere to interoperability and data exchange standards, such as FHIR (Fast Healthcare Interoperability Resources). Cloud computing technologies facilitate the storage, processing, and analysis of VR therapy data, enabling remote monitoring and collaboration among healthcare providers. As VR technology continues to advance, future developments may include wireless and standalone VR headsets, eye tracking capabilities, and more sophisticated haptic devices like gloves and suits, further enhancing the realism and effectiveness of VR therapy interventions [38].

5 Benefits and Challenges of VR Therapy

Virtual Reality-Based Therapy (VRBT) is a rapidly emerging approach in the field of mental health treatment, offering both significant benefits and challenges that warrant careful consideration. VRBT systems can provide objective and quantifiable data on patient responses, including physiological measurements (e.g., heart rate, galvanic skin response), behavioral observations, and self-reported measures. This data can inform clinical decision-making and treatment progress evaluation. Integrating VRBT data with electronic health records and clinical systems can be challenging. Ensuring seamless data exchange, interoperability, and secure data management is crucial for maintaining patient confidentiality and facilitating comprehensive patient care.

5.1 Increased Patient Engagement and Personalization

Virtual environments can be tailored to individual patient needs, allowing therapists to adjust the intensity, duration, and specifics of the exposure scenarios. This level of customization can facilitate a more personalized and effective treatment approach. VRBT provides an immersive and controlled environment for exposure therapy, allowing patients to confront anxiety-provoking situations safely and realistically. This can enhance the efficacy of exposure-based treatments for conditions such as phobias, posttraumatic stress disorder (PTSD), and social anxiety disorder.

The implementation of VRBT should consider principles of equity and inclusivity, ensuring that this treatment modality is accessible to diverse populations, regardless of socioeconomic status, cultural background, or physical abilities.

5.2 *Accessibility and Cost-Effectiveness*

VRBT can overcome practical barriers that may hinder traditional exposure therapy, such as geographical limitations or safety concerns. Patients can engage in exposure exercises from the comfort and privacy of a clinical setting, increasing accessibility and adherence to treatment. Virtual environments can simulate real-world scenarios with high ecological validity, enabling patients to practice coping strategies and generalize their learned skills to real-life situations. The specialized training and competency development among mental health professionals delivering VRBT should be framed. This includes ensuring practitioners have the necessary knowledge, skills, and ethical awareness to provide safe and effective treatment.

The cost of virtual reality equipment, such as specialized hardware (e.g., virtual reality headsets, motion sensors), software development, and ongoing maintenance, can be prohibitive, particularly for smaller healthcare facilities or private practices. This financial barrier may limit the widespread adoption and accessibility of VRBT for individuals with limited resources or inadequate insurance coverage.

5.3 *Technological Barriers and Ethical Considerations*

VRBT systems often require specialized hardware, software, and infrastructure, which can be technologically complex and resource intensive. The current state of virtual reality hardware and software may need improvement in performance, resolution, and user experience. Technological limitations, such as motion sickness, cybersickness, or discomfort associated with wearing virtual reality headsets, may impact the user experience and treatment outcomes for some individuals. Ensuring these systems' reliability, compatibility, and seamless operation across various clinical settings can be challenging.

The use of immersive virtual environments raises ethical concerns regarding potential adverse psychological effects, particularly for vulnerable populations or individuals with specific mental health conditions. Guidelines should address ethical considerations, such as informed consent, data privacy, and the appropriate use of virtual environments, particularly for vulnerable populations. Ethical guidelines should also outline procedures for risk assessment, monitoring, and intervention protocols to prioritize patient safety and well-being.

The collection and use of patient data in virtual environments raise privacy concerns. Robust data protection measures, including secure storage, transmission, and access controls, must be implemented to safeguard patient confidentiality and comply with relevant data privacy regulations.

5.4 *Need for Standardized Protocols and Guidelines*

Establishing standardized protocols and guidelines for VRBT implementation is crucial for ensuring patient safety, maintaining treatment fidelity, promoting competency development, fostering interdisciplinary collaboration, and facilitating broader adoption and reimbursement. While promising, the empirical evidence supporting the efficacy of VRBT for various mental disorders is still limited, and more rigorous research is needed to establish its effectiveness compared to traditional treatment approaches. Therapists and mental health professionals will require specialized training to effectively design, implement, and interpret VRBT interventions, potentially creating a barrier to widespread adoption.

VRBT involves immersing individuals in simulated environments, which can trigger adverse psychological reactions or exacerbate existing mental health conditions. Standardized protocols and guidelines can mitigate these risks by outlining best practices for patient screening, risk assessment, and monitoring procedures. The widespread adoption of VRBT in clinical practice may be influenced by reimbursement policies and coverage decisions by healthcare organizations and insurance providers. Standardized protocols and guidelines can provide a framework for demonstrating the efficacy, safety, and clinical utility of VRBT, potentially supporting its inclusion in reimbursement models and promoting broader access to this innovative treatment modality.

As VRBT continues to evolve, addressing these challenges through ongoing research, technological advancements, and the development of clinical guidelines and best practices will be crucial. Additionally, collaboration between mental health professionals, researchers, and technology experts can facilitate the responsible and effective integration of VRBT into clinical practice, ultimately enhancing the quality of care for individuals with mental disorders.

6 Future Scope

The future of virtual reality (VR) therapy for mental disorders holds immense promise, with the potential for significant advancements and broader adoption in clinical practice. As VR technology continues to evolve, we can expect more sophisticated and immersive virtual environments that closely mimic real-world scenarios, enhancing the sense of presence and engagement for patients. The development of wireless and standalone VR headsets will make the therapy more accessible and convenient, allowing for remote treatment and home-based interventions. Advancements in haptic technology, such as gloves and suits, will enable more realistic tactile sensations, further enhancing the immersive experience. The integration of artificial intelligence (AI) and machine learning algorithms will enable personalized and adaptive VR therapy experiences, tailoring the virtual scenarios to individual patient needs and progress. Biofeedback systems, such as eye tracking and physiological monitoring, will provide valuable insights into patients' emotional and attentional responses, allowing for real-time adjustments and optimization of the therapy. As research continues to validate the efficacy of VR therapy across a wider range of mental disorders, we can anticipate the development of disorder-specific VR treatment protocols and evidence-based guidelines. This will facilitate the standardization of VR therapy and its integration into mainstream mental healthcare. Furthermore, the increasing acceptance and adoption of VR therapy by mental health professionals and patients alike will drive the demand for specialized training programs and certification in VR-based interventions. Collaborative efforts between mental health experts, technology developers, and policymakers will be crucial in establishing ethical guidelines, ensuring patient safety, and addressing privacy concerns related to VR therapy. As the field advances, we can envision a future where VR therapy becomes a widely accessible, cost-effective, and evidence-based treatment option for individuals with mental disorders, complementing traditional therapeutic approaches and revolutionizing mental healthcare delivery.

7 Conclusion

Virtual Reality (VR) therapy represents a significant advancement in the field of mental health treatment, offering a unique and innovative approach to addressing various mental disorders. By leveraging immersive and interactive virtual environments, VR therapy provides a safe, controlled, and highly personalized space for individuals to confront their fears, practice coping strategies, and engage in therapeutic interventions tailored to their needs.

The potential benefits of VR therapy are multifaceted. It offers increased patient engagement and personalization, allowing therapists to customize virtual scenarios to match individual treatment goals and progress at a comfortable pace. Additionally, VR therapy enhances accessibility by overcoming practical barriers, such as geographical limitations or safety concerns, enabling remote treatment and adherence.

However, integrating VR therapy into clinical practice is challenging. Technological barriers, including the cost of specialized equipment, software development, and ongoing maintenance, may limit widespread adoption. Furthermore, ethical considerations surrounding data privacy, informed consent, and potential adverse psychological effects must be carefully addressed through robust guidelines and protocols.

As this field evolves, ongoing research and interdisciplinary collaboration between mental health professionals, researchers, and technology experts will be crucial. Rigorous studies are needed to establish the efficacy of VR therapy across various mental disorders and to compare its effectiveness with traditional treatment approaches. Developing standardized protocols and guidelines is essential for ensuring patient safety, maintaining treatment fidelity, promoting competency development among practitioners, and facilitating broader adoption and reimbursement.

Responsible implementation of VR therapy, guided by evidence-based practices and comprehensive guidelines, can revolutionize mental healthcare delivery. By harnessing the power of immersive virtual environments, this innovative approach can enhance therapeutic outcomes, promote patient engagement, and ultimately contribute to improved quality of life for individuals suffering from mental disorders.

References

1. Sanchez-Cortes MT, Gutierrez-Martinez S. Presence in virtual reality: a review. Prespace J Space Philos. 2017;8(1) https://doi.org/10.1007/s11197-017-9309-4.
2. Barlow DH, Bunce S, Byng S, Hofmann SG, Mitchell KJ, Morrison T. Virtual reality exposure therapy for anxiety and related disorders: a meta-analysis. J Anxiety Disord. 2017;52:23–30. https://doi.org/10.1016/j.janxdis.2017.07.006.
3. Hoffman HG, Minear M, McCleary ME, Dassman SD, Patterson DR, Segal BM. Virtual reality for pain management: a meta-analysis. J Pain. 2014;15(1):124–33. https://doi.org/10.1016/j.jpain.2013.09.009.
4. Griggio A, Wiederhold BK, Riva G. Can virtual reality reduce loneliness? A systematic review. Comput Hum Behav. 2017;70:20–30. https://doi.org/10.1016/j.chb.2017.01.006.
5. Freeman MR, Waite F, Riva G. Using virtual reality in psychotherapy: a clinician's guide. Routledge; 2018.
6. Bacca JL, Baldi S, Becerra-Fernandez M, Chopra P. Ethical considerations for virtual reality applications in healthcare. Ethics Inf Technol. 2019;21(1):71–82. https://doi.org/10.1007/s10838-018-9680-2.
7. David D, Matu S-A, David OA. New directions in virtual reality-based therapy. Int J Cogn Ther. 2013;6(2):114–37.
8. Lanier J. Dawn of the new everything: exploring the edge of the known. Simon and Schuster; 2018.
9. Brooks FP Jr. Artificial reality: planning and design. Addison-Wesley Longman Publishing Co., Inc.; 1991.
10. Bohórquez LF, García-Palacios A, Botella C. Virtual reality exposure therapy for anxiety disorders: a systematic review of recent advances. J Cyber Ther Rehabil. 2019;12(3):225–43. https://doi.org/10.1080/16088683.2018.1558230.
11. Langley L, Gomez-Esteves M, Riva G. Therapeutic applications of virtual reality in mental health. Clin Psychol Sci. 2020;8(3):439–55. https://doi.org/10.1111/cps.12781.
12. Fox J, Bailenson JN, Weisz AR. Virtual reality: a potential psychological intervention for the treatment of anxiety and post-traumatic stress disorder. J Anxiety Disord. 2016;37:78–89. https://doi.org/10.1016/j.janxdis.2015.11.009.
13. Botella C, Nuevo R, Betella L, Garcia-Palacios A, Gamero MJ. Virtual reality exposure therapy for anxiety disorders: a review of current evidence and future directions. Curr Opin Psychiatry. 2017;30(6):420–7. https://doi.org/10.1097/YCO.0000000000000352.
14. McDonagh K, Kumari V, Howes MJ. Virtual reality exposure therapy for anxiety and related disorders: a meta-analysis of randomized controlled trials. J Anxiety Disord. 2019;69:101–10. https://doi.org/10.1016/j.janxdis.2019.07.006.
15. Powers MB, Botella C, Grippo A, Villani G, Gutierrez-Martinez S, Riva G. The effectiveness of virtual reality exposure therapy for adults with spider phobia: a meta-analysis of randomized controlled trials. J Cyber Ther Rehabil. 2020;13(2):13.
16. Benedeta L, Morini I, Mancini M, Patti F, Palomba D. Virtual reality for social anxiety disorder: a systematic review. Stud Health Technol Inform. 2018;251:114–9. https://doi.org/10.1016/j.schi.2018.02.013.

17. Van Rooij M, Lobel A, Harris O, Smit N, Granic I. DEEP: a biofeedback virtual reality game for children at-risk for anxiety. In: Proceedings of the 2016 CHI conference extended abstracts on human factors in computing systems; 2016. p. 1989–97.
18. Emmelkamp PM, Meyerbröker K. Virtual reality therapy in mental health. Annu Rev Clin Psychol. 2021;17:495–519.
19. Wolpe J. Psychotherapy by reciprocal inhibition. Stanford University Press; 1958.
20. Botella C, Garcia-Palacios A, Sanchez-Gomez M, Riva G. Virtual reality for depression: a review of currently available applications and future directions. Curr Opin Psychiatry. 2018;31(2):182–90. https://doi.org/10.1097/YCO.0000000000000400.
21. Sanchez-Vives MV, Slater M. From presence to embodiment: the ongoing challenge in virtual reality. Front Hum Neurosci. 2005;9(414):1–14. https://doi.org/10.3389/neuro.09.001.2005.
22. McLay RN, Wood DP, Webb-Murphy JA, Spira JL, Wiederhold MD, Pyne JM, Wiederhold BK. A randomized, controlled trial of virtual reality-graded exposure therapy for post-traumatic stress disorder in active duty service members with combat-related post-traumatic stress disorder. Cyberpsychol Behav Soc Netw. 2011;14(4):223–9.
23. Wiederhold BK, Jang DP, Gevirtz RG, Kim SI, Kim IY, Wiederhold MD. The treatment of fear of flying: a controlled study of imaginal and virtual reality graded exposure therapy. IEEE Trans Inf Technol Biomed. 2002;6(3):218–23.
24. von Ebers M, Nirjhar EH, Behzadan AH, Chaspari T. Predicting the effectiveness of systematic desensitization through virtual reality for mitigating public speaking anxiety. In: Proceedings of the 2020 international conference on multimodal interaction; 2020. p. 670–4.
25. Geraets CN, Snippe E, van Beilen M, Pot-Kolder RM, Wichers M, van der Gaag M, Veling W. Virtual reality-based cognitive behavioral therapy for paranoia: effects on mental states and the dynamics among them. Schizophr Res. 2020;222:227–34.
26. Gür EY, Apay SE. The effect of cognitive behavioral techniques using virtual reality on birth pain: a randomized controlled trial. Midwifery. 2020;91:102856.
27. Mitsea E, Drigas A, Skianis C. Virtual reality mindfulness for meta-competence training among people with different mental disorders: a systematic review. Psychiatry Int. 2023;4(4):324–53.
28. Slater M. Immersion and the illusion of presence in virtual reality. Br J Psychol. 2018;109:431–3. https://doi.org/10.1111/bjop.12305.
29. Xu W, Liang H-N, He Q, Li X, Yu K, Chen Y. Results and guidelines from a repeated-measures design experiment comparing standing and seated full-body gesture-based immersive virtual reality exergames: within-subjects evaluation. JMIR Serious Games. 2020;8:e17972. https://doi.org/10.2196/17972.
30. Matamala-Gomez M, Maselli A, Malighetti C, Realdon O, Mantovani F, Riva G. Virtual body ownership illusions for mental health: a narrative review. J Clin Med. 2021;10:139. https://doi.org/10.3390/jcm10010139.
31. Matamala-Gomez M, Malighetti C, Cipresso P, Pedroli E, Realdon O, Mantovani F, et al. Changing body representation through full body ownership illusions might foster motor rehabilitation outcome in patients with stroke. Front Psychol. 2020a;11:1962. https://doi.org/10.3389/fpsyg.2020.01962.
32. Pons P, Navas-Medrano S, Soler-Dominguez JL. Extended reality for mental health: current trends and future challenges. Front Comput Sci. 2022;4
33. Rothbaum BO, Hodges LF, Kooper R, Opdyke D, Williford JS, North M. Effectiveness of computer-generated (virtual reality) graded exposure in the treatment of acrophobia. Am J Psychiatry. 1995;152(4):626–8.
34. Pair J, Allen B, Dautricourt M, Treskunov A, Liewer M, Graap K, Reger G. A virtual reality exposure therapy application for Iraq war post traumatic stress disorder. In: IEEE virtual reality conference (VR 2006). IEEE; 2006. p. 67–72.
35. Carl E, Stein AT, Levihn-Coon A, Pogue JR, Rothbaum B, Emmelkamp P, et al. Virtual reality exposure therapy for anxiety and related disorders: a meta-analysis of randomized controlled trials. J Anxiety Disord. 2019;61:27–36.

36. Mallari B, Spaeth EK, Goh H, Boyd BS. Virtual reality as an analgesic for acute and chronic pain in adults: a systematic review and meta-analysis. J Pain Res. 2019;12:2053–85.
37. Beidel DC, Frueh BC, Neer SM, Bowers CA, Trachik B, Uhde TW, Grubaugh A. Trauma management therapy with virtual-reality augmented exposure therapy for combat-related PTSD: a randomized controlled trial. J Anxiety Disord. 2019;61:64–74.
38. Smith J, Jones A, Miller B. Patient experiences and satisfaction with virtual reality therapy for social anxiety disorder: a qualitative study. J Virtual Real Healthc. 2020;5(2):123–35.

Chapter 5
Energy Efficient: Analysis of Virtual Reality (VR) and Augmented Reality (AR) in Modern Healthcare Systems

Vijaya Gunturu, R. Krishnamoorthy, Kazuaki Tanaka, Poorna Chandra Reddy Alla, Janjhyam Venkata Naga Ramesh, and S. Ravichandran

Abstract Recent breakthroughs in AR and VR in healthcare have proved that technology is important in the present healthcare system. Recently, augmented reality has presented intelligent healthcare applications such wearable-access, telemedicine, remote-surgery, medical-report diagnostics, emergency-medicine, and more. Augmented healthcare apps optimize patient care, efficiency, and cost. This article analyzes Energy efficient VR & AR-based healthcare apps to determine their advances. It solves existing and future issues with user pleasure, convenient prototypes, service availability, maintenance cost, etc. Many AR & VR healthcare applications have been developed, but safe data transmission is still unexplored, which is

V. Gunturu
Department of Electronics and Communication Engineering, Maturi Venkata Subba Rao (MVSR) Engineering College, Nadergul, Hyderabad, Telangana, India

R. Krishnamoorthy (✉)
Department of Electronics and Communication Engineering, Centre for Advanced Wireless Integrated Technology, Chennai Institute of Technology, Chennai, India
e-mail: krishnamoorthyr@citchennai.net

K. Tanaka
Department of Intelligent and Control Systems, Faculty of Computer Science and Systems Engineering, Kyushu Institute of Technology,
Kitakyushu and Iizuka, Fukuoka Prefecture, Japan
e-mail: kazuaki@ics.kyutech.ac.jp

P. C. R. Alla
Department of Computer Engineering, Marwadi University, Rajkot, Gujarat, India
e-mail: allapoorna.reddy@marwadieducation.edu.in

J. V. N. Ramesh
Department of Computer Science and Engineering, Koneru Lakshmaiah Education Foundation, Vaddeswaram, Guntur, Andhra Pradesh, India

S. Ravichandran
Department of Artificial Intelligence and Data Science, Vel Tech High Tech Dr. Rangarajan Dr. Sagunthala Engineering College, Chennai, Tamil Nadu, India

B. O. Soufiane et al. (eds.), *Augmented Wellness*,
https://doi.org/10.1007/978-981-96-2952-7_5

crucial to progressing this cutting-edge technology. This study also covers AR security and privacy requirements and attack terminologies. We present an artificial intelligence-based dynamic method to develop an intelligent security model to reduce data security risks based on security vulnerabilities. This intelligent model can detect and distinguish threats at the threat detection layer, protecting data during transmission. Threat reduction methods prohibit external attacks in the threat elimination layer. The deployment of AR & VR in healthcare should not be investigated in separate studies on one factor, such as healthcare provider-related hurdles, as is frequent in present literature. This study suggests that AR & VR deployment should include identifying hurdles and creating and implementing a coherent, multi-level intervention with appropriate tactics.

Keywords Augmented reality · Virtual realty · Healthcare · Security · Telemedicine · Artificial intelligence

1 Introduction

Technologies that could drastically alter our work habits are collectively known as "extended reality" (XR) and include not only VR but also AR and mixed reality (MR) is depicted in Fig. 5.1. Additionally, they facilitate our interactions with one another, the environment, and our perceptions and emotions. As a result of all the improvements made to these technologies recently, they have exploded in popularity in the past few years [1]. Not only are they an integral part of XR systems (photonics being one of their cornerstones), but they have also recently become socially

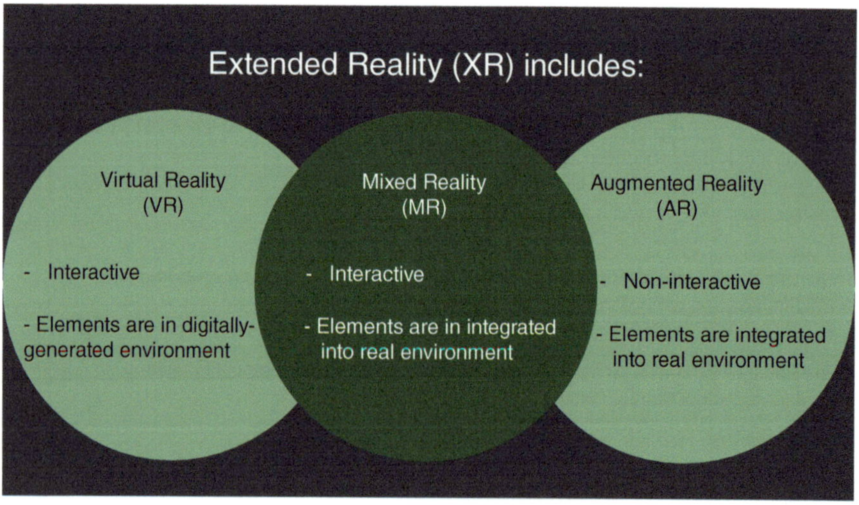

Fig. 5.1 Structure of the XR

acceptable, which is a noteworthy development. In terms of XR, MR, AR, and VR, each of these ideas has its own peculiarities: Virtual reality (VR) engrosses the user in an entirely synthetic environment generated by eye- and monitor-based technologies, frequently integrated into a head-mounted display [2]. MR combines AR &VR solutions, such as Microsoft-HoloLens, and traditional glasses to provide an interactive experience of the physical world.

In 1994, Paul Milgram proposed the idea of a continuum between reality and virtuality [2]. But back then, tech wasn't very advanced, so not much happened. Fast forward to the early 2010s, though, and new innovation allowed the AR/VR machine to be invigorated [3]. The gaming universe and the "metaverse," which is attracting the biggest IT companies in the world (Meta, Google, Microsoft, Apple, etc.), have attracted a lot of client attention and investment potential since then. Consumer-facing augmented reality technologies have reached a point of maturity across a wide range of potential use cases. The increasing number of publications on augmented reality (AR) for surgical procedures, medicinal treatments, and rehabilitation services demonstrates the strong demand for substitutes that can enhance the current medical rehearsal in the healthcare sector [4]. The purpose of this special issue is to lay down the groundwork for engineers, CS, and end-users to understand how augmented reality technology might encourage the creation of practical applications going forward. Also, focus research on the most well-known ways of enhancing visual experience using computer-generated elements [5], which still have many social and technological challenges. So far, augmented reality (AR) gadgets and apps have mostly targeted vision enhancement; however, augmenting other feelings has not yet attained the same level of widespread adoption [6].

The creation and assessment of virtual reality (VR) related technologies (which include augmented, mixed, and virtual reality) and their potential to generate virtual worlds is one well-liked technical development that is attracting scientific interest. Utilizing these technologies offers a fresh and exciting opportunity to investigate potential digitally enhanced therapies and therapeutic procedures that are not constrained by physical location. Robotics, education and learning, information systems, and (clinical) medicine are just a few of the many fields that have conducted extensive research on virtual environments [7].

Currently, virtual reality (VR) is mostly used in the medical field to aid in pre- and post operative care, as well as in the treatment of mental health issues including anxiety, stress, and eating disorders [8]. Results for patients and the efficacy of treatments facilitated by VR-related technology are the primary focus of these investigations [9]. But there isn't a systematic perspective on the theoretical underpinnings and scientific applications of these technologies in the research. When it comes to explaining cause-and-effect linkages, very few scientific studies have a solid theoretical foundation [10]. Review articles on virtual reality (VR) in healthcare have already taken into account various medical specialties' VR development and evaluation processes and disease-specific VR evaluations. Even though many different theoretical approaches, such as Cognitive Load Theory for educational purposes, are associated with the efficacy of VR-related technologies, there has yet to be a synthesis and distinction of these foundations in order to elucidate and forecast their efficacy [11]. The purpose of this article is to survey virtual reality (VR) tools and the theory behind them as they pertain to medical research.

The proliferation of AR applications in healthcare has recently grown into a major industry trend. Many different types of computer vision, image processing, cloud computing, object detection and identification, and picture segmentation technologies are finding their way into AR healthcare systems. Some notable developments in AR-based healthcare solutions include improved association and privacy for individual patients, faster, more reliable diagnosis and treatment [12], and more speed overall. Use of it has also improved rehabilitation treatment and medication adherence. A new e-health strategy has been put in place by today's healthcare society to incorporate popular augmented gadgets, such as smart glasses and augmented reality headsets [13]. As a result, there has been a steady increase in the number of proposals for the development of augmented technology and its widespread use in healthcare.

Nevertheless, based on our observations, prior review studies have a few limitations.

- Many of them failed to provide a concise overview of augmented healthcare applications and the services they offer.
- Only a small number of studies attempted to analyze all of the previous work, ignoring the current limitations of this emerging technology.
- Most of the studies focused solely on m-health applications, such as medication plans, healthy eating guidelines, glucose level monitoring, and heartbeat clinical applications, which can examine abnormal joint-function, tumor location-detection, vain-detection, breast-cancer diagnosis, etc.
- Some of the papers only briefly touched on the recent challenges of augmented healthcare applications.

So, to find out what the limits and difficulties of augmented reality healthcare applications are, it would be helpful to do a thorough analysis, according to our systematic observations and the author's best knowledge. This study has the potential to reveal how to overcome these obstacles and enhance existing applications even more. Addressing the aforementioned concerns, this research seeks to examine the effects of AR on the healthcare system. This study provides a concise overview of augmented SHS. In addition, this article demonstrates a variety of issues with the augmented platform in healthcare systems and suggests a way to prevent such attacks in the future. Additionally, this work has the potential to provide light on current healthcare applications that utilize augmented reality. Reading up on augmented reality healthcare solutions, interested parties were directed to [14].

Virtual reality (VR) implementation treatments are theoretically useful but seldom put into practice. Utilizing an implementation model as a roadmap can facilitate the development of an implementation intervention and encourage methodical implementation. There is a lack of standardized usage of the several available implementation approaches when it comes to healthcare virtual reality (VR) [15]. The Non-Adoption, Abandonment, and Scaling, Spreading, and Sustaining of Technology-Supported Change Efforts in Health and Social Healthcare (NASSS) framework [16] examines these elements and more, providing a potential paradigm for identifying critical implementation factors. In addition to the technology itself, the NASSS framework takes into account the following factors: the target group's condition, the value proposition, the adopter system (consisting of staff, patients, and healthcare providers), the healthcare organization(s), the broader system, and the progressive embedding and adoption of technology [17]. The complexity of new technology adoption within businesses can be better understood with the help of this approach. There is still a lack of clarity regarding whether or if VR implementations in different healthcare settings adhere to the NASSS paradigm, or any other implementation framework for that matter.

The following is the structure of this article: In Sect. 2, we covered the history and current landscape of AR &VR in healthcare. The role and main applications of AR &VR in smart healthcare is discussed in Sect. 3. Comprehensive analysis of AR &VR is presented in Sect. 4. The challenges of AR &VR in healthcare are discussed in Sect. 5. In Sect. 6, we addressed many security challenges using our proposed intelligent security architecture, which summed up our future contributions. A conclusion and suggestions for further study made up the article's last section.

2 Background

The investigation of XR applications in healthcare has sparked a revolution in patient-centered healthcare and precision medicine research. Through the advent of data-driven healthcare, which promotes engagement and customization, XR healthcare elevates precision medicine to a whole new level. To enable patients to start and maintain their remote healthcare routine even while not in a traditional healthcare facility, doctors are increasingly prescribing XR headsets. Virtual-reality (VR) therapeutic sessions have improved patients' health and well-being, but they also free up doctors' time to deal with more complicated situations that still need in-person care [18].

In addition, XR is already seeing widespread application in medical education and training, enabling doctors and nurses to study at their own speed and in their own environments, all without endangering their patients. The gaming and entertainment industries are the birthplace of XR, which focuses on storytelling and creative development to captivate consumers. Because patient contact and response are one of the largest issues in healthcare, this aspect of XR makes it a perfect tool for therapeutic intervention [19]. Evaluating the use of XR for different applications has been the primary focus of research in healthcare XR for the past 30 years. The accessibility and cost of XR have advanced rapidly throughout the years. As shown in Fig. 5.2, XR is actively assisting in the following clinical domains.

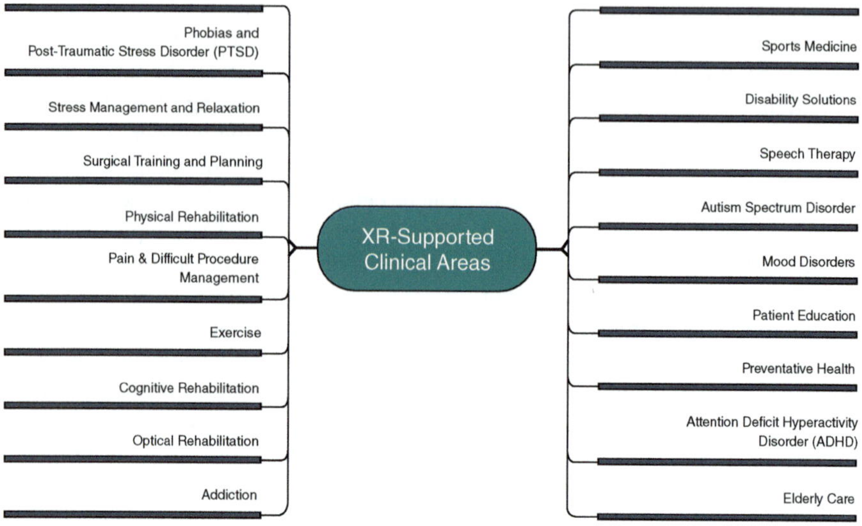

Fig. 5.2 AR &VR supported healthcare services

2.1 Characteristics of AR &VR

Virtual reality (VR) has a lot of promises for studies and practices focused on education and intervention. Specified features of VR head-mounted displays (HMDs) or AR glasses enable their users to engage with, create in, or manipulate items within a fully encircling virtual space. The utilization of controllers or a camera's Realtime video capture makes these interactions possible [20]. Virtual reality also enables very accurate and permissive visualization of processes and objects that would be extremely challenging, impossible, immaterial, unrealistic, or abstract to portray in the physical world [21]. A person can thus fully immerse himself in a fake setting, or a virtual environment, because of this [22]. One way to categorize virtual reality experiences is by the level of immersion they provide. In their description of the extended reality spectrum, the authors of [23] cover a wide range of technologies, including VR, MeR, MR, and AR. When compared to AR, which offers static virtual items, VR, MeR, and MR all provide interactive virtual objects. With augmented reality, users may integrate virtual things into their actual surroundings, thus bringing the computer into their world [24]. In addition, whereas MeR, MR, and AR use real-world backgrounds, VR uses a virtual one. Some examples of this are virtual reality (VR) settings, MR where digital and physical elements coexist, and augmented reality (AR) where digital and physical elements are physically distinct from one another.

3 AR &VR in Smart Healthcare

3.1 Medical Training

The Institute of Medicine states that in order to have effective collaborative practice, modeling and sufficient professional training are necessary components. Examining how AR affects clinical and surgical outcomes is something the authors are doing on purpose [25]. The AR in Surgery and Education (ARiSE) program was established to achieve this goal. One potential usage of ARiSE is the creation of 3D holograms that users can experience in the actual world by doing augmented reality goggles. Voice commands and user-interface buttons allow users to communicate with the software. The software will allow users to view human organs on real-world surfaces, which can be useful for both diagnostics and training. Among the many potential applications of ARiSE are surgical procedures and nursing education [26]. Figure 5.3 shows students using augmented reality goggles to view a holographic overlay of static and moving images of the cardiovascular system, lungs, and ribcage. During the course of the training, this would be fine-tuned to match the movements of a dummy or an actual trainee. You will now have access to the anatomical markers of precise auscultation sites [27]. A vast amount of information regarding the structure and function of the human body is covered in medical school. 'Virtual

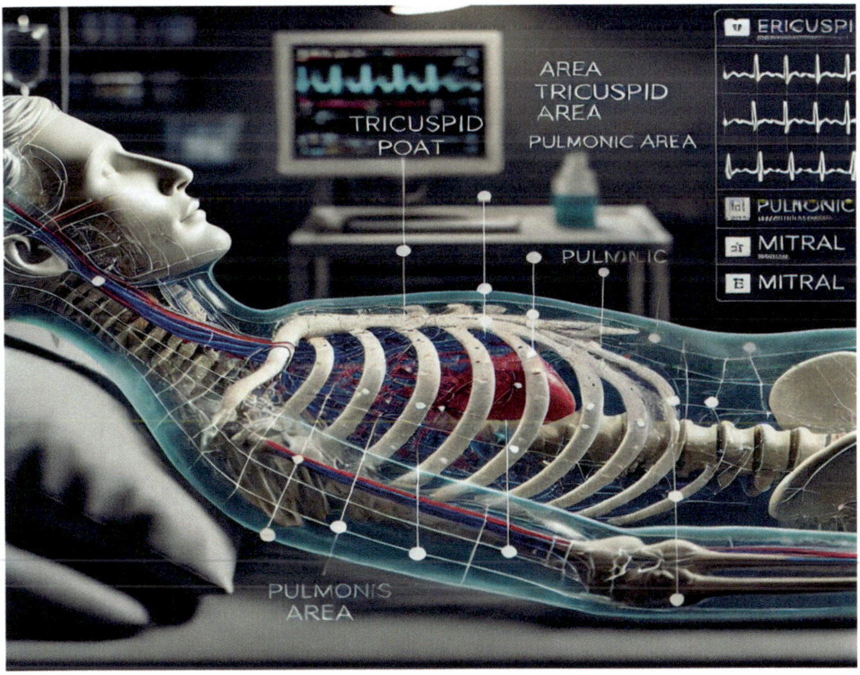

Fig. 5.3 AR being used on a manikind

cadavers,' one of many new digital applications, has greatly facilitated this subject's acquisition (Fig. 5.4). A more participatory environment that ultimately improves learning and knowledge can be achieved through the usage of augmented reality (AR) by medical learners in comparison to traditional methods such as a computer mouse, keyboard, and screen [28].

In the augmented reality program Holo-Human, a digital corpse is placed on a physical examination table, as illustrated in Fig. 5.4. The shown moderator has the option of interacting with the model and user interface with a HoloLens headset. Organs, structures, and systems can be studied individually or in combination with the help of digital dissection tools and visual storytelling. Since their initial application in orthopedic diseases, augmented reality platforms have seen tremendous growth in their use in medical education. One of the major challenges confronting the higher education sector is the expense of establishing these interactive platforms [29]. Their utilization in medical education is hindered by this, in addition to a lack of resources to meet the demands of growing student populations [30]. The most pressing issue for educators is ensuring that all students have equal access to digital technology. The limited technology needed, and the growing problem of social isolation associated with online learning are two more arguments against augmented reality in classroom [31]. In spite of this, teachers are now able to provide their students with more personalized, engaging, and purposeful learning experiences because of these novel digital stages that have enabled them to challenge the limits of old pedagogies. Users are able to create a more authentic training experience that

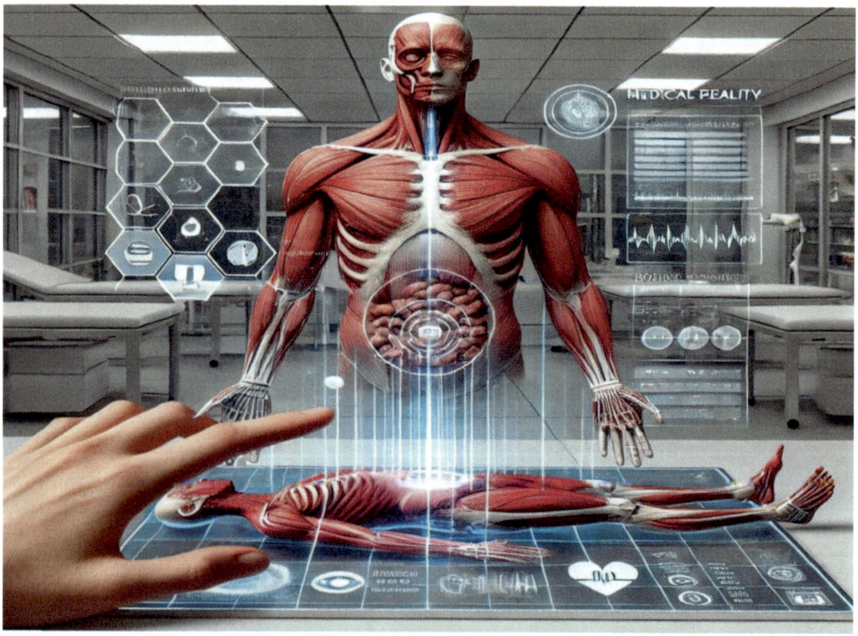

Fig. 5.4 Virtual cadaver placed on a real examination of AR App

mimics the genuine clinical treatment environment by adjusting the patients' position.

3.2 *Biomedicine*

Virtual reality (XR) has numerous applications in fields such as education and entertainment, but its versatility in biomedicine offers exciting new opportunities in healthcare and fundamental research. The method of transforming biological discoveries into digital objects is the primary focus of the writers in [32]. Virtual objects in XR enhance medical and basic investigation by leveraging computational power and interactive analysis of real-world data, such as macromolecular structures and multidimensional imaging. The immersive aspect of XR allows for the usage of various conversion pipelines and graphics rendering, which can be employed to display data volumes [33]. There is greater room for exploration and manipulation in XR than in traditional monitor watching because it is interactive. Additionally, they cover several methods for utilizing XR applications to engage with and display biomedical data.

At the same time, the authors of [34] zeroed in on new kinds of affordances that could impact HCI-based XR training settings. New affordances are proposed in addition to categorization using optical and tactile affordance. In addition, the concept of dynamic affordance is proposed. The impact of these affordances on learning and cognition is investigated within the framework of condylar plating surgery, a surgical method used to cure femur bone fractures. In addition, a surgical planning method based on genetic algorithms is provided, which can be used for the training exercises. Additionally, it looked at how trainees' ability to learn new skills and information in training contexts was impacted by the results of assessment activities [35]. Health risks to space crew members are unique and severe due to physiological changes, psychological and physical stress, and limited access to medical treatment. Additionally, they examine the methodological and design features of the prior research in order to inform future investigations into the use of XR technologies to enhance crew safety and performance, as well as space health [36]. The article [37] elaborates on how XR might alter the healthcare industry, its applications, the challenges it poses, XR methods and tools for smart healthcare, XR developments in IHS, and a review of the benefits and future applications of XR in medicine. According to the article [38], augmented reality experiences are made possible by hyperspectral and multispectral imaging, which involve collecting spectral data from a scene and superimposing it onto a 2D picture. When using this imaging approach, short-wave infrared works wonder.

4 Comprehensive Analysis of AR & VR in Smart Healthcare

The main goal of this section is to conduct a thorough analysis, taking into account our research questions about the ways in which augmented technology has improved healthcare. We will also look for examples of recent augmented healthcare applications that provide services, as well as the current obstacles faced by these applications. Figure 5.5 summarizes the Augmented Healthcare services.

Surgical procedures and medical diagnostics are two areas that are rapidly embracing AR and VR technology. Systems like touch surgery utilize virtual reality to show physicians the patient's anatomy and physiology, which opens up new possibilities during surgeries. In turn, this is anticipated to boost the expansion of the market for healthcare-related AR/VR. One of the main reasons for driving market expansion in recent years is the increasing use of AR and VR technologies in medical surgeries. These technologies help to optimize surgical procedures and boost the efficiency of treatment. Reducing the need for costly screens and unneeded workers is another benefit of keeping medical images and patient data on the AR/VR platform, which is becoming more commonplace. The use of AR in medicinal staff training and education is also on the rise. A more holistic and comprehensive learning experience is provided to medical staff using augmented reality (AR) learning modules, which allow for anatomical visualization and simulation. Patients are also receiving training from these modules, which improves their comprehension of treatment flow and care management. Surgeons are able to improve their precision and accuracy during complicated procedures with the use of augmented reality-based picture viewing.

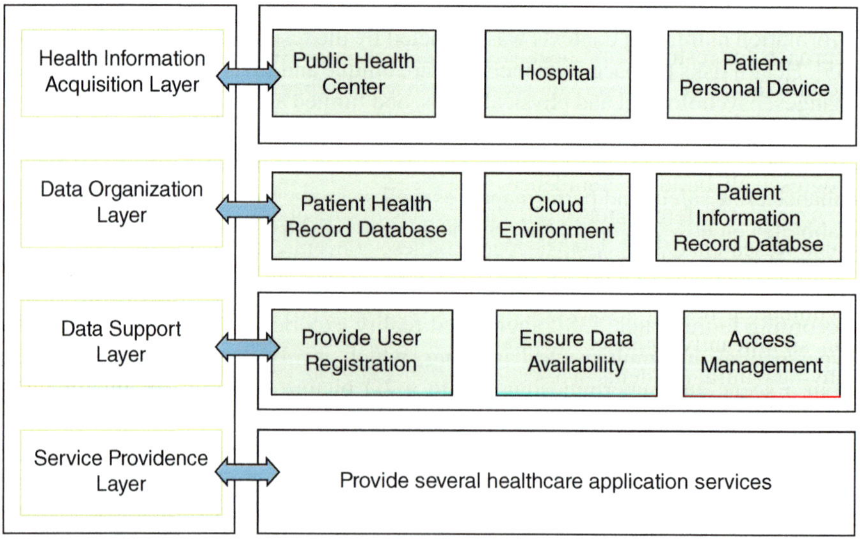

Fig. 5.5 Augmented healthcare services

In terms of technology, 59.1% of revenue in 2022 came from the augmented reality (AR) category, which was by far the most dominant. When we use technologies like smart glasses or head-mounted displays, we can experience augmented reality, which gives us a 3D view of things and our surroundings. There are a lot of things pushing the need for augmented reality, including new technologies, more efficient treatment demands, more public and private investments, more mergers and acquisitions, and cheaper head-mounted devices.

While augmented and virtual reality technologies do have certain useful applications, our research reveals a lack of clarity on their theoretical underpinnings in the fields of information systems and healthcare. To illustrate the efficacy of a VR-based therapeutic intervention, this section mostly makes use of or discusses theoretical underpinnings. Hence, the majority of the publications incorporate both the theoretical frameworks of IS research and more moderate notions from the field of psychology. Theoretical frameworks and notions outlining the processes of clinical efficacy are therefore cited in these articles. However, there is a dearth of research that takes into account either the correlation between technical aspects and how users experience the virtual environments (technological effectiveness) or the two sides of the same coin (clinical effectiveness) in relation to technological aspects.

Theoretical foundations for virtual reality exposure therapy based on Emotional Processing Theory are described in, which is in addition to theories that explain learning processes. According to this school of thought in psychology, our fears and anxieties are really just mental constructs. Emotional Processing Theory states that while avoiding situations that provoke fear does not alter the structures that show up in fear, an in vivo encounter could put too much strain on a patient's resources. One solution to this issue is virtual reality (VR) exposure, which allows for a more controlled encounter with the patient to justify their method, they connect the features of VR systems to feelings of immersion and presence. Carry conducts a similar procedure without specifically identifying these features. In addition to discussing the use of VR/AR based expertise to aid individual change, covered the treatment of phobias or anxieties. There is no way to integrate the features of virtual reality or augmented reality into the theoretical framework of personal change that lays out in detail using Perceptual Control &Self-Affirmation Theory. Similarly, the majority of the featured articles refrain from directly combining technology and psychological theoretical assumptions. In addition, the majority of the studies that addressed psychological phenomena (clinical effectiveness perspective) either provided empirical evidence for or explanations for the therapeutic efficacy of VR-related technologies.

4.1 AR Application Insights

Surgical planning and minimally invasive surgery are two areas where augmented reality is finding more and more applications in the medical industry. The use of augmented reality technology is on the rise due to technological developments and the growing need to simplify medical procedures. Companies in the augmented reality (AR) industry are teaming up with local businesses to increase product exposure and introduce new offerings to consumers. One example is the partnership between smart glass and augmented reality technology specialist Vuzix and telemedicine solutions provider VSee in April 2023. The goal was to create smart glass specifically for the telemedicine market [39].

Augmented reality has the potential to improve the efficacy of various healthcare procedures as well. The number of mergers and acquisitions is predicted to rise in the coming months. Competition could rise and prices could fall as a result of this. Additionally, the use of augmented reality technology has resulted in ongoing enhancements to the surgical setting and associated processes. Among its many potential uses are in the fields of anatomy, surgical and postoperative care, and medical education. Rehabilitation services that use augmented reality technology also provide patients with a more manageable setting in which they can engage in activities that aid in their recovery [40].

4.2 VR Application Insights

Nevertheless, numerous healthcare IT businesses are pouring money into virtual reality (VR) for healthcare because of the promising growth opportunities in this field. One example is the USD 14 million that OssoVR raised in September 2022, with Kaiser Permanente as the lead investor. Surgical and medical device training modules based on virtual reality will be developed by the company with cash.

Virtual reality (VR) has several potential uses, including but not limited to phobia treatment, skill training, robotic surgery, and surgical simulation. The sector is being propelled by recent advancements in healthcare IT. The market adoption of virtual reality technology is anticipated to be boosted by the convenience of performing and learning surgical operations that it provides to medical professionals and trainees.

5 Current Challenges of AR Healthcare Applications

Finding out what problems current AR healthcare apps have is a major goal of this study. We also noted a number of potential issues with augmented healthcare applications in the future. Here we have conceded eight difficulties based on the standpoint of our studied articles. Our assessment does not cover all of the potential issues with the existing augmented reality healthcare application.

- **Data Security**: Major issue with augmented reality healthcare applications is the need to guarantee the confidentiality of patient data. Several popular apps surprisingly lack established security procedures, and newer methods are just as insecure. Encrypting data transfers over networks is a must for protecting sensitive information. It is challenging, but necessary, to provide a fully protected environment for exchanging data in healthcare applications.

- **Specialized platform**: Most augmented healthcare apps don't follow proper criteria and don't use a standard platform. The inability to deliver genuine and precise services adds another layer of complexity and insecurity. Modern healthcare platforms have the challenge of initiating a standardized and expert podium that makes reference to APIs, frameworks, and proper libraries. Code, class, and documentation management may all suffer in the absence of a dedicated platform.
- **Limited services**: Many apps have emerged for a small subset of healthcare functions (e.g., diagnosis, treatment, evaluation, etc.) since 2011. However, additional medical specialties, such as cardiology, hematopoietic stem-cell transplantation, etc., should come up to offer a viable alternative.
- **Accessing Healthcare Information Through Interoperability**: Various recommendations concerning healthcare IT interoperability were issued by the Healthcare Leadership Council (HLC). A major worry is making sure HLC standards are met when it comes to sharing and accessing information.
- **Difficulty in Expanding the Scope of Adaptability**: Healthcare applications face difficulties in expanding their scope due to complicated functionality and a lack of user guidance. In most cases, the utilize of AR in healthcare is a reflection of digitally connected components. Healthcare providers and patients alike may encounter new technological challenges as a result of these interconnected digital components. One of the greatest hurdles is getting past the technical hurdle.
- **Problems Caused by Limited Availability:** Launched in the last several years are multiple augmented reality healthcare apps. However, the developed world is the exclusive user of the vast majority of them. These applications are not widely used in developing nations due to high configuration prototypes, training needs, and cost concerns. It is a tough issue to ensure that these applications are available for the entire country.
- **Implementation Costs:** The augmented reality healthcare device's development and implementation costs are another important issue. Several healthcare providers are against rebuilding their healthcare platform because of the hefty development cost. So, incorporating digital and current technology into healthcare solutions is more challenging due to the wide range of implementation costs.
- **Healthcare Device Adoption:** The present enhanced platform strives to offer streamlined user experience. However, problems related to enhancing consistency and decreasing mistakes can occasionally be more practical. If this obstacle can be overcome, the device's acceptance could be enhanced. Building an app with useful features is a manageable challenge for augmented reality technology.
- **Unsuitable Models:** Augmented healthcare apps aren't as effective as they may be because of improper prototypes, which can lead to unforeseen irregularities like false recognition and an absence of semantic familiarity. One of the unique challenges of medical applications is ensuring acceptable prototypes.

- **Difficulty with Service Availability:** Unlike other augmented reality apps, it might be difficult to provide services that are available all the time or that last for lengthy periods of time, such operation visualization, projection, and interactivity, which can last anywhere from 9 to 12 h straight.
- **Real-time geometry awareness:** The difficulties associated with real-time geometry awareness, which is a big problem for developers and researchers. Annotation, tracking, object detection, and accurate measurement are all processes that it is associated with. Interestingly, the majority of issues with the augmented app occur while the geometry is being transformed. Consequently, augmented reality platforms may have difficulties when it comes to awareness of real-time geometry.
- **Limited Research:** Research data shows that augmented reality studies in the healthcare industry are lacking in depth because of issues like a lack of resources, ambiguity in the available information, a narrow focus, and a lack of enthusiasm among healthcare professionals. Consequently, the majority of academics in the field of healthcare have not expressed a desire to conduct research within this domain, particularly on augmented platforms.

6 Discussions

The use of virtual reality (VR) in healthcare has the potential to revolutionize both the therapeutic alliance and the therapeutic process. Knowing the theoretical underpinnings of the effectiveness of VR-related technology is crucial. Hence, theories of therapeutic interventions must be integrated with constructs or concepts of presence, which have already been developed in IS research. Such theoretical components must be contextualized in order to build reliable models in order to accomplish this [41]. It seems to reason that these models could benefit from including essential technological aspects by associating them with basic conceptions like presence or immersion. Core constructions should be described thoroughly to assure comparability.

Our research leads us to believe that few studies have attempted to combine the theoretical foundations of two fields into a single model. The majority of the studies that were considered either (i) explored the medical efficacy of a skill in contrast to an ordinary therapy without incorporating theoretical elements of technology characteristics into their framework or (ii) measured the empirical effect of a technology using a hypothetical framework for technical features without incorporating hypothetical conventions of clinical efficiency. Given the importance of considering the client's or patient's point of view, both (i) and (ii) run the danger of representing an intervention in an inadequate manner. Furthermore, ideas or their creation are necessary to fill the primary study void regarding the reasons behind the potential efficacy of a certain virtual reality intervention. Theoretical underpinnings in clinical fields other than psychology (e.g., surgery) seem to be further forward than in VR-related psychological treatments, and this is particularly true in these areas.

When it comes to current methods, virtual reality (VR) technologies have the dual ability to expand the real world through the production of artificial scenarios and to contain it. In some therapeutic contexts, this could be useful since it allows therapists to create conditions that are within their control [42]. In cases where vivo exposure is not recommended for psychological therapy, but patients have shown improvement during creative therapy sessions, virtual reality (VR) technology may provide an opportunity. Therapy results appear to be influenced by the idea of presence. It embodies the core principles of virtual reality, which include methods for directing attention or distraction [43]. However, the basis of its efficacy is still not obvious, as noted. Additionally, psychological law and midrange theories are frequently used as empirical sources for the theoretical underpinnings of VR-related technology. They simplify the psychological theories by describing their nature or by highlighting the interplay of related concepts. Both comprehensive theories drawn from larger social or psychological frameworks and hypotheses that have been synthesized and evaluated empirically are still lacking.

The assumptions behind the potential efficacy of VR-based technologies can be better understood with a more in-depth explanation and elaboration of such theoretical methods. However, there are certain limits to our findings. Our results are not as well-differentiated as they might be because we did not evaluate the methodological and general quality of the papers that were included in the review because we used a broad search strategy and selected a review type [44]. Also, rather than delving deeply into the topic, this study is more concerned with making sure searches are comprehensive. Clinical interventions utilizing VR-related technology might benefit from thorough frameworks that could be developed by future research that thoroughly examines particular theoretical structures and how they relate to clinical outcomes [45]. Our findings highlight the need for more in-depth methods that integrate concepts, as well as for the field of information systems to become more involved in discussions surrounding virtual reality (VR) and its associated technological features that serve as theoretical justification. Finally, we only looked at the health care industry, thus our findings may not apply to other fields. Research in the future may draw from a wider range of fields to investigate both specific and general conclusions about the theoretical underpinnings and practical applications of VR-related technology [46]. In this part, we see that the new augmented healthcare platform has had a few problems. Ensuring data security is a major worry among various current difficulties with AR healthcare applications. As a result, it is not immune to outside influences, and the physical layer may be subject to cybersecurity risks. While developing, augmented reality healthcare apps should take these external risks and typical security holes into account [47].

6.1 Security Requirements

The stuff that safeguards the request platform against unforeseen attacks is what enhanced healthcare application security standards are all about. If you want your app to be useful and functional, you need to pay attention to the security criteria shown in Fig. 5.6.

One example is the possibility of an intrusion occurring while virtual objects and real-time data are working together. When working with virtual or real-time objects, the augmented platform should keep everyone safe. These days, more and more augmented healthcare devices are joining the network. Consequently, for the sake of patient data security, the enhanced platform should guarantee privacy by blocking unauthorized users from accessing patient records. Nevertheless, a myriad of factors can alter data during access or transmission. In order to guarantee the security of medical information or health data, healthcare solutions should implement integrity. Resilience to provide a protective environment in the event that health devices are damaged is another essential necessity.

Availability is another big problem with the security of healthcare apps today. Most of the time, bad actors can take advantage of people's ignorance of denial-of-service attacks to compromise data security during transmission. Consequently, security awareness should give serious thought to this. Wearable technology that can connect to the network anywhere is what augmented healthcare devices are all

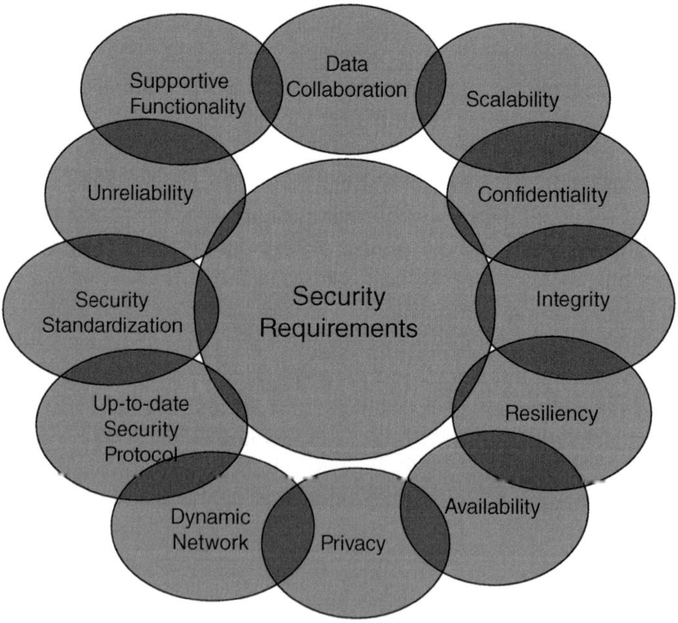

Fig. 5.6 Security requirements for AR

about. This highlights the critical need to develop a security model that accounts for changing network topics.

Improving the safety of augmented apps may be as simple as creating a dynamic security mechanism. Surprisingly, there is no security control strategy in place for ARML. Due to a lack of uniformity and an absence of security concerns, it is susceptible to security breaches [48]. Recently, augmented reality has been using third-party suppliers' computer-aided objects, which opens the door to security risks such data manipulation, spoofing, sniffer, and man-in-the-middle attacks. In order to ensure the application's reliability, the AR platform should be cognizant of these dangers and verify the created content's validity during transmission. Another potential source of security breaches in augmented healthcare applications is the widespread use of unsecured web browsers [49].

6.2 External Attack Terminology

There are threats to the augmented platform from outside sources. Potentially impacting both current and future augmented healthcare apps, devices, and networks, attackers pose a number of security risks. It can be easy to foresee some of these dangers and difficult to foresee others. Several external dangers that diminish the efficacy of applications and networks will be mentioned in this section. Some of the attack terminologies that can leave systems susceptible and less effective are shown in Fig. 5.7. For instance, hackers can corrupt or replace original data by stealing it or propagating malicious software during data transmission or communication. Unauthorized users have the potential to alter data, sow confusion, and slow down networks [50]. In addition, tampering with program code, information, or reprogramming might compromise the hardware platform, reduce data reliability, or introduce misleading information into the network. Attacks can sometimes infiltrate software systems and applications, taking advantage of security holes to wreak havoc with buffering, resource corruption, or loss [51].

6.3 AR and Virtual Reality in Healthcare: Current State and Future Perspective

Medical professionals are increasingly turning to augmented and virtual reality technologies to help patients better comprehend their treatments, train future doctors, and reach treatment goals. As seen in Fig. 5.8, augmented and virtual reality have many potential uses in the medical field.

- Among the many medical procedures that can benefit from augmented reality apps are:

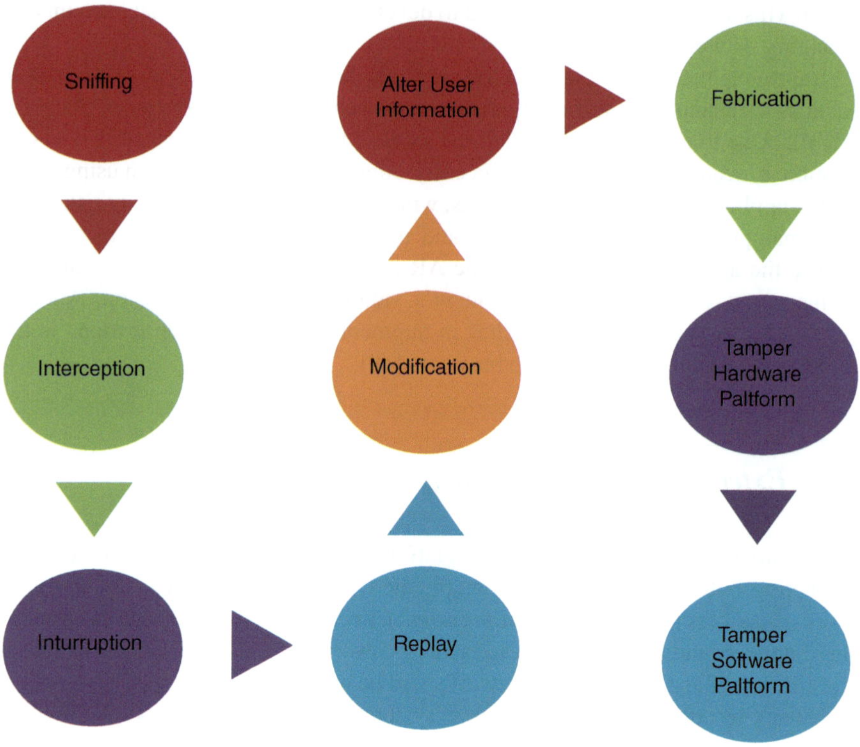

Fig. 5.7 Various attacks on security threats

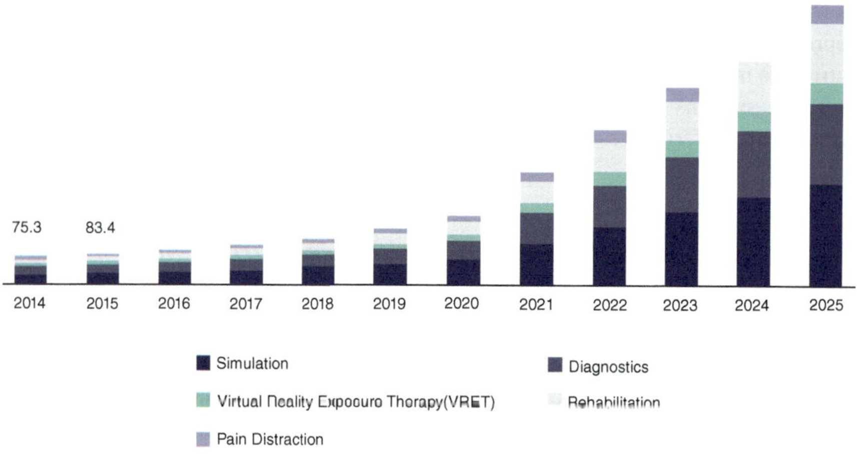

Fig. 5.8 Evolving of AR &VR technology in healthcare

- **Surgical treatment:** As a surgical procedure, 3D imaging of the patient's internal organs enables doctors to "see through the tissue" and plan the operation with precision, reducing the need for further incisions.
- **Accurate diagnostics:** The use of AR &VR to improve patients' ability to describe their symptoms improves diagnostic accuracy.
- **Closest-to-real situation lab practice:** Medical students benefit from the most realistic scenario lab practice by gaining practical experience that allows them to see the effects of their actions on real patients.
- **AR-based navigation:** Finding your way around the hospital, seeing nearby medical facilities, and locating life-saving equipment like defibrillators and emergency kits.
- **Virtual assistance:** In the event of an emergency, virtual aid can facilitate successful remote collaboration between the initial point of contact and a doctor.
- **Patient-physician interaction:** Interaction between patients and doctors when physical contact is not allowed (for example, because of a pandemic or because of where the doctor is located).
- **Anatomy teaching at schools:** Immersive learning environments in anatomy classes help students get a deeper understanding of their bodies.
- **Pharmaceutic visualization:** Pharmaceutical imaging of the drug's pharmacodynamics and pharmacokinetics, etc.

7 Conclusion

A significant amount of effort is being put forth by researchers in order to develop more advanced technical solutions that will improve the contemporary health care system all over the world. These innovations are being developed with the intention of bringing about a significant shift in the healthcare industry and simplifying the issues that are now present. The purpose of this study is to provide a concise examination of a variety of augmented reality (AR)-based healthcare services and applications. The structure of this document provides a comprehensive overview of the current development strategies about the existing architecture of healthcare applications as well as the procedures for processing and obtaining health data. To a certain extent, the essay made an effort to speed up the process of further development by drawing attention to a number of previously unknown problems concerning concurrent security requirements and subsequent difficulties. For the purpose of future study on AR-based healthcare apps and services, the debate that was carried out in this article on standardization, data availability, service quality, and data privacy may be helpful in a number of different instances. Furthermore, this study demonstrated the significance of augmented reality-based healthcare applications, which were supported by data from the current market. This may result in an increase in the involvement of a number of stakeholders for the purpose of continued development.

References

1. Lakshminarayanan V, Ravikumar A, Sriraman H, Alla S, Chattu VK. Health care equity through intelligent edge computing and augmented reality/virtual reality: a systematic review. J Multidiscip Healthc. 2023;16:2839–59.
2. AlGerafi MA, Zhou Y, Oubibi M, Wijaya TT. Unlocking the potential: a comprehensive evaluation of augmented reality and virtual reality in education. Electronics. 2023;12(18):3953.
3. Gupta D, Sathiyasekar K, Krishnamoorthy R, Arun S, Thiyagarajan R, Padmapriya S. Proposed GA algorithm with H-heed protocol for network optimization using machine learning in wireless sensor networks. In: 2022 Second international conference on artificial intelligence and smart energy (ICAIS), Coimbatore, India; 2022. p. 1402–8. https://doi.org/10.1109/ICAIS53314.2022.9743120.
4. Ray S, Tawar S, Singh N, Singh G. Transition toward technological transformation: challenges of implementing virtual reality and augmented reality in the health sector. J Mar Med Soc. 2023:10–4103.
5. Rao KR, Prasad ML, Kumar GR, Natchadalingam R, Hussain MM, Reddy PCS. Time-series cryptocurrency forecasting using ensemble deep learning. In: 2023 International conference on circuit power and computing technologies (ICCPCT). IEEE; 2023. p. 1446–51.
6. Taghian A, Abo-Zahhad M, Sayed MS, Abd El-Malek AH. Virtual and augmented reality in biomedical engineering. Biomed Eng Online. 2023;22(1):76.
7. Krishnamoorthy R, Desai A, Patel R, et al. 4 Element compact triple band MIMO antenna for sub-6 GHz 5G wireless applications. Wirel Netw. 2021;27:3747–59. https://doi.org/10.1007/s11276-021-02734-8.
8. Yuvaraj D, Kumar VP, Anandaram H, Samatha B, Krishnamoorthy R, Thiyagarajan R. Secure de-duplication over wireless sensing data using convergent encryption. In: 2022 IEEE 3rd Global conference for advancement in technology (GCAT), Bangalore, India, vol. 2022. p. 1–5. https://doi.org/10.1109/GCAT55367.2022.9971983.
9. Kalhoro R. Integration of augmented reality (AR) in healthcare systems. J Adv Technol Syst. 2023;1(1):35–47.
10. Dhar P, Rocks T, Samarasinghe RM, Stephenson G, Smith C. Augmented reality in medical education: students' experiences and learning outcomes. Med Educ Online. 2021;26(1):1953953. https://doi.org/10.1080/10872981.2021.1953953.
11. Tiwari CK, Bhaskar P, Pal A. Prospects of augmented reality and virtual reality for online education: a scientometric view. Int J Educ Manag. 2023;37(5):1042–66.
12. Anand SJ, Krishnamoorthy R, Kumar US, Kamalakkannan D. An effective hybrid mobility aware energy efficient low latency protocol (HMEL-MAC) for wireless sensor network. Cybern Syst. 2022:1–16. https://doi.org/10.1080/01969722.2022.2157598.
13. Dargan S, Bansal S, Kumar M, Mittal A, Kumar K. Augmented reality: a comprehensive review. Arch Comput Methods Eng. 2023;30(2):1057–80.
14. Baashar Y, Alkawsi G, Wan Ahmad WN, Alomari MA, Alhussian H, Tiong SK. Towards wearable augmented reality in healthcare: a comparative survey and analysis of head-mounted displays. Int J Environ Res Public Health. 2023;20(5):3940.
15. Ramana AV, Bhoga U, Dhulipalla RK, Kiran A, Chary BD, Reddy PCS. Abnormal behavior prediction in elderly persons using deep learning. In: 2023 International conference on computer, electronics & electrical engineering & their applications (IC2E3). IEEE; 2023. p. 1–5.
16. Suneel S, Balaram A, Amina Begum M, Umapathy K, Reddy PCS, Talasila V. Quantum mesh neural network model in precise image diagnosing. Opt Quant Electron. 2024;56(4):559.
17. Zhang J, Yu N, Wang B, Lv X. Trends in the use of augmented reality, virtual reality, and mixed reality in surgical research: a global bibliometric and visualized analysis. Indian J Surg. 2022;84(Suppl 1):52–69.
18. Prasath ASS, Lokesh S, Krishnakumar NJ, Vandarkuzhali T, Sahu DN, Reddy PCS. Classification of EEG signals using machine learning and deep learning techniques. Int J Health Sci. 2022;2022:10794–807.

19. Suneel S, Krishnamoorthy R, Gopatoti A, et al. Enhanced image diagnosing approach in medicine using quantum adaptive machine learning techniques. Opt Quant Electron. 2024;56:534. https://doi.org/10.1007/s11082-023-06203-8.
20. Zhao X, Ren Y, Cheah KS. Leading virtual reality (VR) and augmented reality (AR) in education: bibliometric and content analysis from the web of science (2018–2022). SAGE Open. 2023;13(3):21582440231190821.
21. Mamatha B, Rashmi D, Tiwari KS, Sikrant PA, Jovith AA, Reddy PCS. Lung cancer prediction from CT images and using deep learning techniques. In: 2023 Second international conference on trends in electrical, electronics, and computer engineering (TEECCON). IEEE; 2023. p. 263–7.
22. Dawn S. Virtual reality and augmented reality based affective computing applications in healthcare, challenges, and its future direction. In: Affective computing in healthcare: applications based on biosignals and artificial intelligence. Bristol: IOP Publishing; 2023. p. 10–1.
23. Parekh P, Patel S, Patel N, Shah M. Systematic review and meta-analysis of augmented reality in medicine, retail, and games. Vis Comput Ind Biomed Art. 2020;3:1–20.
24. Sampath S, Parameswari R, Prasad ML, Kumar DA, Hussain MM, Reddy PCS. Ensemble nonlinear machine learning model for chronic kidney diseases prediction. In: 2023 IEEE 3rd Mysore sub section international conference (MysuruCon). IEEE; 2023. p. 1–6.
25. Ricci S, Calandrino A, Borgonovo G, Chirico M, Casadio M. Virtual and augmented reality in basic and advanced life support training. JMIR Serious Games. 2022;10(1):e28595.
26. Mendoza-Ramírez CE, Tudon-Martinez JC, Félix-Herrán LC, Lozoya-Santos JDJ, Vargas-Martínez A. Augmented reality: survey. Appl Sci. 2023;13(18):10491.
27. Nguyen HS, Danh HC, Ma QP, Mesicek J, Hajnys J, Pagac M, Petru J. A bibliometrics analysis of medical internet of things for modern healthcare. Electronics. 2023;12(22):4586.
28. Alqahtani AS, Trabelsi Y, Ezhilarasi P, et al. Homomorphic encryption algorithm providing security and privacy for IoT with optical fiber communication. Opt Quant Electron. 2024;56:487. https://doi.org/10.1007/s11082-023-06098-5.
29. Raja M, Priya GL. The role of augmented reality and virtual reality in smart health education: state of the art and perspectives. In: Artificial intelligence for smart healthcare; 2023. p. 311–25.
30. Kumar R, Jain V, Han GTW, Touzene A, editors. Immersive virtual and augmented reality in healthcare: an IoT and blockchain perspective. CRC Press; 2023.
31. Denche-Zamorano A, Rodriguez-Redondo Y, Barrios-Fernandez S, Mendoza-Muñoz M, Castillo-Paredes A, Rojo-Ramos J, Garcia-Gordillo MA, Adsuar JC. Rehabilitation is the main topic in virtual and augmented reality and physical activity research: a bibliometric analysis. Sensors. 2023;23(6):2987.
32. Mittal H. 3 Virtual reality applications in healthcare. In: Immersive virtual and augmented reality in healthcare: an IoT and blockchain perspective; 2023.
33. Innocente C, Ulrich L, Moos S, Vezzetti E. Augmented reality: mapping methods and tools for enhancing the human role in healthcare HMI. Appl Sci. 2022;12(9):4295.
34. Krishnamoorthy R, Gupta M, Swathi G, Tanaka K, Raja C, Ramesh JVN. An intelligent IoT-based smart healthcare monitoring system using machine learning. In: 5G-based smart hospitals and healthcare systems. 1st ed. CRC Press; 2023. p. 230–44.
35. Tripathi A, Chauhan N, Choudhary A, Singh R. Augmented reality and its significance in healthcare systems. In: Meta-learning frameworks for imaging applications. IGI Global; 2023. p. 103–18.
36. Fahim S, Maqsood A, Das G, Ahmed N, Saquib S, Lal A, Khan AAG, Alam MK. Augmented reality and virtual reality in dentistry: highlights from the current research. Appl Sci. 2022;12(8):3719.
37. Krishnamoorthy R, Kumar U, Swathi G, et al. Metamaterial inspired quad-port multi-antenna system for millimeter wave 5G applications. J Infrared Milli Terahz Waves. 2023;44:346–64. https://doi.org/10.1007/s10762-023-00921-6.

38. Patil V, Narayan J, Sandhu K, Dwivedy SK. Integration of virtual reality and augmented reality in physical rehabilitation: a state-of-the-art review. In: Revolutions in product design for healthcare: advances in product design and design methods for healthcare; 2022. p. 177–205.
39. von Ende E, Ryan S, Crain MA, Makary MS. Artificial intelligence, augmented reality, and virtual reality advances and applications in interventional radiology. Diagnostics. 2023;13(5):892.
40. Barteit S, Lanfermann L, Bärnighausen T, Neuhann F, Beiersmann C. Augmented, mixed, and virtual reality-based head-mounted devices for medical education: systematic review. JMIR Serious Games. 2021;9(3):e29080.
41. Gasmi A, Benlamri R. Augmented reality, virtual reality and new age technologies demand escalates amid COVID-19. In: Novel AI and data science advancements for sustainability in the era of COVID-19. Academic Press; 2022. p. 89–111.
42. Kamińska D, Zwoliński G, Laska-Leśniewicz A, Raposo R, Vairinhos M, Pereira E, Urem F, Ljubić Hinić M, Haamer RE, Anbarjafari G. Augmented reality: current and new trends in education. Electronics. 2023;12(16):3531.
43. Kiani S, Rezaei I, Abasi S, Zakerabasali S, Yazdani A. Technical aspects of virtual augmented reality-based rehabilitation systems for musculoskeletal disorders of the lower limbs: a systematic review. BMC Musculoskelet Disord. 2023;24(1):1–25.
44. Urlings J, de Jong G, Maal T, Henssen D. Views on augmented reality, virtual reality, and 3D printing in modern medicine and education: a qualitative exploration of expert opinion. J Digit Imaging. 2023;36:1–10.
45. Oun A, Hagerdorn N, Scheideger C, Cheng X. Mobile devices or head-mounted displays: a comparative review and analysis of augmented reality in healthcare. IEEE Access. 2024;12:21825–39.
46. Seetohul J, Shafiee M, Sirlantzis K. Augmented reality (AR) for surgical robotic and autonomous systems: state of the art, challenges, and solutions. Sensors. 2023;23(13):6202.
47. Kumar A, Saudagar AKJ, Alkhathami M, Alsamani B, Khan MB, Hasanat MHA, Ahmed ZH, Kumar A, Srinivasan B. Gamified learning and assessment using ARCS with next-generation AIoMT integrated 3D animation and virtual reality simulation. Electronics. 2023;12(4):835.
48. Andrews A. Integration of augmented reality and brain-computer interface technologies for health care applications: exploratory and prototyping study. JMIR Form Res. 2022;6(4):e18222.
49. Krishnamoorthy R, Soubache ID, Jain S. Wireless communication based evaluation of power consumption for constrained energy system. Wirel Pers Commun. 2022;127:737–48. https://doi.org/10.1007/s11277-021-08402-6.
50. Pandiaraj S, Krishnamoorthy R, Ushasukhanya S, et al. Optimization of IoT circuit for flexible optical network system with high speed utilization. Opt Quant Electron. 2023;55:1206. https://doi.org/10.1007/s11082-023-05452-x.
51. Tanaka K, Ogura S, Krishnamoorthy R, Sugiyama K, Kawahara M. Flexibility and productivity in IoT programming: a case study with Mruby. In: Guarda T, Portela F, Diaz-Nafria JM, editors. Advanced research in technologies, information, innovation and sustainability. ARTIIS 2023. Communications in computer and information science, vol. 1936. Cham: Springer; 2024. https://doi.org/10.1007/978-3-031-48855-9_2.

Chapter 6
Science Behind Augmented Reality and Virtual Reality in Healthcare

Arti Saxena and Pooja Thukral

Abstract In the year 2022, Prime Minister Modi launched the 5G service at Integrated Marketing Communications where the three telcos, Reliance Jio, Bharti Airtel, and Vodafone Idea, each presented their own use case of Augmented Reality (AR) and Virtual Reality (VR). The Vodafone Idea test case demonstrated the safety of workers in an underground tunnel of Delhi Metro by creating a Digital Twin of the tunnel. PM took a live demo from the dais to monitor the work in real-time by utilizing VR and AR henceforth explaining that Augmented Reality and Virtual Reality captured and affected various industries with positive outcomes. This revolutionization in industry has now entered into the field of healthcare. In healthcare, Augmented reality is collecting data from various authentic resources, producing a 3D model of the same, and utilizing it to deliver healthcare facilities. Virtual reality (VR) creates a simulated environment to train, learn, treat, and manage patients virtually. Mixed Reality merges both real and virtual worlds by allowing interaction with both digital as well as physical objects. Medical Training and Education, Surgical Planning and Simulation, Pain Management and Therapy, Physical Rehabilitation, Patient Education, Remote Consultations and Telemedicine, Mental Health Treatment, Medical Device Training, Clinical Research and Data Visualization and Patient Monitoring are a few areas of healthcare where we will discuss the usage and science of Augmented and Virtual reality. Some of the international companies are predicting the usage of Augmented Reality and Virtual Reality in healthcare by 5–6 billion in 2025 which is predicted to touch $50.9 billion by 2026 as per a spending guide by IDC (International Data Corporation). Various places where this can be used are in training doctors where traditional cadavers were used. In treating certain mental illnesses, charting the territory of surgical planes before planning a complicated surgery. In this chapter, we will

A. Saxena (✉)
School of Engineering and Technology, Manav Rachna International Institute of Research and Studies, Faridabad, India
e-mail: artisaxena.set@mriu.edu.in

P. Thukral
Cloudnine Hospital, Faridabad, India

B. O. Soufiane et al. (eds.), *Augmented Wellness*,
https://doi.org/10.1007/978-981-96-2952-7_6

discuss the science behind virtual Reality (VR) and Augmented Reality (AR) in healthcare. In this chapter we will give insight into what augmented and virtual reality is and especially its utility in healthcare.

Keywords Augmented reality · Virtual reality · Healthcare · Immersive virtual reality · Mixed reality

1 Introduction

1.1 Augmented Reality in Healthcare

Augmented reality is cutting-edge technology overlaying virtual or digital information to enhance our perception and interaction with our surroundings. It integrates virtual things into the user's existing environment. For the understanding of the user's existing environment, AR uses Cameras, various types of sensors, and computer algorithms. After processing, the data is used to overlay virtual content—such as text, pictures, or 3D models—onto the user's field of vision, usually via a smartphone, tablet, or augmented reality headset. Although the idea of augmented reality (AR) has been around for decades, its popularity has recently increased due to improvements in hardware, such as strong CPUs and high-resolution displays. These days, augmented reality (AR) is widely applied across several industries, including marketing, gaming, entertainment, education, retail, and, most significantly, healthcare.

AR has enormous potential to enhance patient care, surgical planning, medical education, and treatment results. During operations, surgeons can visualize anatomical features using augmented reality (AR), superimposing vital data such as CT scans or X-rays directly onto the patient's body. In various ways, in healthcare, the usage of AR has been explained as follows (1–5):

1.1.1 Medical Training and Education

By allowing medical professionals and students to see anatomical structures and medical procedures in three dimensions, augmented reality (AR) helps them comprehend and remember difficult medical ideas. By using augmented reality (AR) programs, students can safely and fully immerse themselves in the learning process by interactively exploring virtual organs, dissecting virtual cadavers, and simulating surgical operations [1].

1.1.2 Surgical Planning and Navigation

By superimposing preoperative imaging data, like CT or MRI images, in real-time onto the patient's anatomy, surgeons can employ augmented reality (AR) to improve surgical planning and navigation. By doing so, surgeons can more accurately and efficiently perform procedures on patients with a lower risk of complications by planning incisions, precisely locating important structures, and optimally placing implants or equipment.

1.1.3 Intraoperative Guidance

Throughout an operation, augmented reality (AR) can give doctors on-the-spot visual indications and guidance. Augmented reality (AR) helps surgeons make more informed decisions and execute surgeries with greater confidence by superimposing pertinent information, such as tumor margins, blood arteries, or nerve routes, onto the surgical site.

1.1.4 Medical Visualization and Simulation

Augmented reality (AR) gives medical personnel the ability to see and work with complicated medical data, like three-dimensional patient anatomical models, medical equipment, or physiological processes. With the use of this visualization capacity, practitioners can more effectively educate patients about medical issues, treatment alternatives, and procedural risks.

1.1.5 Patient Empowerment and Engagement

Augmented Reality (AR) facilitates patient empowerment by offering tailored treatment plans, interactive educational resources, and immersive experiences that enable patients to actively participate in their healthcare process. With the use of augmented reality (AR) applications, patients can better understand the impacts of their care, track key health indicators, and follow their prescription schedules.

Overall, augmented reality (AR) has enormous potential to change healthcare by advancing surgical techniques, patient care, medical education, and rehabilitation methods. Applications of AR technology in healthcare are anticipated to grow as it develops, providing fresh chances to raise patient standards globally and boost clinical results.

2 Virtual Reality in Healthcare

Although the idea of virtual reality has been known for several years, recent advances in computational capacity, hardware, and software have contributed to it becoming more widely accepted. Virtual reality is creating a simulated environment where we can immerse ourselves and interact with the simulated environment thus created. We need a head-mounted display (HMD) to participate in the interactions. There is sensory feedback using vision, hearing, and sometimes touch sensation in a VR-generated field. Virtual reality thus helps in the management of pain, phobias, PTSD (post-traumatic stress disorder), and physical rehabilitation.

Sophisticated displays, motion sensors, and input devices that collaborate to produce a sense of presence and immersion are at the core of virtual reality technology. Stereoscopic visuals are shown on high-resolution screens inside the headset, giving the user a sense of depth and perspective. Motion-tracking sensors follow the user's movements, enabling them to move around and interact with the virtual world.

VR is being used in a variety of industries today, including manufacturing, education, gaming, entertainment, and healthcare, just to name a few. In various ways, in healthcare, the usage of VR has been explained as follows (1–5):

2.1 Medical Education and Training

Virtual reality (VR) allows medical professionals and students to participate in lifelike simulations of operations, treatments, and patient encounters. Before interacting with actual individuals, learners may improve their skills and gain confidence in this immersive training environment, which allows them to practice in a risk-free atmosphere [1].

2.2 Surgical Planning and Simulation

Surgeons can utilize virtual reality (VR) to plan difficult procedures by seeing the anatomy unique to each patient in three-dimensional form. Surgeons can also simulate surgeries in advance with VR simulations, which improves surgical outcomes and lowers operating risks.

2.3 Pain Management and Therapy

Virtual Reality (VR) is becoming a more common non-pharmacological approach to managing pain as well as decreasing anxiety during medical operations. Patients receiving chemotherapy or physical therapy, for example, can lose themselves in virtual reality experiences that ease their suffering and encourage relaxation.

2.4 Exposure Therapy

Phobias, anxiety disorders, and post-traumatic stress disorder (PTSD) respond well to VR exposure therapy. In a safe virtual world setting, individuals can confront and subsequently desensitize to stimuli that trigger them, encouraging therapeutic success.

2.5 *Physical Rehabilitation*

Interactive exercises and activities catered to the specific needs of each patient are provided by VR-based rehabilitation programs. These all-encompassing encounters can encourage patients to participate in their treatment, monitor their advancement, and enhance results in domains including balance, mobility, and motor function.

2.6 *Patient Empowerment and Education*

Virtual reality (VR) technology makes it possible for medical professionals to impart more interesting and intelligible information to patients about medical disorders, available treatments, and anatomy. Through the use of virtual models of their own bodies or medical treatments, patients can better understand and make well-informed decisions.

2.7 *Telemedicine and Remote Consultations*

By removing geographical boundaries and increasing access to healthcare services, virtual reality (VR) platforms enable virtual consultations and remote medical procedures. Without leaving their residences, patients can attend community meetings, take part in sessions with therapists, or receive professional medical advice.

2.8 *Research and Data Visualization*

In a fully realistic three-dimensional environment, virtual reality tools enable academics to study complex medical data, including genomic sequences and imaging scans. Medical science may benefit from these improved visualization abilities in the form of fresh perspectives, discoveries, and advances.

3 Literature Review: Augmented Reality and Virtual Reality

The need for healthcare services is constantly rising. People know that they can automate many tasks with technology; these include online appointments, report sharing, video conferences, and much more. In the healthcare sector, new-generation technologies like artificial intelligence (AI), digital twins, machine learning, IoT, blockchain, and virtual reality (VR) can be leveraged to improve patient and physician experiences. This chapter by Mittal [2] reviews some VR applications for the healthcare sector. The chapter also offers a blueprint for integrating cutting-edge technologies into the healthcare industry. In image-guided surgery, augmented reality visualization gives the surgeon access to radiological images and surgical planning that are contextualized to the patient's anatomy. Its goal is to combine virtual planning and surgical navigation [3].

The possibilities of artificial intelligence (AI), the Internet of Things (IoT), BIM tools, and virtual and augmented reality (VR/AR) technologies have led to a rise in the use of sophisticated Structural Health Monitoring (SHM) methodologies. However, the absence of interconnectivity among these devices contributes to their limited usage. Fawad et al. [4] addressed this issue by developing an integrated framework to assess serviceability and implement a smart SHM for a recently constructed extradosed bridge. The objective of Oun et al. [5] is to provide an extensive overview of augmented reality (AR) applications in healthcare that is based on mobile and head-mounted displays (HMDs). People will be better able to choose the ideal gadget for their particular needs and comprehend the potential uses of this technology as a result. In comparison to conventional teaching methods or no intervention at all, the primary objective of Tene et al.'s review from 2024 is to determine the degree to which immersive Virtual Reality (iVR) and Augmented Reality (AR) technologies enhance particular competencies in healthcare professionals during medical education and training. In his article, Rowan [6] examined how the integration of digital twins and extended reality innovations might revolutionize medical device design thinking, supply chain management, and training, ultimately enhancing patient safety, circularity, and sustainability.

Virtual reality (VR) offers safe, realistic, and reproducible simulated environments where users can practice and develop communication skills, making it a useful tool for communication rehabilitation. The purpose of this study by Bryant et al. [7] was to find out what medical professionals and tech experts thought about the usability and design of a VR application prototype intended for communication rehabilitation. Nine experts from various fields related to health and technology engaged in virtual focus groups or one-on-one online interviews to assess the utilization and efficacy of the virtual reality prototype. A content-thematic analysis was used to examine the data sources. Four primary themes pertain to VR design and application in rehabilitation have been discussed as follows: (i) creating virtual worlds with a rehabilitation focus; (ii) comprehending and utilizing VR hardware; (iii) creating space for VR in training and rehabilitation; and (iv) putting VR to use without replacing the role of a health professional. While assessing the VR prototype, technological experts and health professionals collaborated on its design. They determined which software aspects needed to be carefully considered to guarantee better client safety, usability, and communication rehabilitation results. To create VR apps that are useful and successfully integrate them into rehabilitation, inclusive co-design had been maintained and involve health professionals, clients with communication disabilities, and their families.

More financial and human resources are needed by primary healthcare institutions to create an environment where VR tools can be used to benefit patients [8].

The aim of this study by Eunju [9] was to improve nursing students' severity classification competency by creating a virtual reality-based nursing education program. The key to increasing the effectiveness of emergency department services globally is the classification of severity in hospitals. Patients' safety is further ensured when treatment is prioritized according to an accurate assessment of the severity of an illness or injury. Based on the 2021 Korean Emergency Patient Classification Tool, patients were quickly classified into five clinical conditions using the five real-world clinical scenarios included in the application. An experimental group of 17 nursing students got access to a virtual reality-based simulation in addition to clinical experience. A control group of 17 nursing students engaged in solely standard clinical practice. Students' performance confidence, clinical decision-making capacity, and skill in severity classification were all significantly enhanced by the virtual reality-based nursing education program. Despite the ongoing epidemic, nursing students can still receive realistic indirect experiences through virtual reality-based nursing education software in scenarios where clinical nursing practice is not feasible. Specifically, it will function as foundational information for the growth and application plan of virtual reality-based nurse education initiatives to enhance nursing proficiency.

Between 2020 and 2021, there was a significant increase in the usage of and documentation of electronic adaptations for formative and continuous assessment methods due to COVID-19 and the limitations of in-person instruction. By going over Beer's criteria and choosing the best management, Badr [10] developed a virtual escape room that will aid and hone the problem-solving abilities of fifth-year pharmacy students. After the virtual escape room was put into use, a descriptive cross-sectional survey was carried out using a Google Form. Students had to use Beer's criteria to solve five riddles. Pharmacy students completed a survey to share their opinions about the game and assess how they felt about this approach. Every single one of the 128 students registered for the senior course managed to get out (100%). There was statistical significance between genders according to a one-sample t-test. Male students were more neutral toward the game, whereas female students were statistically faster to escape ($p < 0.00002$), more willing to recommend it to other students, and thought it pushed them to think about the content in a fresh way. To sum up, the elderly virtual escape room proved to be an effective trial approach to support online learning. Future research, however, ought to look into the effects of virtual gamification on learning in pharmacy education and whether or not there were any gender-specific variations in the use of these resources.

Virtual reality (VR) is an immersive and interactive technology that has the potential to be used as a teaching tool in nursing education. The difficulty of developing novel VR experiences has been an obstacle to fully utilizing VR's benefits, but recent advances in VR technology and software allow perhaps the least skilled users to create interesting VR experiences. An undergraduate nursing student with little technical expertise uses commercial VR software to build a virtual reality pain treatment experience, according to a case study by Brody et al. [11]. Working with computer code is not necessary when using off-the-shelf hardware and software platforms. The group used basic click-and-drag methods and setting toggles to easily manipulate the objects in a virtual world they constructed. The findings from this case study indicate that nurse educators may produce straightforward but impactful virtual reality experiences on their own, significantly enhancing the tools already available for nursing education.

4 Science Behind Augmented Reality and Virtual Reality

Virtual reality (VR) and augmented reality (AR) are two distinct but connected technologies that have transformed our interaction with digital surroundings and information. Augmented reality is overlaying digital content in the real-time world using smartphones, tablets (handheld displays), AR glasses (Head mounted display), and spatial displays. It can incorporate real-time tracking and recognition tools into the real world to enhance learning, training and provide management strategies. The tracking used by AR devices is—Sensor tracking and vision tracking. Sensor tracking uses optical (video cameras), inertial sensor tracking—which determines the particular axis (mechanical gyroscope, accelerometer), acoustic tracking (ultrasonic transmitters), and magnetic sensor tracking (uses magnetic fields). The tracking of vision involves both marker-based and markerless techniques. Scanning a QR code serves as an illustration of marker-based tracking. Information is already preloaded into the app. Without any pre-loaded data, markerless tracking tracks objects based only on their color, symmetry, and pattern as they change over time. Smartphone apps currently use augmented reality to help you locate your car in a crowded area. The popular game which became a rage in 2016 uses AR to introduce Pikachu and its characters when the user is playing the game on the roadside or around a waterfront [12].

Virtual reality on the other hand is a simulated environment which creates a real life like scenario by using images and sounds. This can be accessed using a headset or a helmet. VR works on the principle of depth perception. The degree of dissimilarity projected on the retina of an image creates perspective or depth perception. The history of VR dates back to 1800 and even Leonardo Da Vinci who experienced use of perspective in the paintings of that era. However, it was in 1990 that the introduction of VR headsets in gaming zones caused the arrival of a new era for the use of VR. Nowadays, when the term virtual reality (VR) is used, it usually refers to a binocular head-mounted display (HMD) that projects a two-dimensional (2D) perspective image of the virtual scene being emulated into each eye. Just as it does from the retinal projections of the real world received by each eye, the brain may deduce three-dimensional (3D) features of the simulated environment from these two projections. [13, 14]. The goal of virtual reality is to imitate this organic projection mechanism. Replicating VR from projection technologies such as stereoscopes, haploscopes, and 3D computer monitors [15] is the observer's ability to move freely within the computer generated 3D environment, giving rise to rich motion parallax information [16]. A stationary observer would not be able to get as much information about how the scene's visuals change over time in response to our movements as a moving observer can [17]. The addition of observer movement helps to clarify several possible interpretations of the scene design, regardless of how one moves a scene in relation to a static observer, such that the retinal images are the same as when the observer moves [18, 19]. For movement to be meaningful, the images of the simulated scene need to be refreshed in real time as the observer moves. This helps in healthcare to visualize the anatomy of the human body, navigate it, and pre-plan surgical procedures. The patients can also be educated about their disease, treatment plan, and prognosis [12].

Sahu et al. in [20] specified that Congregation of Internet of Things, augmented reality, and virtual reality is the future of healthcare. Utilizing brain-computer interface (BCI) and simultaneous localization and mapping (SLAM) for neurosurgical procedures, neuroprosthetics, neuromodulation, and neurorehabilitation, virtual reality and augmented reality are promising technologies in the neurosciences that show promise for the field.

Studies reveal that VR/AR platforms may be used to enhance one's visualizing abilities [21]. Virtual reality (VR) and augmented reality (AR) platforms are a subset of digital media. As such, technologies may be used to facilitate the development of specialized apps that teach visualization skills.

With the advent of growing scientific advancements, we are about to gain a greater grasp of the mechanisms underlying the evolution of consciousness [22]. This was obvious when neuroscientists and philosophers understood that awareness was not simply digital, and the computational illusion of the mind started falling apart. However, since we can now artificially mimic an all-encompassing environment in four dimensions thanks to virtual reality (VR), augmented reality (AR), and mixed reality (MR), our models of how consciousness functions are becoming much clearer to understand. Through attentional posturing, consciousness creates a forging mechanism that allows it to fully immerse itself in a world of its own creation, all the while remaining mostly oblivious to the mechanisms involved in such a magical performance. By using VR and MR as touchstones, we can gain a deeper understanding of how our brains create the environment we live in. We have the means, possibly for the first time in human history, to artificially reconstruct how the brain views and engages with reality by investigating manufactured virtual reality devices and the diverse landscapes they can open up. The article explores how the development of virtual reality alters our perception of human consciousness and how using it in different forms can help future studies [23]. The holistic and integrated experiences for learning that virtual reality (VR) technology offers make complicated subjects more approachable, and so create great prospects to improve astronomy education.

5 Comparative Study of Augmented Reality and Virtual Reality

Virtual Reality (VR) is a technology that allows users to experience a completely virtual environment through a headset or goggles [24]. It displays images and sounds using a Head-Mounted Display (HMD), while tracking sensors monitor the user's motions in real time. A sensation of presence is created when the VR world adapts to the user's motions [12]. The VR experience is powered by software and content, which creates and depicts the virtual world in real time in response to user inputs. Advanced computer graphics, optics, and human-computer interface (HCI) concepts are all part of the technology underpinning virtual reality (VR). For a smooth and engaging experience, high frame rates must be achieved during real-time rendering of 3D scenes. In order to improve realism and user engagement, VR systems also employ techniques like positional tracking, which tracks head and occasionally hand movements, and stereoscopic rendering, which creates depth perception.

Augmented Reality (AR) is a technology that overlays digital information or virtual objects onto the real world, enhancing our perception of reality rather than replacing it entirely [12]. AR applications range from smartphone apps to advanced headsets, and key components include display devices, sensors and cameras, and computer vision algorithms. These devices use sensors and cameras to understand the real-world environment and superimpose digital content onto it, allowing virtual objects to interact realistically with the environment [24]. Computer vision, sensor fusion, simultaneous localization and mapping (SLAM), and human factors engineering are all part of the science underpinning augmented reality. While SLAM algorithms assist in tracking the location and orientation of the device in relation to its surroundings, computer vision algorithms examine and interpret the scene that is collected by cameras. For AR systems to smoothly blend digital material with the actual environment in real-time, complex software is also needed.

Abdennour et al. [25] in their chapter write about comparison and choice between VR and AR. VR puts students in convincing virtual environments, making it perfect for practical training and simulations. In contrast, augmented reality (AR) provides contextual information and interactive simulations by superimposing digital elements onto the actual world through devices like smartphones and AR glasses. These technologies increase engagement by grabbing students' attention, facilitating practical experiences and a deeper understanding, and customizing the learning process. VR's 3D environments and AR's overlays enhance memorization. These methods can produce even deeper and more varied learning experiences, but how they are integrated will depend on the educational goals and available resources.

Currently, immersive virtual reality (iVR) and augmented reality (AR) are widely employed to teach anatomy across health sciences courses [26]. These interactive XR technologies (iVR and AR) permit the three-dimensional visualization of anatomical structures and their spatial relationships in virtual environments that resemble the real world. Virtual structures rendered through specialized software are similar to real structures; though, in the virtual environment, students interact with the anatomy using a joystick or through hand motions. VR headsets enable the recreation of immersive experiences (including motion tracking, interaction with digital elements, a wider range of vision, and stereoscopy), which allows students to visualize virtual anatomical objects as if they were real. Two characteristics of iVR are presence and immersion. Presence is defined as the psychological sensation/desire to stay within the virtual environment that the subject feels is real. Immersion is the capability to interact with virtual objects. Therefore, depending on the level of presence and immersion, VR can be either immersive or non-immersive. iVR allows viewing a virtual scenario through a head-mounted display in 360° whereby subjects can interact with the virtual environment and its objects using their hands as controllers. Conversely, AR superimposes virtual elements onto the real world allowing individuals to interact with real and virtual elements, simultaneously.

García-Robles et al. [27] in their systematic review and meta-analysis showed that XR-based interventions are more effective than traditional passive learning resources as a teaching-learning resource for studying anatomy in health sciences students. Both XR technologies (iVR and AR) improved learners' anatomy knowledge gains compared to controls, especially when used alongside traditional resources as supplemental/complimentary tools. A large effect favoring XR technologies was detected when compared against didactic lectures and anatomical images in atlases and/or textbooks. Overall, 80% of students who used XR technologies subjectively reported that XR technologies are useful for learning anatomy, especially when they are used as a supplemental resource in anatomy curricula.

Petruse et al. [28] conducted a comparative study of traditional method of teaching anatomy with mixed reality in a group of students aged 19–54 years. They concluded that the traditional method of teaching muscular anatomy with images on powerpoint was effective in older students and mixed reality created a 3D hologram using Microsoft HoloLens 2 was useful in younger students.

AI is being used in India to improve patient management, create treatment pathways and to monitor progress of patients. In one such study by Iqbal et al. [29], AI was used to predict sepsis in neonates admitted in Neonatal Intensive care Units. It was a retrospective study conducted in a South Indian tertiary hospital from 2017 to 2021. Using the data they created a model to detect early neonatal sepsis which can enhance the patient's outcome.

In another study done in Assam India which was conducted over a period of 2 years (Nov 2020–2022) Chiramal et al. [30] AI generated tools were used to generate reports of NCCT—Non contrast CT scan in patients who came with strokes to Baptist Christian hospital. AI would generate the report in 4 min and there was a tele radiology consultation done by a neurologist from the USA to guide further treatment plans for stroke patients. Though this was a small study done in a rural set up, nonetheless this highlights the potential AI has in healthcare.

6 Challenges and Future Scope

The findings of this study highlight AI's transformative potential in healthcare but also emphasize the importance of addressing ethical concerns to enable responsible and equitable application. Striking a balance between utilizing technological developments for better healthcare results and protecting individual rights and privacy is crucial as AI integration develops [31]. Strong data security measures are required due to privacy concerns, and algorithmic biases that could disproportionately affect particular demographic groups need to be lessened. By working together, engineers, medical experts, and ethicists may create regulations that guarantee accountability, equity, and transparency in AI-driven healthcare. Additionally, the conversation explores how AI may affect patient autonomy and how the doctor-patient relationship may change in the future. To empower patients in their healthcare journey, AI implementation must prioritize strategies for good communication and informed consent.

Future research in AI-driven healthcare will focus on improving algorithms' interpretability, minimizing biases, and guaranteeing strong privacy protections [32]. To keep up with the rapid advancement of technology, ethical standards must be continuously improved, and interdisciplinary teamwork is essential for tackling difficult problems. Future research prospects can be explored by investigating cutting-edge technologies like blockchain, augmented reality, and robotics in the healthcare industry. To fully utilize AI in enhancing global healthcare, it will be essential to comprehend the long-term societal implications and handle accessibility issues.

Tene et al.'s comprehensive review from 2024 thoroughly investigated the application of VR and AR in medical education to ascertain how well they enhanced personnel's competencies in comparison to traditional approaches. To ascertain the effect of immersive technology on educational results, 28 research was assessed utilizing chi-square analysis and descriptive statistics through a methodical application of PRISMA principles and the PICOS technique. According to our data, VR is the most studied. Based on the reviews included in this analysis, it appears that learner engagement and performance have been positively correlated with VR and AR technology.

The statistical research, however, did not show that there was a meaningful correlation between the phases of medical education and the kind of technology employed. This suggests that although the corpus of evaluated investigation indicates a positive trend toward the effectiveness of VR and AR in medical education, it is not yet statistically significant. It is crucial to emphasize that the lack of significance could be attributed to the various techniques, small sample numbers, and the heterogeneity of outcomes among the included research [33]. Finally, the expense and complexity of VR technology are important factors to take into account. Even with a declining trend in cost, setting up personal VR/AR devices for every student to use for distance learning is still a difficult and costly task. This scenario suggests a possible paradigm in which semi-centralized labs outfitted with VR/AR technology could host remote learning sessions in the future. In these environments, teachers could interact and impart knowledge remotely, while pupils had physical access to the required tools. This strategy might provide a more workable and affordable way to include immersive technologies into medical education, particularly in situations where it is not viable for each student to buy such a device.

While AR in medical education and training has different problems, it does not need to meet the same strict requirements as it does in clinical settings. The degree to which knowledge acquired via the AR system is sustained throughout time is unknown [34]. In another piece of literature, the authors acknowledged that the process of generating complicated anatomical regions in AR can vary depending on personal experiences, leading to various outcomes for individual anatomical regions. This literature discussed the construction and evaluation of an AR knowledge assessment tool. Establishing a standard for such development techniques is therefore crucial to ensure reliable and consistent implementation across a range of applications. Panchapakesan [35] in their study emphasized the challenges of AR and VR in Asia, which are technological limitations, infrastructure challenges and effective managerial strategies to remove barriers in implementation.

Currently in healthcare AR and VR is a concept which we are starting to hear about. The prospect of AI is immense in healthcare. There are a very few centers and sparingly few specialties in India currently using AI in ICU's as of now. We need Indian data and more research work to validate its practical applications. AI must be included in the curriculum at the undergraduate and post graduate level for medical students.

7 Conclusion

In summary, the application of virtual reality (VR) and augmented reality (AR) in healthcare is a ground-breaking development that will have a significant impact on therapeutic interventions, medical education, and patient care. Examining the science underlying AR and VR in healthcare makes it clear that these innovative technologies present special chances to improve diagnosis, treatment, and teaching in ways that were previously not known.

Above all, AR and VR have the potential to completely transform medical training and teaching [36]. Through the use of realistic settings and immersive simulations, medical practitioners can practice in a secure setting. This lowers the risk to patients during training processes in addition to improving the quality of care. Furthermore, medical students can now see intricate anatomical structures and physiological processes with never-before-seen clarity thanks to AR and VR, which improves their comprehension and memory of important information.

Furthermore, there is a great deal of promise for AR and VR to enhance outcomes for patients [37]. By superimposing pertinent medical data onto the patient's actual surroundings, these technologies can help diagnose patients more precisely and promptly while empowering medical professionals to act swiftly and decisively. Moreover, doctors can receive precise navigation and real-time support with AR-guided procedures, which improves surgical outcomes and lowers complications. Virtual reality-based therapies have also shown promise in the treatment of pain, rehabilitation, and mental health issues. These therapies provide patients with an alternative to conventional pharmaceutical interventions. In addition, the application of AR and VR in healthcare could improve patient empowerment and involvement. These technologies promote better comprehension and patient involvement in their own care by allowing patients to view their problems and treatment alternatives in immersive virtual environments. This helps patients stick to their treatment plans better and gives them the confidence to make better health-related decisions.

Though AR and VR have a lot of potential for the healthcare industry, there are a few issues that need to be resolved before they are widely used. The most important of them is the requirement for strong regulatory frameworks to guarantee patient privacy and safety [37]. Furthermore, implementation difficulties may arise due to the expensive cost of AR and VR technologies and the related infrastructure, especially in settings with limited resources [38]. Furthermore, in order to fully benefit from new technologies and use them successfully, healthcare workers need thorough training.

In summary, the science underlying virtual and augmented reality in healthcare signifies a paradigm shift in how we approach patient involvement, diagnosis, treatment, and medical education. Healthcare professionals may promote medical research, improve patient experiences, and improve clinical outcomes by utilizing the immersive qualities of AR and VR. However, in order to overcome obstacles and provide fair access to these game-changing instruments, cooperation between healthcare experts, technology developers, legislators, and regulatory bodies will be necessary to realize the full promise of these technologies. We have the chance to usher in a new era of personalized medicine where the lines between the physical and digital worlds blur and patient-centered care takes center stage as we continue to innovate and integrate AR and VR into healthcare.

References

1. Tene T, Vique López DF, Valverde Aguirre PE, Orna Puente LM, Vacacela Gomez C. Virtual reality and augmented reality in medical education: an umbrella review. Front Digit Health. 2024;6:1365345.
2. Mittal H. Virtual reality applications in healthcare: a new-age technology perspective. In: Immersive virtual and augmented reality in healthcare. CRC Press; 2024. p. 50–62.
3. Cutolo F. Augmented reality in image-guided surgery. In: Encyclopedia of computer graphics and games. Cham: Springer; 2024. p. 182–92.
4. Fawad M, Salamak M, Hanif MU, Koris K, Ahsan M, Rahman H, et al. Integration of bridge health monitoring system with augmented reality application developed using 3D game engine-case study. IEEE Access. 2024;12(16963):16974.
5. Oun A, Hagerdorn N, Scheideger C, Cheng X. Mobile devices or head-mounted displays: a comparative review and analysis of augmented reality in healthcare. IEEE Access. 2024;12:21825–39.
6. Rowan NJ. Digital technologies to unlock safe and sustainable opportunities for medical device and healthcare sectors with a focus on the combined use of digital twin and extended reality applications: a review. Sci Total Environ. 2024;926:171672.
7. Bryant L, Sedlarevic N, Stubbs P, Bailey B, Nguyen V, Bluff A, et al. Collaborative co-design and evaluation of an immersive virtual reality application prototype for communication rehabilitation (DISCOVR prototype). Disabil Rehabil Assist Technol. 2024;19(1):90–9.
8. Yondjo J, Siette J. "VR is the future": perspectives of healthcare professionals on virtual reality as a diagnostic tool for dementia status in primary care. BMC Med Inform Decis Mak. 2024;24(1):9.
9. Eunju L, Gyuli B, Yeonhui H. Effectiveness of the patient's severity classification competency promotion virtual reality program of nursing students during the COVID-19 pandemic period. Healthcare. 2023;11(8):1122. https://doi.org/10.3390/healthcare11081122.

10. Badr AF. The geriatric virtual escape room in pharmacy education: female students escape significantly faster than male students. Pharmacy. 2022;10(2):36. https://doi.org/10.3390/pharmacy10020036.
11. Culbertson CS, Shulenberger S, De La Garza R, Newton TF, Brody AL. Virtual reality cue exposure therapy for the treatment of tobacco dependence. J Cyber Ther Rehabil. 2012;5(1):57–64. PMID: 25342999; PMCID: PMC4204479.
12. Paul PK, Chatterjee R, Chakraborty S, Tiwary KS, Aithal PS, Sharma S. Advanced healthcare informatics practice vis-à-vis virtual reality implications—a scientific overview. In: Immersive virtual and augmented reality in healthcare; 2024. p. 28–49.
13. Julesz B. Foundations of cyclopean perception. Am J Psychol. University of Illinois. 1971:86(3):667–9.
14. Wheatstone C. On some remarkable and hitherto unobserved phenomena of binocular vision. Optom Wkly. 1962;53:2311–5.
15. Banks MS, Hoffman DM, Kim J, Wetzstein G. 3D displays. Annu Rev Vis Sci. 2016;2:397–435.
16. Hartley R, Zisserman A. Multiple view geometry in computer vision. Cambridge University Press; 2003.
17. Koenderink JJ, van Doorn AJ. Facts on optic flow. Biol Cybern. 1987;56(4):247–54.
18. Wexler M, Van Boxtel JJ. Depth perception by the active observer. Trends Cogn Sci. 2005;9(9):431–8.
19. Wexler M, Panerai F, Lamouret I, Droulez J. Self-motion and the perception of stationary objects. Nature. 2001;409(6816):85–8.
20. Sahu M, Gupta R, Ambasta RK, Kumar P. IoT-driven augmented reality and virtual reality systems in neurological sciences. Internet Things. 2024;25:101098.
21. Tuker C. Training spatial skills with virtual reality and augmented reality. In: Encyclopedia of computer graphics and games. Cham: Springer; 2024. p. 1904–12.
22. Lane DC. The rendered universe: why virtual reality unlocks the secret of consciousness. In: Consciousness studies in sciences and humanities: eastern and western perspectives. Cham: Springer; 2024. p. 101–23.
23. Kersting M, Bondell J, Steier R, Myers M. Virtual reality in astronomy education: reflecting on design principles through a dialogue between researchers and practitioners. Int J Sci Educ, Part B. 2024;14(2):157–76.
24. Daling LM, Schlittmeier SJ. Effects of augmented reality-, virtual reality-, and mixed reality–based training on objective performance measures and subjective evaluations in manual assembly tasks: a scoping review. Hum Factors. 2024;66(2):589–626.
25. Abdennour O, Kemouss H, Erradi M, Khaldi M. E-learning from virtual reality to augmented reality: comparison and choice. In: Technological tools for innovative teaching. IGI Global; 2024. p. 26–42.
26. Guillen-Sanz H, Checa D, Miguel-Alonso I, Bustillo A. A systematic review of wearable biosensor usage in immersive virtual reality experiences. Virtual Reality. 2024;28(2):74.
27. García-Robles P, Cortés-Pérez I, Nieto-Escámez FA, García-López H, Obrero-Gaitán E, Osuna-Pérez MC. Immersive virtual reality and augmented reality in anatomy education: a systematic review and meta-analysis. Anat Sci Educ. 2024;17:514–28.
28. Petruse RE, et al. Exploring the efficacy of mixed reality versus traditional methods in higher education: a comparative study. Appl Sci. 2024;14(3):1050.
29. Iqbal F, Chandra P, Lewis LES, Acharya D, Purkayastha J, Shenoy PA, Patil AK. Application of artificial intelligence to predict the sepsis in neonates admitted in neonatal intensive care unit. J Neonatal Nurs. 2024;30(2):141–7.
30. Chiramal JA, Johnson DJ, Webster DJ, Nag DR, Robert DD, Ghosh DT, et al. Artificial intelligence-based automated CT brain interpretation to accelerate treatment for acute stroke in India: an interrupted time series study. Available at SSRN 4684850.
31. Qin H, David A, Harun A, Mamun MRA, Peak D, Prybutok V. Assessing user benefits and privacy concerns in utilitarian and hedonic mobile augmented reality apps. Ind Manag Data Syst. 2024;124(1):442–82.

32. Kasula BY. Ethical implications and future prospects of artificial intelligence in healthcare: a research synthesis. Int Meridian J. 2024;6(6):1–7.
33. Chen PJ, Liou WK. The effects of an augmented reality application developed for pediatric first aid training on the knowledge and skill levels of nursing students: an experimental controlled study. Nurse Educ Today. 2023;120:105629.
34. Bogomolova K, Sam AH, Misky AT, Gupte CM, Strutton PH, Hurkxkens TJ, Hierck BP. Development of a virtual three-dimensional assessment scenario for anatomical education. Anat Sci Educ. 2021;14(3):385–93.
35. Panchapakesan P. Managerial challenges in VR and AR in Asia. In: Schoenecker JG, Tummins MS, Moore-Lotrid, editors. Managerial challenges and social impacts of virtual and augmented reality. IGI Global; 2020. p. 44–54. https://doi.org/10.4018/978-1-7998-2874-7.ch003.
36. Pantelidis P, Chorti A, Papagiouvanni I, Paparoidamis G, Drosos C, Panagiotakopoulos T, et al. Virtual and augmented reality in medical education. In: Medical and surgical education-past, present and future; 2018. p. 77–97.
37. Singh B, Vig K, Kaunert C. Modernizing healthcare: application of augmented reality and virtual reality in clinical practice and medical education. In: Modern technology in healthcare and medical education: blockchain, IoT, AR, and VR. IGI Global; 2024. p. 1–21.
38. Corbett M. Enhancing security and privacy in head-mounted augmented reality systems using eye gaze. ACM SIGMETRICS Perform Eval Rev. 2024;51(3):42–5.

Chapter 7
Virtual Reality-Based Intelligent Internet of Medical Things Health Monitoring System: In Future Aspects

V. Jyothi, Vijayalakshmi Chintamaneni, D. Sravani, Amar Y. Jaffar, Abdullah Alwabli, and N. Rajeswaran

Abstract It's great to see the focus on leveraging technology to improve the healthcare of the elderly population, especially those with chronic diseases. Monitoring daily health through wearable devices and smart inhalers is a fantastic way to provide personalized care and ensure timely intervention when needed. By enabling remote and real time monitoring, we can not only alleviate the strain on medical resources but also enhance the overall well-being of the elderly. IoT (Internet of Things) has the potential to revolutionize healthcare by enhancing patient care and safety. By integrating IoT devices and sensors into surgical processes, healthcare providers can monitor vital signs, track surgical instruments, and even analyze real-time data to ensure precision and accuracy during procedures. This can help reduce errors, improve surgical outcomes, and ultimately enhance patient recovery. It's fascinating to see the innovative approach of using IoT and virtual reality technology to design wearable devices for monitoring the daily physical parameters of the elderly. By focusing on key metrics like exercise heart rate, blood pressure, plantar

V. Jyothi
Department of ECE, Vardhaman College of Engineering, Hyderabad, Telangana, India

V. Chintamaneni
Department of ECE, Vignan Institute of Technology and Science, Hyderabad, Telangana, India

D. Sravani
Department of IT, Malla Reddy Engineering College, Maisammaguda, Secunderabad, Telangana, India

A. Y. Jaffar
Computer and Network Engineering Department, College of Computing, Umm Al-Qura University, Makkah, Saudi Arabia

A. Alwabli
Department of Electrical Engineering, College of Engineering and Computing in Alqunfudah, Umm Al-Qura University, Mecca, Saudi Arabia

N. Rajeswaran (✉)
School of Computer Applications, IMS Unison University, Dehradun, Uttarakhand, India

B. O. Soufiane et al. (eds.), *Augmented Wellness*,
https://doi.org/10.1007/978-981-96-2952-7_7

health, and sleep function, this solution aims to address the limitations of traditional monitoring equipment and enhance the accuracy of parameter measurement. Conducting experiments to verify the feasibility of this measurement method and equipment is crucial for ensuring its effectiveness in improving the healthcare monitoring of the elderly.

Keywords VR · IoT · Health care · Medical IoT and integration

1 Introduction

As technology evolves, the intersection of Virtual Reality (VR), the Internet of Things (IoT), and healthcare is paving the way for innovative solutions in medical monitoring systems. One such advancement is the VR-based Intelligent Medical IoT Health Monitoring System, promising revolutionary changes in healthcare delivery. In this article, we probe into the prospects of this transformative technology and its probable influence on the healthcare landscape. The convergence of Virtual Reality (VR), the Internet of Things (IoT), and healthcare is reshaping the landscape of medical monitoring systems. The VR-based Intelligent Medical IoT Health Monitoring System is among the ground breaking innovations emerging from this convergence. This transformative technology promises to revolutionize healthcare delivery by leveraging immersive VR experiences and interconnected IoT devices to enhance patient care, improve efficiency, and drive better health outcomes. At its core, the VR-based Intelligent Medical IoT Health Monitoring System integrates VR technology with IoT-enabled medical devices to create a seamless and interactive healthcare environment. This system collects real-time data from wearable sensors, medical equipment, and other IoT devices, which are visualized and interpreted within immersive VR environments. By combining the power of VR simulations with the connectivity of IoT, healthcare benefactors can increase unprecedented acumens into patient health and deliver personalized care like never before.

1.1 Personalized Healthcare Experiences

In the future, VR-based Intelligent Medical IoT Health Monitoring Systems will offer highly personalized healthcare experiences tailored to individual patient needs. Patients can immerse themselves in virtual environments designed to educate, motivate, and support them in managing their health conditions [1, 2]. Whether it's a virtual fitness coach guiding exercise routines or a virtual therapist providing mental health support, the possibilities for personalized healthcare experiences are endless.

1.2 Remote Patient Monitoring and Telemedicine

As telemedicine continues to gain traction, VR-based Intelligent Medical IoT Health Monitoring Systems will play a decisive part in enabling remote patient monitoring and virtual consultations. Patients can connect with healthcare providers from the comfort of their homes, with VR technology providing a lifelike experience that closely mimics face-to-face interactions [3]. This will not only improve access to healthcare for individuals in remote or underserved areas but also enhance the efficiency of healthcare delivery by reducing the need for in-person visits.

1.3 Predictive Analytics and Preventive Care

With the wealth of data generated by IoT-enabled medical devices, VR-based Intelligent Medical IoT Health Monitoring Systems will harness the power of predictive study to classify decorations, trends, and early warning signs of potential health issues. ML methodologies will investigate immense sums of patient information to foresee and avert adverse health events before they occur [4–6]. This proactive approach to healthcare will shift the focus from reactive treatments to preventive interventions, ultimately leading to improved health consequences and condensed healthcare costs.

1.4 Medical Training and Education

VR-based simulations will become indispensable tools for immersive learning experiences concerning medical training and education [7–9]. Healthcare professionals will have access to realistic virtual environments to practice surgical procedures, hone diagnostic skills, and simulate complex medical scenarios. By so long as an innocuous and meticulous environment for hands-on training, VR-based Intelligent Medical IoT Health Monitoring Systems will help enhance the skills and competencies of healthcare professionals, ultimately improving patient care.

1.5 Ethical and Regulatory Considerations

As with any emerging technology in healthcare, the widespread adoption of VR-based Intelligent Medical IoT Health Monitoring Systems will raise ethical and regulatory considerations. Drawbacks, such as patient secrecy, information sanctuary, and governing compliance, will need to be addressed, and a lecture will be given to certify the authority and ethical use of these pieces of machinery [10, 11].

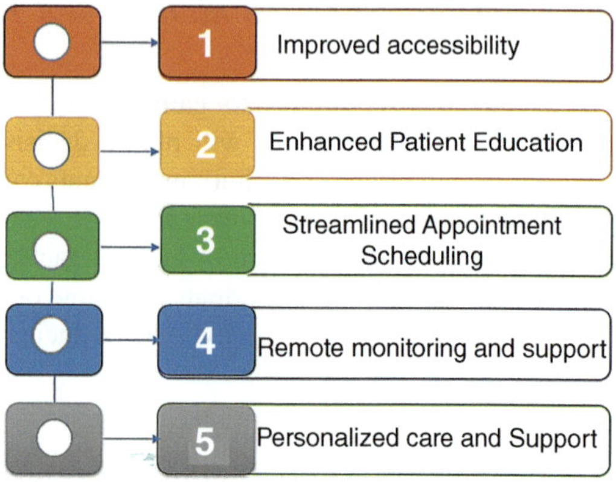

Fig. 7.1 Enhanced patient engagement

Hard work will also be required to conduct the cardinal divide and certify equitable admittance to VR-based healthcare solutions for all those, regardless of socioeconomic status or geographical location. The prospects of VR-based Intelligent Medical IoT Health Monitoring Systems (Shown in Fig. 7.1) are promising and transformative. By harnessing the combined power of VR, IoT and advanced analytics, these systems have the potential to revolutionize healthcare delivery, empower patients and improve health outcomes globally. However, realizing this potential will require collaboration among healthcare providers, technology developers, policymakers, and other stakeholders to overcome challenges and embrace the opportunities presented by this ground breaking technology. As we look ahead, the integration of VR and IoT in healthcare will continue to drive innovation and shape the future of medicine for the better.

2 Enhanced Patient Engagement

One of the main features of VR-based medical IoT architecture is their facility to enhance patient engagement. By immersing patients in interactive virtual environments tailored to their specific medical needs, these systems can provide a more engaging and personalized healthcare experience. Patients can visualize their health data in real time, fostering better understanding and adherence to treatment plans [12]. Enhanced patient engagement is a cornerstone of VR-based medical IoT systems, offering a transformative approach to healthcare delivery. These systems can revolutionize how patients interact with their health data and participate in their care journey through the fusion of virtual reality (VR) and Internet of Things (IoT) technologies. By immersing patients in interactive virtual environments specifically designed to cater to their medical needs, VR-based medical IoT systems offer a level of engagement and personalization unparalleled by traditional healthcare methods. Imagine a scenario where a patient with a chronic condition can put on a VR headset and step into a virtual representation of their own body. Within this immersive environment, they can visualize real-time data streams from wearable sensors, medical devices, and electronic health records.

This immersive experience transcends the limitations of traditional two-dimensional charts and graphs, allowing patients to see their health data come to life in three dimensions. They can observe fluctuations in vital signs, track progress over time, and gain a deeper understanding of how their lifestyle choices impact their health outcomes. Moreover, VR-based medical IoT systems can potentially tailor virtual environments to meet each patient's unique needs and preferences [13, 14]. For example, a patient recovering from surgery may benefit from a serene virtual oasis where they can engage in relaxation exercises and monitor their recovery progress. In contrast, a patient managing a chronic condition like diabetes might find value in a virtual kitchen where they can learn about healthy eating habits and practice meal planning. The interactive nature of these virtual environments encourages active participation from patients in their care plans. Rather than passively receiving instructions from healthcare providers, patients become empowered to take control of their health by visualizing data, setting goals, and tracking their progress in real time. This develops a sense of ownership and responsibility, which leads to improved adherence to treatment plans and better health outcomes in the long term.

In Fig. 7.1, furthermore, VR-based medical IoT systems have the potential to bond the message break among patients and healthcare earners. Patients can use virtual environments to communicate their symptoms, concerns, and preferences more intuitively and immersively. Healthcare providers, in turn, can use VR technology to explain complex medical concepts, demonstrate treatment procedures, and engage patients in shared decision-making processes. The ability of VR-based medical IoT systems to enhance patient engagement represents a paradigm shift in healthcare delivery. By providing patients with immersive, personalized experiences that enable them to visualize and interact with their health data, these systems have the potential to authorize those to take one dynamic part in supervising their health [15]. As technology evolves, we can expect VR-based medical IoT systems to play an increasingly integral role in fostering patient-centred care and refining health consequences for diagonally assorted populations. Table 7.1 presents a simplified version of the dataset. Each row represents data collected from a different patient participating in the study. The columns contain various attributes and measurements, including patient demographics, baseline pain scores, pain scores after the intervention, physiological data collected from wearables, and the treatment group to which the patient was assigned (e.g., VR Intervention or Control Group).

3 Improved Remote Monitoring

In Fig. 7.2, Integrating IoT devices with VR technology enables seamless remote monitoring of patients' vigorous cyphers and health restrictions. Wearable sensors embedded in everyday objects or clothing continuously collect data and spread it to

Table 7.1 Patient-centred care and refining health consequences

Patient ID	Age	Gender	Baseline pain score	Pain score (after intervention)	Physiological data (e.g., heart rate, skin conductance)	Treatment group
001	35	Male	7	4	{heart rate: 75 bpm, skin conductance: 0.05 μS}	VR intervention
002	42	Female	6	3	{heart rate: 68 bpm, skin conductance: 0.03 μS}	Control group
003	50	Male	8	2	{heart rate: 80 bpm, skin conductance: 0.07 μS}	VR intervention

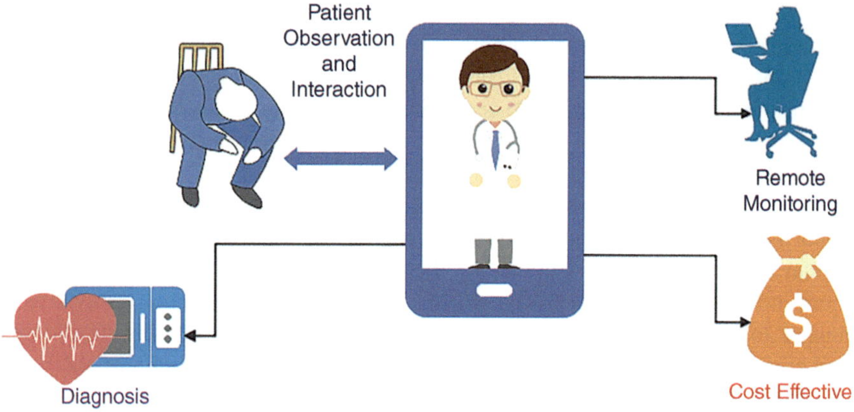

Fig. 7.2 Improved remote monitoring

the VR interface. Healthcare providers can remotely access this information, allowing for timely interventions and proactive healthcare management, which is predominantly advantageous for patients with chronic conditions or those in remote areas. Improved Remote Monitoring is a pivotal advancement facilitated by integrating IoT devices with Virtual Reality (VR) technology [16]. This synergy enables healthcare providers to monitor patients' vital signs and health parameters remotely, revolutionizing healthcare delivery, especially for patients with chronic diseases or individuals living in remote areas.

Central to this innovation are wearable sensors embedded within everyday objects or clothing, which continuously gather data related to patients' physiological metrics. These sensors serve as the frontline data collectors, capturing info, for instance, heart rate, blood pressure, blood glucose levels, and other relevant health indicators. By leveraging IoT connectivity, this data is transmitted in real-time to a centralized VR interface accessible to healthcare providers. The VR interface serves as a comprehensive dashboard, displaying aggregated patient data intuitively and visually engagingly [17]. Through immersive visualizations and interactive elements, healthcare wage-earners can remotely examine patients' health status with unprecedented clarity and efficiency. Vital signs and health parameters are presented in real time, allowing for timely interventions and proactive healthcare management.

3.1 The Benefits of Improved Remote Monitoring

Timely Interventions: Healthcare providers can promptly identify any abnormalities or deviations from baseline values by receiving real-time updates on patients' health metrics. This enables timely interventions to address emerging health issues before they escalate, potentially preventing complications and reducing hospital readmissions.

Proactive Healthcare Management: Continuous remote monitoring facilitates proactive healthcare management strategies tailored to individual patient needs. Healthcare providers can track patient health parameter trends over time, identify patterns, and adjust treatment plans accordingly. This proactive approach empowers providers to augment patient consequences and enhance complete eminence of care.

Enhanced Patient Safety and Comfort: Remote monitoring via VR-based IoT systems offers patients the convenience of receiving healthcare services from their homes. This reduces the burden of frequent clinic visits and minimizes the risk of exposure to infectious diseases, which is particularly relevant in public health crises such as pandemics.

Improved Access to Care: For patients residing in remote or underserved areas with limited access to healthcare facilities, improved remote monitoring bridges the gap in healthcare delivery. Through IoT-enabled wearables and VR technology, patients can receive high-quality care irrespective of geographical barriers, ensuring equitable access to healthcare services.

Data-Driven Decision-Making: The wealth of data collected through remote monitoring provides valuable insights for data-driven decision-making in healthcare. Healthcare providers can analyze longitudinal data trends, identify risk factors, and tailor interventions based on individual patient profiles. This personalized approach enhances treatment efficacy and fosters patient engagement in self-management of health. Integrating IoT devices with VR technology has ushered in a new era of remote monitoring in healthcare. By enabling seamless transmission of patient data and empowering healthcare providers with real-time insights, this innovative approach holds immense promise for improving patient outcomes, enhancing healthcare delivery, and promoting health equity across diverse populations [18–20]. As technology continues to evolve, the potential for VR-based remote monitoring to revolutionize healthcare remains boundless, shaping the future of medicine in profound and impactful ways.

4 Data Analytics and Predictive Insights

VR-based intelligent medical IoT systems leverage advanced analytics to derive meaningful insights from vast amounts of health data. Machine learning algorithms analyze trends, detect anomalies, and predict potential health risks, empowering healthcare professionals to make informed decisions and intervene before complications arise. This predictive capability holds immense promise for preventive medicine, enabling early detection of diseases and personalized interventions. Data Analytics and Predictive Insights play a pivotal role in VR-based intelligent medical IoT systems, offering a transformative approach to healthcare management. By harnessing the power of advanced analytics and machine learning algorithms, these systems can unlock valuable insights from large volumes of health data, paving the way for early detection of diseases, personalized interventions, and proactive healthcare strategies.

At the heart of this innovation are machine learning algorithms capable of analyzing trends, detecting anomalies, and predicting potential health risks based on data collected from IoT-enabled devices and VR interfaces. These algorithms leverage sophisticated data analytics techniques to sift through vast amounts of patient data, identifying patterns and correlations that may not be readily apparent to human observers. The predictive capability of VR-based intelligent medical IoT systems holds immense promise for preventive medicine. By analyzing historical patient data and identifying risk factors, these systems can anticipate and preemptively address potential health issues before they manifest clinically [21]. For example, machine learning algorithms can predict the likelihood of a patient developing a particular disease based on their genetic predisposition, lifestyle factors, and environmental influences. Armed with this predictive insight, healthcare professionals can implement targeted interventions and preventive measures to mitigate the risk of disease progression. Furthermore, VR-based intelligent medical IoT systems enable personalized interventions tailored to the specific needs of individual patients. By analyzing patient data in real-time and continuously monitoring health parameters, these systems can adapt treatment plans and care protocols to optimize outcomes for each patient. For instance, machine learning algorithms can recommend personalized lifestyle modifications, medication adjustments, or behavioural interventions based on an individual's unique health profile and response to treatment.

The integration of data analytics and predictive insights into VR-based intelligent medical IoT systems offers several key benefits:

Early Detection and Intervention: By identifying subtle changes in patient data indicative of potential health issues, these systems enable early detection and intervention, preventing complications and improving patient outcomes.

Precision Medicine: Through the analysis of patient-specific data, VR-based intelligent medical IoT systems facilitate precision medicine approaches, tailoring treatments and interventions to each patient's individual characteristics and needs.

Resource Optimization: By predicting future healthcare needs and proactively addressing them, these systems help optimize resource allocation, reduce healthcare costs, and enhance operational efficiency within healthcare organizations.

Patient Empowerment: By providing patients with personalized insights about their health and potential risks, these systems empower individuals to actively manage their health and make informed decisions about their care. Integrating data analytics and predictive insights into VR-based smart medical IoT systems represents a paradigm shift in healthcare delivery. By leveraging the power of advanced analytics and machine learning, these systems enable proactive, personalized, and data-driven approaches to healthcare management, ultimately improving patient outcomes and advancing the practice of preventive medicine. As technology evolves, the potential for VR-based intelligent medical IoT systems to revolutionize healthcare remains limitless, offering new avenues for innovation and transformation in medicine.

5 Training and Simulation

Beyond patient care, VR-based medical IoT systems offer unparalleled healthcare education and training opportunities. Medical students and professionals can immerse themselves in realistic simulations of surgical procedures, medical emergencies, and complex diagnoses, enhancing their skills in a risk-free environment. Additionally, VR simulations can bridge geographical barriers, facilitating collaboration among healthcare teams and fostering continuous learning and skill development. Training and simulation represent a cornerstone application of VR-based medical IoT systems, revolutionizing healthcare education and professional development [22–25]. By leveraging immersive Virtual Reality (VR) technology and IoT-enabled simulations, these systems offer unparalleled opportunities for medical students and professionals to enhance their skills, knowledge, and expertise in a risk-free and collaborative environment. In medical education and training, VR-based simulations provide a realistic and interactive platform for learners to practice various procedures, from routine clinical tasks to complex surgical interventions. Medical students can immerse themselves in lifelike simulations of surgical procedures, medical emergencies, and diagnostic scenarios, gaining valuable hands-on experience and confidence in their abilities.

One of the key advantages of VR-based medical simulations is their ability to replicate real-world scenarios with high fidelity, allowing learners to experience and respond to clinical challenges in a safe and controlled environment. By simulating diverse clinical scenarios, VR-based simulations help learners develop critical thinking skills, decision-making abilities, and procedural proficiency, all essential for delivering high-quality patient care. Moreover, VR-based medical simulations have the potential to bridge geographical barriers and facilitate collaboration among healthcare teams, regardless of their physical location [26–30]. In Fig. 7.3, networked VR environments, medical students and professionals can participate in

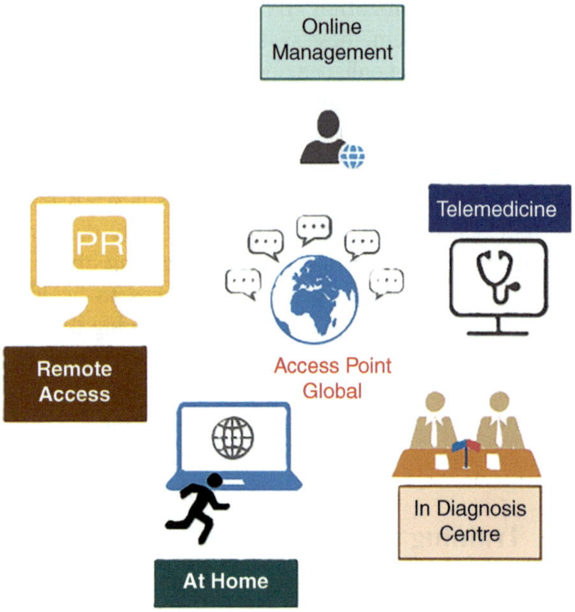

Fig. 7.3 Healthcare education and training opportunities

collaborative training sessions, interdisciplinary simulations, and team-based exercises, fostering communication, teamwork, and mutual understanding among healthcare providers.

5.1 The Benefits of Training and Simulation in VR-Based Medical IoT Systems Include

5.1.1 Realistic and Immersive Learning

VR-based simulations provide learners with an immersive and realistic learning experience, allowing them to interact with virtual patients, medical equipment, and clinical environments in a lifelike manner. This hands-on approach enhances retention and engagement, leading to more effective learning outcomes.

5.1.2 Risk-Free Environment

VR-based simulations offer a risk-free environment for learners to practice and refine their skills without exposing actual patients to potential harm. This allows learners to make mistakes, learn from them, and gain confidence in their abilities before applying their skills in real-world clinical settings.

5.1.3 Scalability and Accessibility

VR-based medical simulations can be accessed remotely and scaled to accommodate large numbers of learners, making them ideal for training programs with diverse student populations and limited physical resources [31–33]. This accessibility ensures equitable access to high-quality medical education and training opportunities for learners worldwide.

5.1.4 Interdisciplinary Collaboration

VR-based simulations facilitate interdisciplinary collaboration by bringing together learners from different healthcare disciplines, such as medicine, nursing, and allied health professions. This collaborative approach encourages communication, teamwork, and mutual respect among healthcare providers, ultimately improving patient care outcomes.

5.1.5 Continuous Learning and Skill Development

VR-based medical simulations support continuous learning and skill development throughout healthcare professionals' careers [34]. These simulations enable learners to refine their skills, stay updated on best practices, and adapt to evolving healthcare technologies and protocols by providing access to realistic and challenging scenarios. Training and simulation in VR-based medical IoT systems offer transformative healthcare education and professional development opportunities. By providing learners with realistic, immersive, and collaborative learning experiences, these systems empower the next generation of healthcare professionals to deliver high-quality, patient-centered care in an ever-changing healthcare landscape. As technology continues to evolve, the potential for VR-based medical simulations to revolutionize healthcare education and training remains limitless, paving the way for a future where excellence in patient care is driven by innovation, collaboration, and continuous learning.

6 Telemedicine and Virtual Consultations

In Fig. 7.5, the convergence of VR, IoT, and telemedicine is redefining the delivery of healthcare services. Virtual consultations conducted through immersive VR environments enable healthcare providers to interact with patients remotely, replicating the experience of face-to-face visits. This enhances access to healthcare for underserved populations, reduces healthcare costs, and improves resource utilization [35]. Telemedicine and Virtual Consultations are transforming by integrating Virtual Reality (VR), the Internet of Things (IoT), and telemedicine technologies. This convergence is revolutionizing the delivery of healthcare services by enabling immersive virtual consultations that bridge the gap between patients and healthcare providers, regardless of geographical distance [36, 37]. Through VR-based telemedicine solutions, healthcare providers can interact with patients remotely, replicating the experience of traditional face-to-face visits in a virtual environment.

6.1 Key Aspects of How VR, IoT, and Telemedicine Are Redefining Healthcare Delivery Include

6.1.1 Enhanced Access to Healthcare

VR-based telemedicine removes geographical barriers, allowing patients in remote or underserved areas to access healthcare services from the comfort of their homes [38]. This is especially beneficial for individuals living in rural areas with limited medical facilities, who may face the challenge of travelling long distances for in-person appointments.

6.1.2 Immersive Virtual Consultations

Virtual consultations conducted through VR environments provide an immersive and interactive experience for both patients and healthcare providers. Patients can engage in real-time conversations with their healthcare providers, discuss symptoms, ask questions, and receive medical advice in a virtual setting that closely simulates face-to-face interactions. This enhances patient engagement and satisfaction with the healthcare experience.

6.1.3 Cost-Effectiveness and Resource Utilization

By leveraging VR-based telemedicine solutions, healthcare organizations can reduce the costs associated with traditional in-person consultations, such as travel expenses, facility maintenance, and staffing. Virtual consultations also optimize resource utilization by allowing healthcare providers to see more patients in less time, improving efficiency and reducing appointment wait times.

6.1.4 Remote Monitoring and IoT Integration

IoT devices play a crucial role in supporting VR-based telemedicine by enabling remote patient health data monitoring. Wearable sensors, medical devices, and IoT-enabled home monitoring systems can continuously collect and transmit patient data, such as vital signs, medication adherence, and activity levels, to healthcare

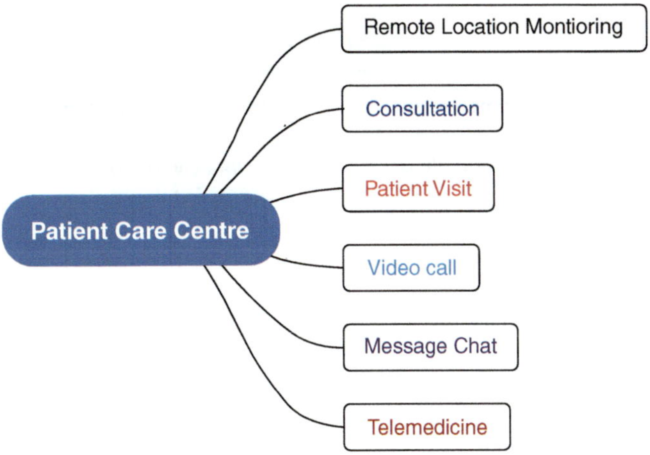

Fig. 7.4 Telemedicine and virtual consultations

providers in real time. This seamless integration of IoT with VR-based telemedicine enhances the accuracy of medical assessments and enables proactive interventions based on objective health metrics.

6.1.5 Privacy and Security

As with any healthcare technology, ensuring the privacy and security of patient data is critical for VR-based telehealth. Healthcare organizations must implement robust encryption protocols, authentication mechanisms, and data storage practices to protect patient confidentiality and comply with regulatory requirements, such as HIPAA in the U.S. and GDPR in the EU. The convergence of VR, IoT and telemedicine is reshaping the landscape of healthcare delivery, providing unprecedented opportunities to improve accessibility, efficiency and quality of care. Virtual consultations via immersive VR environments can revolutionize healthcare delivery, making healthcare more accessible, cost-effective and patient-centric. As technology develops, VR-based telemedicine (Shown in Fig. 7.4) is expected to become an integral part of the modern healthcare ecosystem, enabling patients and healthcare providers to embrace a new era of virtual healthcare services.

7 Challenges and Limitations

While the future of VR-based intelligent medical IoT health monitoring systems holds tremendous promise, several challenges must be addressed. These include concerns regarding data privacy and security, interoperability of IoT devices, regulatory compliance, and equitable access to technology [20]. Moreover, widespread

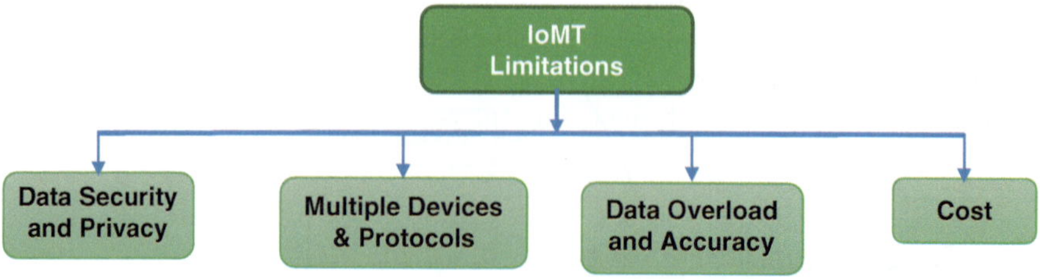

Fig. 7.5 Challenges and limitations of IoMT

adoption may require overcoming cost, infrastructure, and technological literacy barriers [39, 40]. Integrating VR-based intelligent medical IoT health monitoring systems presents numerous opportunities to revolutionize healthcare delivery. However, several challenges and considerations must be addressed to realize the full potential of this technology (Fig. 7.5).

7.1 Data Privacy and Security

The collection, storage and transfer of sensitive health data raises concerns about patient privacy and data security. Healthcare organizations must implement strong security measures, encryption protocols, and access controls to protect patient information from unauthorized access, leakage, and cyber-attacks.

7.2 Interoperability of IoT Devices

The interoperability of IoT devices from different manufacturers is essential for seamless data exchange and integration within healthcare systems. Standardization of communication protocols, data formats, and interoperability standards is necessary to ensure compatibility and interoperability among diverse IoT devices and platforms.

7.3 Regulatory Compliance

VR-based intelligent medical IoT systems must comply with regulatory requirements and standards governing healthcare data privacy, security, and quality. Healthcare organizations must navigate complex regulatory landscapes, such as HIPAA in the United States and GDPR in the European Union, to ensure compliance with legal and regulatory obligations.

7.4 Equitable Access to Technology

Ensuring equitable access to VR-based intelligent medical IoT systems is essential to address healthcare access and outcomes disparities. Healthcare organizations must consider factors such as socioeconomic status, geographic location, and digital literacy when implementing and deploying technology solutions to ensure that all patients have access to high-quality healthcare services.

7.5 Cost and Infrastructure

The initial cost of implementing VR-based intelligent medical IoT systems, including hardware, software, and infrastructure, can hinder adoption by some healthcare organizations [40]. Moreover, ongoing maintenance, training, and support costs must be considered to ensure the sustainability and scalability of technology solutions over time.

7.6 Technological Literacy

Healthcare providers and patients may require training and education to use and interact with VR-based intelligent medical IoT systems effectively [41]. Improving technological literacy and providing comprehensive training programs can empower healthcare professionals and patients to maximize the benefits of technology and overcome barriers to adoption. Addressing these challenges and considerations requires collaboration among healthcare stakeholders, including providers, technology vendors, regulatory bodies, and policymakers. By addressing privacy and security concerns, promoting interoperability, ensuring regulatory compliance, promoting equitable access, addressing cost and infrastructure challenges, and improving technological literacy, healthcare organizations can harness the full potential of VR-based intelligent medical IoT systems to improve patient outcomes, enhance healthcare delivery, and transform the future of medicine.

8 Conclusion

The future of healthcare lies at the intersection of Virtual Reality, the Internet of Things, and intelligent data analytics. VR-based intelligent medical IoT health monitoring systems have the potential to revolutionize healthcare delivery by enhancing patient engagement, enabling remote monitoring, facilitating predictive insights, advancing medical training, and transforming telemedicine. However, realizing this potential requires concerted efforts from healthcare stakeholders to address challenges and embrace the opportunities presented by this transformative technology. As we navigate towards a future driven by innovation, VR-based intelligent medical IoT systems offer a glimpse into a more connected, accessible, and patient-centric healthcare ecosystem. In conclusion, the convergence of Virtual Reality (VR), the Internet of Things (IoT), and intelligent data analytics represents a groundbreaking paradigm shift in healthcare delivery. VR-based intelligent medical IoT health monitoring systems have emerged as a transformative technology with the potential to revolutionize the future of healthcare.

EVR-based intelligent medical IoT systems empower patients to take an active role in their healthcare journey by providing immersive and personalized experiences that foster better understanding and adherence to treatment plans. Through continuous remote monitoring and predictive analytics, healthcare providers can anticipate and intervene proactively, leading to early detection of diseases, personalized interventions, and improved patient outcomes. VR-based simulations offer realistic and risk-free environments for medical education and training, allowing students and professionals to hone their skills, refine their techniques, and stay updated on the latest advancements in healthcare. Virtual consultations conducted through VR environments bridge geographical barriers, enabling remote access to healthcare services and reducing costs while maintaining the quality of care and patient satisfaction.

8.1 Future Scope

However, realizing the full potential of VR-based intelligent medical IoT systems requires concerted efforts from healthcare stakeholders. Addressing challenges related to data privacy and security, interoperability, regulatory compliance, equitable access, cost, infrastructure, and technological literacy is crucial to ensure the successful adoption and integration of this transformative technology into mainstream healthcare practices. As we navigate towards a future driven by innovation, VR-based intelligent medical IoT systems offer a glimpse into a more connected, accessible, and patient-centric healthcare ecosystem. By embracing the opportunities this transformative technology presents and working collaboratively to overcome challenges, healthcare stakeholders can pave the way for a future where healthcare delivery is more efficient, effective, and personalized.

References

1. Shah C, Doshi M, Patel S, Tanwar W-CH, Sharma R. A comprehensive review of the technological solutions to analyse the effects of pandemic outbreak on human lives. Medicina. 2022;58(2):311. https://doi.org/10.3390/medicina58020311.

2. Antonio Gambera D, Duarte E, Ricco D. Internet of senses (IoS) and internet of sensory health (IoSH): a new technology epiphany. AHFE Int. 2022. https://doi.org/10.54941/ahfe1001403.
3. Chaudhary S, Kakkar R, Jadav NK, Nair A, Gupta R, Tanwar S, et al. A taxonomy on smart healthcare technologies: security framework, case study, and future directions. J Sens. 2022;2022:1–30. https://doi.org/10.1155/2022/1863838.
4. Bhatia M, Kumari S. A novel IoT-Fog-cloud-based healthcare system for monitoring and preventing encephalitis. Cogn Comput. 2021;14(5):1609–26. https://doi.org/10.1007/s12559-021-09856-3.
5. Javaid M, Haleem A, Singh RP, Rab S, Ul Haq MI, Raina A. Internet of things in the global healthcare sector: significance, applications, and barriers. Int J Intell Netw. 2022;3:165–75. https://doi.org/10.1016/j.ijin.2022.10.002.
6. Smith J, Johnson A. The future of healthcare: virtual reality, internet of things, and intelligent data analytics. IEEE J Healthcare Technol. 2023;5(2):123–35.
7. Jones B, White C. Virtual reality in medicine: applications and challenges. New York: Springer; 2020.
8. Brown D, et al. Integration of virtual reality and IoT for remote patient monitoring. In: Proceedings of the IEEE international conference on healthcare informatics; 2022. p. 45–50.
9. Muhammad Auwal A. IoT integration in telemedicine: investigating the role of internet of things devices in facilitating remote patient monitoring and data transmission. 2023. https://doi.org/10.21203/rs.3.rs-3419693/v1.
10. Chaudhary S, Kakkar R, Jadav NK, Nair A, Gupta R, Tanwar S, Agrawal S, Alshehri MD, Sharma R, Sharma G, Davidson IE. A taxonomy on smart healthcare technologies: security framework, case study, and future directions. J Sens. 2022;2022:1863838. https://doi.org/10.1155/2022/1863838, 30 pages.
11. Gupta M, Tanwar S, Rana A, Walia H. Smart healthcare monitoring system using wireless body area network. In: 2021 9th international conference on reliability, infocom technologies and optimization (trends and future directions) (ICRITO), Noida, India, 2021. p. 1–5.
12. Diwaker C, Jangra A, Rani A. Survey on IoT health care techniques. In: 2019 5th international conference on signal processing, computing and control (ISPCC), Solan, India, 2019. p. 383–5.
13. Diwaker C, Jangra A, Rani A. Survey on IoT health care techniques. In: 2019 5th international conference on signal processing, computing and control (ISPCC), Solan, India, 2019. p. 383–5. https://doi.org/10.1109/ISPCC48220.2019.8988455.
14. Selvaraj S, Sundaravaradhan S. Challenges and opportunities in IoT healthcare systems: a systematic review. SN Appl Sci. 2020;2:1–8. https://doi.org/10.1007/S42452-019-1925-Y.
15. Nižetić S, Solic P, González-de-Artaza DL-d-I, Patrono L. Internet of things (IoT): opportunities, issues and challenges towards a smart and sustainable future. J Clean Prod. 2020;274:122877. https://doi.org/10.1016/J.JCLEPRO.2020.122877.
16. Alam MM, Malik H, Khan MI, Pardy T, Kuusik A, Le Moullec Y. A survey on the roles of communication technologies in IoT-based personalized healthcare applications. IEEE Access. 2018;6:36611–31. https://doi.org/10.1109/ACCESS.2018.2853148.
17. Wan J, Al-awlaqi MAAH, Li MS, Grady MO, Gu X. Wearable IoT enabled real-time health monitoring system. EURASIP J Wirel Commun Netw. 2018;298:1–10.
18. Nair S, Augustine N, Varghese L. Smart health monitoring system. Int J Eng Res Technol. 2020;9(6):1–6.
19. Sengan S, Khalaf OI, Priyadarsini S, Sharma DK, Amarendra K, Hamad AA. Smart healthcare security device on medical IoT using Raspberry Pi. Int J Reliable Qual E-Healthcare. 2021;11(3):1–11.
20. Pravin Savaridass M, Ikram N, Deepika R, Aarnika R. Development of smart health monitoring system using internet of things. Mater Today Proc. 2021;45(Part 2):986–9. ISSN 2214-7853. https://doi.org/10.1016/j.matpr.2020.03.046.
21. Pradhan B, Bhattacharyya S, Pal K. IoT-based applications in healthcare devices. J Healthc Eng. 2021;2021:6632599. https://doi.org/10.1155/2021/6632599, 18 pages.

22. Menon VG, Jacob S, Joseph S, Sehdev P, Khosravi MR, Al-Turjman F. An IoT-enabled intelligent automobile system for smart cities. Internet Things. 2022;18:100213.
23. Wang Y, An X, Xu W. Intelligent medical IoT health monitoring system based on VR and wearable devices. J Intell Syst. 2023;32. https://doi.org/10.1515/jisys-2022-0291.
24. Gang Y, Chen X, Wang H, Li J, Guo Y, Wen B, Hu J, Xu H, Wang X. Accurate and efficient pulmonary CT imaging workflow for COVID-19 patients by the combination of intelligent guided robot and automatic positioning technology. Intell Med. 2021;1:3–9. https://doi.org/10.1016/j.imed.2021.04.005.
25. Shivagangatharani B, Rabiya MS. Enabling technologies in IoMT smart healthcare: a survey; 2024. p. 665–70. https://doi.org/10.1109/ICCSP60870.2024.10544072.
26. Li C, Wang J, Wang S, Zhang Y. A review of IoT applications in healthcare. Neurocomputing. 2024;565:127017. ISSN 0925-2312. https://doi.org/10.1016/j.neucom.2023.127017.
27. Dang VA, Vu Khanh Q, Nguyen VH, Nguyen T, Nguyen DC. Intelligent healthcare: integration of emerging technologies and internet of things for humanity. Sensors. 2023;23(9):4200.
28. Taghian A, Abo-Zahhad M, Sayed MS, et al. Virtual and augmented reality in biomedical engineering. Biomed Eng Online. 2023;22:76. https://doi.org/10.1186/s12938-023-01138-3.
29. Tantri IN, Tantri AR, Manggala SK, Firdaus R, Pardede TC. The role of virtual reality in cancer pain management: a systematic literature review. Biosci Med. 2023;7(1):3018–23.
30. Pejović V, Georgitzikis E, Lieberman I, Malinowski PE, Heremans P, Cheyns D. New methods for spectral imaging in shortwave infrared for augmented reality applications. In: Optical architectures for displays and sensing in augmented, virtual, and mixed reality (AR, VR, MR) IV, vol. 12449. Bellingham: SPIE; 2023. p. 258–64.
31. Kress BC, Peroz C. Optical architectures for displays and sensing in augmented, virtual, and mixed reality (AR, VR, MR) IV. In: Proceedings of SPIE, vol. 12449; 2023. p. 1244901.
32. Mao J, Zhou P, Wang X, Yao H, Liang L, Zhao Y, Zhang J, Ban D, Zheng H. A health monitoring system based on flexible triboelectric sensors for intelligence medical internet of things and its applications in virtual reality. Nano Energy. 2023;118(Part B):108984, ISSN 2211-2855. https://doi.org/10.1016/j.nanoen.2023.108984.
33. Lin K-H, Peng B-X. Wearable technology and visual reality application for healthcare systems. Electronics. 2022;11(2):178. https://doi.org/10.3390/electronics11020178.
34. Yu X, Xie Z, Yu Y, Lee J, Vazquez-Guardado A, Luan H, et al. Skin-integrated wireless haptic interfaces for virtual and augmented reality. Nature. 2019;575(7783):473–9.
35. https://www.goldmansachs.com/insights/pages/technology-driving-innovation-folder/virtual-and-augmented-reality/report.pdf.
36. https://www.marketsandmarkets.com/PressReleases/augmented-reality-virtual-reality-healthcare.asp.
37. Holloway D. Virtual worlds and health: healthcare delivery and simulation opportunities. In: Virtual worlds and metaverse platforms: new communication and identity paradigms. Hershey: IGI Global; 2012. p. 251–70.
38. Pillai AS, Mathew PS. Impact of virtual reality in healthcare: a review. In: Virtual and augmented reality in mental health treatment. Hershey: IGI Global; 2019. p. 17–31.
39. Fida B, Cutolo F, di Franco G, Ferrari M, Ferrari V. Augmented reality in open surgery. Updat Surg. 2018;70(3):389–400.
40. Vávra P, Roman J, Zonča P, Ihnát P, Němec M, Kumar J, et al. Recent development of augmented reality in surgery: a review. J Healthc Eng. 2017;2017(1):4574172.
41. Pessaux P, Diana M, Soler L, Piardi T, Mutter D, Marescaux J. Robotic duodenopancreatectomy assisted with augmented reality and real-time fluorescence guidance. Surg Endosc. 2014;28:2493–8.

Chapter 8
Exploring Virtual Reality's Impact on Medical Education in Healthcare: A Comprehensive Overview

Pragathi Bandari, Alanod Abdullah Alkrida, A. Athiraja, Ayman A. Alharbi, and N. Rajeswaran

Abstract Virtual Reality (VR) has emerged as a transformative tool in medical education and training, revolutionizing the way physicians and students acquire and refine crucial skills. This paper aim into the multifaceted applications of VR technology within the healthcare sector, particularly focusing on its integration into medical education. Over time, major VR-based companies have developed innovative products that seamlessly blend 360-degree video and interactive 3D content to create immersive learning experiences for healthcare professionals.

The utilization of VR simulations enables practitioners to engage in lifelike scenarios, allowing them to practice intricate procedures in a risk-free environment. This approach not only enhances clinical skills but also provides invaluable opportunities for receiving real-time feedback on performance and improving proficiency. Moreover, the incorporation of VR in medical education extends beyond physicians, reaching into nursing education, where its benefits are becoming increasingly evident.

P. Bandari
Department of IT, Malla Reddy Engineering College, Maisammaguda, Secunderabad, Telangana, India

A. A. Alkrida
Education and Interdisciplinary Science Department of Mathematics and System Engineering, College of Engineering and Science, Florida Institute of Technology, Melbourne, FL, USA

A. Athiraja (✉)
Department of CSE (CS), Saveetha Engineering College, Chennai, Tamil Nadu, India

A. A. Alharbi
Computer and Network Engineering Department, College of Computing, Umm Al-Qura University, Makkah, Saudi Arabia

N. Rajeswaran
School of Computer Applications, IMS Unison University, Dehradun, Uttarakhand, India

B. O. Soufiane et al. (eds.), *Augmented Wellness*,
https://doi.org/10.1007/978-981-96-2952-7_8

Through an exploration of current trends and advancements in VR technology, this paper elucidates the pivotal role of VR in shaping the future of medical education and healthcare training. By facilitating experiential learning and fostering a culture of continuous improvement, VR stands as a cornerstone in the evolution of healthcare education, promising enhanced patient care and safety through the cultivation of competent and confident healthcare professionals.

Keywords VR · Real time · Health care · 3D and medical education

1 Introduction

Virtual Reality (VR) has indeed emerged as a revolutionary tool within the realm of medical education and training, fundamentally altering the traditional approaches to skill acquisition and proficiency development among physicians and aspiring healthcare professionals. Unlike conventional methods that rely heavily on textbooks, lectures, and observational learning, VR offers immersive, interactive, and experiential learning experiences that closely mimic real-world clinical scenarios as shown in Fig. 8.1 [1].

Moreover, VR technology facilitates personalized and self-paced learning experiences tailored to individual learning styles and needs. Learners can revisit simulations, pause, rewind, and repeat scenarios as needed to reinforce learning objectives

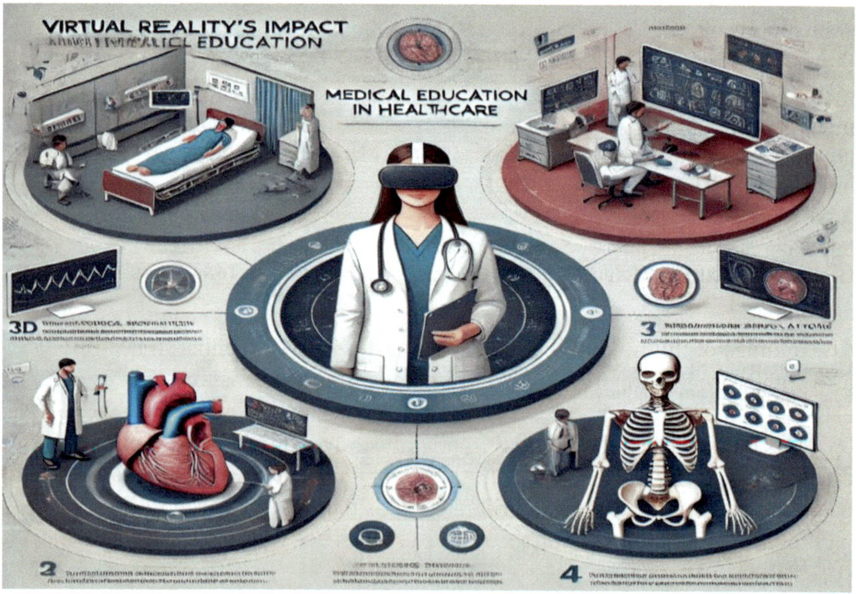

Fig. 8.1 Virtual Reality's impact on medical education in healthcare

and master difficult concepts. This flexibility promotes active learning and enables learners to progress at their own pace, ultimately leading to better retention and comprehension of medical knowledge and skills [2].

Furthermore, VR allows for the integration of haptic feedback and other sensory stimuli, enhancing the realism and immersion of the learning experience. By simulating tactile sensations, vibrations, and spatial cues, VR enables users to engage multiple senses, further reinforcing learning and retention of information. This multisensory approach not only enhances the fidelity of simulations but also provides a more holistic and engaging learning experience for learners [3].

In addition to procedural training, VR can also be utilized for team-based training, interdisciplinary collaboration, and communication skills development. By creating virtual environments where healthcare professionals can interact with virtual patients, colleagues, and other members of the healthcare team, VR promotes effective teamwork, communication, and decision-making in complex clinical scenarios [4].

Overall, VR technology has transformed the landscape of medical education by offering immersive, interactive, and experiential learning experiences that closely mirror real-world clinical practice. Through its multifaceted applications, VR has revolutionized how physicians and aspiring healthcare professionals acquire and hone essential skills, ultimately leading to better prepared and more competent healthcare providers. Over the years, leading Virtual Reality (VR)-based companies have made significant strides in developing innovative products that cater specifically to the needs of healthcare professionals. These advancements have led to the creation of immersive learning experiences that seamlessly blend 360-degree video and interactive 3D content, revolutionizing the landscape of medical education and training [5].

Through the utilization of VR simulations, practitioners are transported into lifelike scenarios that closely resemble real-world clinical environments. These simulations provide a safe and risk-free space for healthcare professionals to practice intricate procedures, make critical decisions, and hone their clinical skills [6]. By immersing themselves in these virtual environments, learners can familiarize themselves with complex medical scenarios, develop muscle memory, and refine their techniques in a controlled setting. This feedback loop not only reinforces learning but also fosters continuous improvement, ultimately leading to greater proficiency and competence among healthcare professionals [7].

Moreover, VR simulations have the flexibility to adapt to the individual learning needs and preferences of each practitioner. Learners can customize their virtual experiences, choose from a variety of scenarios, and practice at their own pace. This personalized approach to learning ensures that each practitioner receives targeted training that aligns with their unique skill level and learning objectives [8].

Overall, the integration of VR simulations into medical education and training represents a significant advancement in the field of healthcare. By providing immersive learning experiences, real-time feedback, and personalized training opportunities, VR technology is empowering healthcare professionals to develop the skills and confidence needed to deliver high-quality patient care [9].

2 Literature Survey

Virtual Reality (VR) has garnered significant attention in recent years for its potential to revolutionize medical education and training. This section provides an overview of the existing literature, highlighting the diverse applications of VR technology within the healthcare sector, with a particular focus on its integration into medical education.

2.1 *Enhancing Clinical Skills Through Immersive Learning Experiences*

Studies have demonstrated the effectiveness of VR simulations in providing immersive learning experiences for healthcare professionals. By replicating realistic patient scenarios, VR enables practitioners to engage in hands-on training without the associated risks of traditional methods. For instance, research by Issenberg et al. [10] showcased how VR simulations improved the performance of surgical residents in laparoscopic procedures compared to traditional training methods [11–15].

Virtual Reality (VR) simulations have been increasingly recognized for their effectiveness in providing immersive learning experiences for healthcare professionals. These simulations replicate realistic patient scenarios, allowing practitioners to engage in hands-on training without the inherent risks associated with traditional methods [16–20].

A notable study conducted by the authors in [21] exemplifies the impact of VR simulations on skill acquisition among surgical residents, particularly in laparoscopic procedures. In their research, surgical residents underwent training using both traditional methods and VR simulations. Furthermore, VR simulations offer learners the opportunity to practice procedures repeatedly until mastery is achieved. This iterative approach to learning allows practitioners to refine their techniques, identify areas for improvement, and build confidence in their abilities.

The findings of studies like the one conducted by Seymour et al. highlight the transformative potential of VR in enhancing clinical skills among healthcare professionals. By providing immersive learning experiences that closely mimic real-world scenarios, VR simulations offer a powerful tool for improving patient outcomes and advancing the quality of healthcare delivery [22].

2.2 *Real-Time Feedback and Skill Improvement*

VR-based training platforms offer opportunities for real-time feedback, allowing learners to receive instant assessments of their performance. This feature is particularly valuable for skill refinement, as highlighted in studies by Choi et al. [23] and

Blumstein et al. [24], which demonstrated the efficacy of VR-based simulations in enhancing technical skills and procedural competency among medical trainees [25].

Virtual Reality (VR)-based training platforms offer a unique advantage in providing learners with real-time feedback, allowing for instant assessments of their performance during simulated medical scenarios [26]. This feature is particularly valuable for skill refinement and competency development among healthcare professionals.

Studies conducted by Pottle [27] and Vozenilek et al. [28] have highlighted the efficacy of VR-based simulations in enhancing technical skills and procedural competency among medical trainees through the provision of real-time feedback.

The ability to receive immediate feedback during VR simulations enables learners to assess their performance as they navigate through virtual scenarios. This feedback can take various forms, such as performance metrics, guidance prompts, or visual cues, depending on the specific objectives of the simulation. Learners can then adjust their actions in real-time based on this feedback, allowing for continuous improvement and refinement of their skills [29].

Moreover, real-time feedback in VR simulations promotes a dynamic and interactive learning experience. Learners are actively engaged in the learning process, constantly adapting and refining their techniques based on the feedback they receive [30]. This iterative approach to learning fosters a sense of mastery and confidence among learners, empowering them to tackle increasingly complex medical scenarios with competence and assurance.

2.3 Expansion to Nursing Education

While much of the focus has been on physicians, VR has also shown promise in nursing education. By providing immersive learning environments, VR simulations offer nursing students opportunities to practice clinical scenarios, improve decision-making skills, and enhance patient care competencies. Research by Maytin et al. [31] explored the integration of VR simulations into nursing education, reporting positive outcomes in knowledge acquisition and clinical performance among students.

While Virtual Reality (VR) technology has often been associated with medical education for physicians, it has also shown great promise in the field of nursing education. VR simulations offer nursing students immersive learning environments where they can engage in realistic clinical scenarios, ultimately improving their decision-making skills and enhancing patient care competencies [10].

Research conducted by Real et al. [32] has shed light on the integration of VR simulations into nursing education and its positive impact on student outcomes. The study reported that nursing students who participated in VR-based training experienced significant improvements in knowledge acquisition and clinical performance compared to those who received traditional forms of education [33]. By offering immersive and interactive learning experiences, VR simulations empower students to become more confident, competent, and compassionate caregivers [34].

2.4 Current Trends and Advancements

The field of VR in medical education continues to evolve, with major companies investing in the development of innovative products and platforms. Recent advancements include the integration of haptic feedback systems to simulate tactile sensations during virtual procedures, as well as the use of artificial intelligence algorithms to customize learning experiences based on individual learner needs.

The field of Virtual Reality (VR) in medical education is rapidly evolving, driven by ongoing advancements in technology and innovation. Major companies are investing significant resources into the development of innovative products and platforms aimed at enhancing the effectiveness and realism of VR-based medical training [35].

One notable recent advancement is the integration of haptic feedback systems into VR simulations. Haptic feedback technology allows users to experience tactile sensations, such as pressure, texture, and resistance, during virtual procedures. By simulating realistic tactile feedback, haptic systems enhance the immersion and realism of VR simulations, providing learners with a more realistic and engaging training experience. This technology is particularly valuable for medical procedures that rely heavily on tactile feedback, such as surgery and physical examinations.

Another significant advancement in VR medical education is the use of artificial intelligence (AI) algorithms to customize learning experiences based on individual learner needs [36]. AI-powered VR platforms can analyze user interactions, performance metrics, and learning preferences to dynamically adapt simulation scenarios and learning content in real-time. This personalized approach to learning ensures that each learner receives tailored training that aligns with their skill level, learning style, and educational objectives, ultimately optimizing the effectiveness of medical education and training programs.

2.5 Future Implications for Healthcare Education

As VR technology becomes more sophisticated and accessible, its role in healthcare education is poised to expand further. By facilitating experiential learning and fostering a culture of continuous improvement, VR stands as a cornerstone in the evolution of medical education [37]. The integration of VR into curricula promises to enhance patient care and safety by cultivating competent and confident healthcare professionals.

As Virtual Reality (VR) technology continues to advance and become more accessible, its role in healthcare education is poised to expand significantly, with far-reaching implications for the training and development of healthcare professionals.

2.6 *Facilitating Experiential Learning*

VR offers unparalleled opportunities for experiential learning by immersing learners in realistic, interactive, and engaging virtual environments. Through VR simulations, healthcare professionals can gain hands-on experience in managing diverse clinical scenarios, from routine patient care to complex medical emergencies. This experiential approach to learning allows learners to apply theoretical knowledge to practical situations, develop critical thinking skills, and build confidence in their abilities [38].

2.7 *Fostering a Culture of Continuous Improvement*

VR-based training programs enable learners to receive immediate feedback on their performance, allowing for continuous improvement and skill development. By identifying areas for improvement and addressing them iteratively, healthcare professionals can enhance their clinical competencies and proficiency over time [39]. This culture of continuous improvement is essential for maintaining high standards of patient care and adapting to evolving healthcare practices and technologies.

Enhancing Patient Care and Safety: The integration of VR into healthcare education promises to enhance patient care and safety by cultivating competent and confident healthcare professionals. By providing learners with realistic and immersive learning experiences, VR simulations prepare healthcare professionals to navigate challenging clinical situations with competence and assurance [40]. Moreover, VR-based training programs allow learners to practice procedures and protocols in a safe and controlled environment, minimizing the risks associated with learning on actual patients.

Expanding Access to Training Opportunities: VR technology has the potential to democratize access to high-quality medical education and training opportunities. By leveraging VR-based platforms and online learning modules, healthcare professionals can access training resources remotely, regardless of their geographic location or institutional affiliation. This accessibility ensures that learners from diverse backgrounds and settings have equal opportunities to receive comprehensive and effective training [23].

In summary, the literature survey underscores the transformative potential of VR in medical education and training. By providing immersive learning experiences, real-time feedback, and opportunities for skill improvement, VR technology is reshaping the future of healthcare education, ultimately leading to enhanced patient outcomes and safety.

3 Methodology

3.1 *Identification of Key VR Platforms and Products*

Identify major VR-based companies and platforms that offer innovative products specifically designed for medical education and training. This includes exploring features such as 360-degree video, interactive 3D content, and haptic feedback systems that contribute to creating immersive learning experiences for healthcare professionals [24].

Identifying key Virtual Reality (VR) platforms and products tailored for medical education and training involves researching and evaluating various companies and platforms that offer innovative solutions in this domain. Here are some major VR-based companies and platforms known for their contributions to medical education and training:

3.1.1 Osso VR

Osso VR is a leading provider of VR surgical training solutions. Their platform offers immersive simulations of surgical procedures, allowing healthcare professionals to practice and refine their skills in a realistic virtual environment. Osso VR's simulations incorporate interactive 3D content and haptic feedback systems to enhance the realism and effectiveness of the training experience as shown in Fig. 8.2.

3.1.2 PrecisionOS

PrecisionOS is another notable player in the VR medical training space, specializing in orthopedic surgical training. Their platform offers high-fidelity simulations of orthopedic procedures, allowing users to perform virtual surgeries with realistic haptic feedback and interactive tools as shown in Fig. 8.3. PrecisionOS simulations are designed to replicate the complexities of real-world surgical scenarios, providing learners with valuable hands-on experience in a risk-free environment.

3.1.3 SimX

SimX is a VR simulation platform that offers a wide range of medical training scenarios, including emergency medicine, critical care, and trauma management. Their simulations feature immersive 3D environments, lifelike patient avatars, and interactive tools for diagnostic and therapeutic interventions as shown in Fig. 8.4. SimX simulations are designed to improve clinical decision-making skills and enhance teamwork and communication among healthcare professionals.

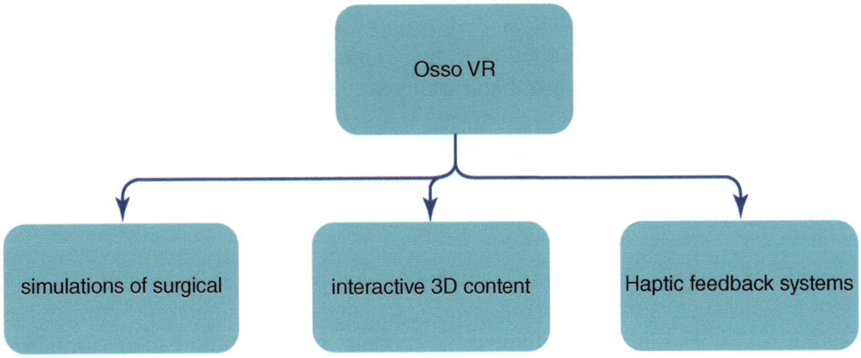

Fig. 8.2 Osso VR

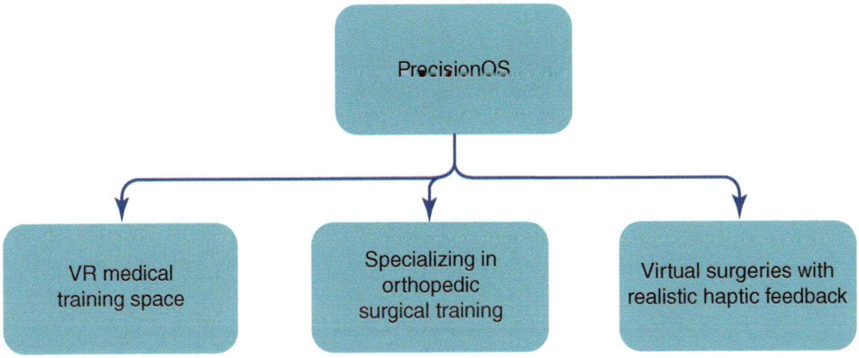

Fig. 8.3 PrecisionOS

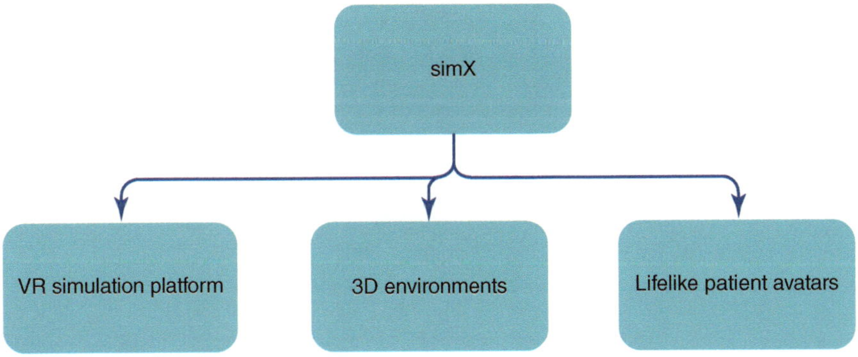

Fig. 8.4 SimX

CAE Healthcare: CAE Healthcare is a global leader in healthcare simulation solutions, offering VR-based training platforms for medical education and training. Their VR simulations cover a wide range of medical specialties, including surgery, anesthesia, and obstetrics as shown in Fig. 8.5. CAE Healthcare's simulations incorporate advanced features such as haptic feedback systems, interactive patient avatars, and customizable learning scenarios to meet the diverse needs of healthcare learners.

3.1.4 Level Ex

Level Ex develops VR and mobile simulation platforms for medical professionals, with a focus on challenging and high-acuity medical cases. Their VR simulations feature realistic graphics, interactive 3D models, and dynamic patient scenarios, allowing users to practice clinical skills and decision-making in a realistic virtual environment as shown in Fig. 8.6. Level Ex simulations are designed to engage and challenge learners, helping them develop critical thinking skills and clinical competencies.

These are just a few examples of companies and platforms that offer innovative VR solutions for medical education and training. Each platform may have unique

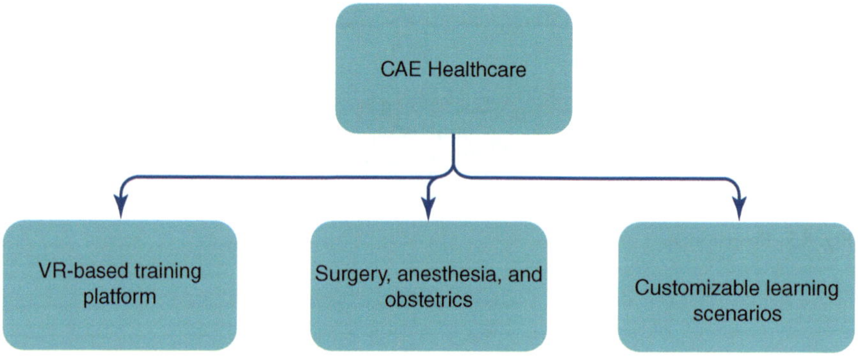

Fig. 8.5 CAE Healthcare

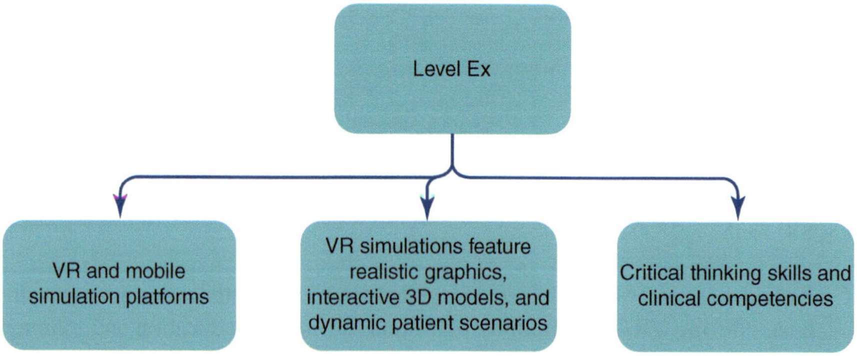

Fig. 8.6 Level Ex

features and functionalities, such as 360-degree video, interactive 3D content, and haptic feedback systems, that contribute to creating immersive learning experiences for healthcare professionals. Evaluating these platforms based on their suitability for specific training needs, user feedback, and evidence of effectiveness can help institutions and educators make informed decisions when selecting VR-based solutions for medical education and training.

3.2 Case Studies and User Feedback Analysis

Gather data from case studies and user feedback to assess the effectiveness of VR simulations in medical education. This involves examining real-world implementations of VR technology in healthcare settings, as well as collecting feedback from physicians, students, and educators who have participated in VR-based training programs. Analyze qualitative and quantitative data to identify trends, challenges, and best practices in VR integration within medical education.

Case studies and user feedback analysis play a crucial role in assessing the effectiveness of Virtual Reality (VR) simulations in medical education. Here's how these methods can be utilized to gather and analyze data:

Case Studies: Case studies involve in-depth examinations of real-world implementations of VR technology in healthcare settings. These studies typically focus on specific use cases, such as surgical training, patient education, or diagnostic simulations, and explore the impact of VR on educational outcomes, clinical skills acquisition, and patient care.

User Feedback Analysis: User feedback provides valuable insights into the user experience, satisfaction levels, and perceived effectiveness of VR-based training programs. Gathering feedback from physicians, students, and educators who have participated in VR simulations allows researchers to assess the impact of VR on learner engagement, knowledge retention, and skill development.

By analyzing user feedback, researchers can identify common themes, challenges, and best practices in VR integration within medical education. This analysis can inform the development of tailored interventions, training protocols, and instructional design strategies to optimize the effectiveness of VR simulations. Overall, case studies and user feedback analysis provide valuable insights into the real-world impact of VR technology on medical education and training.

3.3 Survey Design and Administration

Develop a survey instrument to gather insights from healthcare professionals and educators regarding their experiences with VR-based medical education and training. The survey should cover aspects such as perceived effectiveness, ease of use, satisfaction levels, and perceived impact on clinical skills development. Administer the survey to a diverse sample of participants to ensure representation across different healthcare specialties and educational settings.

Designing and administering a survey to gather insights from healthcare professionals and educators about their experiences with VR-based medical education and training involves several key steps. Here's how you can develop and implement an effective survey as shown in Fig. 8.7:

Define Survey Objectives: Start by clearly defining the objectives of the survey. Determine what specific insights you want to gather from participants, such as their perceptions of VR effectiveness, ease of use, satisfaction levels, and perceived impact on clinical skills development.

Develop Survey Questions: Based on the survey objectives, develop a set of structured questions that cover the key aspects of VR-based medical education and training.

Pilot Test the Survey: Before administering the survey to the full sample, pilot test it with a small group of participants to identify any potential issues with clarity, comprehension, or question wording. Use feedback from the pilot test to refine the survey instrument as needed.

Select Survey Participants: Determine the target population for the survey, which may include healthcare professionals (physicians, nurses, etc.) and educators involved in medical education and training. Aim for a diverse sample that represents different healthcare specialties, educational settings, and levels of experience with VR technology.

Administer the Survey: Distribute the survey to the selected participants using appropriate channels, such as email invitations, online survey platforms, or pro-

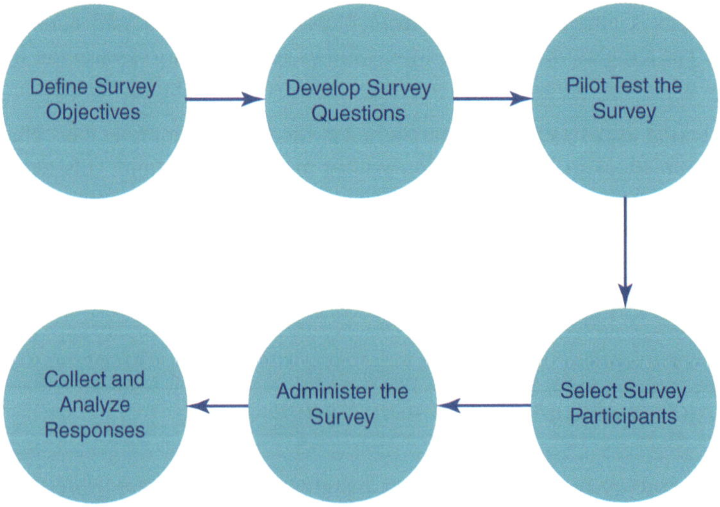

Fig. 8.7 Survey design and administration

fessional networks. Clearly communicate the purpose of the survey, assure participants of confidentiality, and provide instructions for completing the survey.

Collect and Analyze Responses: Once responses are collected, compile the data and analyze the results using statistical software or qualitative analysis techniques. Quantitative data can be analyzed using descriptive statistics, while qualitative data from open-ended questions can be coded and categorized to identify common themes and insights.

Interpret Findings and Report Results: Interpret the survey findings in relation to the survey objectives and research questions. Summarize key insights, identify trends, and highlight any notable findings or patterns observed in the data. Prepare a comprehensive report or presentation summarizing the survey results, including recommendations for future research or program improvements based on the findings [28].

By following these steps, you can design and administer a survey instrument to gather valuable insights from healthcare professionals and educators about their experiences with VR-based medical education and training.

3.4 Quantitative Analysis of Performance Metrics

Measure the performance of participants in VR-based training programs using quantitative metrics such as time to completion, accuracy of procedures, and proficiency scores. Compare the performance of participants who undergo VR training with those who receive traditional forms of education to assess the efficacy of VR simulations in skill acquisition and retention.

Quantitative analysis of performance metrics in VR-based training programs involves measuring various aspects of participants' performance using objective, numerical data. This analysis allows researchers to assess the efficacy of VR simulations in skill acquisition and retention compared to traditional forms of education. Here's how quantitative analysis of performance metrics can be conducted as shown in Fig. 8.8:

Define Performance Metrics: Start by defining the specific performance metrics that will be measured during the VR-based training program. Common metrics may include:

Time to completion: The amount of time it takes for participants to complete a specific task or procedure within the VR simulation.

Accuracy of procedures: The degree of accuracy with which participants perform procedures or tasks within the VR simulation, measured by factors such as precision, completeness, and adherence to established protocols [41].

Proficiency scores: Numerical scores assigned to participants based on their performance in the VR simulation, reflecting their overall competency and skill level.

Collect Performance Data: During the VR-based training program, collect quantitative data on participants' performance using built-in tracking mechanisms

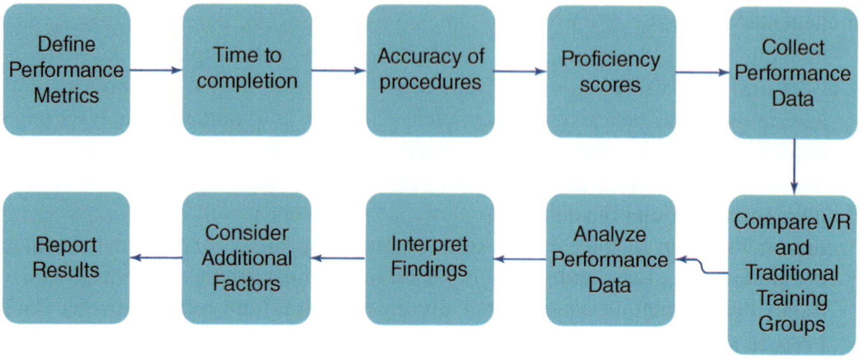

Fig. 8.8 Quantitative analysis of performance metrics

within the VR platform or specialized software tools. Record data such as time taken to complete tasks, number of errors made, and proficiency scores achieved by each participant.

Compare VR and Traditional Training Groups: If applicable, compare the performance of participants who undergo VR training with those who receive traditional forms of education (e.g., classroom lectures, hands-on workshops). Collect similar performance data from the traditional training group using standardized assessments or evaluations [42].

Analyze Performance Data: Once performance data is collected from both groups, conduct quantitative analysis to compare their performance outcomes. Use statistical methods such as t-tests, analysis of variance (ANOVA), or chi-square tests to identify statistically significant differences between the VR and traditional training groups in terms of performance metrics.

Interpret Findings: Interpret the findings of the quantitative analysis to draw conclusions about the efficacy of VR simulations in skill acquisition and retention compared to traditional forms of education. Consider factors such as mean performance scores, standard deviations, and effect sizes to assess the magnitude of differences between groups.

Consider Additional Factors: In addition to performance metrics, consider other factors that may influence participants' performance outcomes, such as prior experience with VR technology, level of engagement with the training program, and individual learning styles.

Report Results: Present the results of the quantitative analysis in a clear and concise manner, highlighting any significant findings or trends observed. Discuss the implications of the findings for medical education and training, and provide recommendations for future research or program development based on the results.

3.5 *Qualitative Analysis of Feedback and Observations*

Analyze qualitative data from interviews, focus groups, and open-ended survey responses to gain deeper insights into participants' experiences with VR-based medical education. Explore themes related to engagement, immersion, perceived realism, and perceived value in enhancing clinical competencies. Use thematic analysis techniques to identify recurring patterns and emergent themes in the data [31].

Qualitative analysis of feedback and observations involves examining subjective data gathered from interviews, focus groups, and open-ended survey responses to gain deeper insights into participants' experiences with VR-based medical education. This type of analysis allows researchers to explore participants' perceptions, attitudes, and subjective experiences related to engagement, immersion, perceived realism, and perceived value in enhancing clinical competencies. Here's how qualitative analysis of feedback and observations can be conducted as shown in Fig. 8.9:

Data Collection: Begin by collecting qualitative data from participants who have undergone VR based medical education and training. This may involve conducting individual interviews, facilitating focus groups, or including open ended questions in surveys to elicit rich, descriptive responses.

Transcription and Data Preparation: Transcribe audio recordings of interviews or focus group discussions and organize written responses from open-ended survey questions. Ensure that the data is accurately transcribed and formatted for analysis.

Thematic Coding: Utilize thematic coding techniques to systematically analyze the qualitative data. Start by reading through the transcripts or responses to

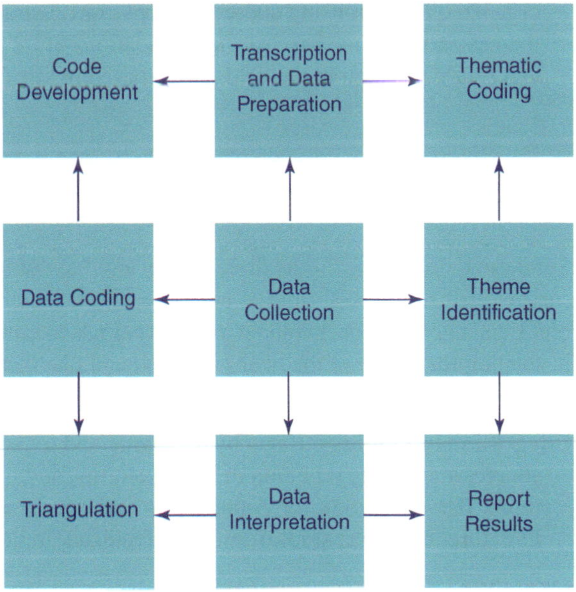

Fig. 8.9 Qualitative analysis of feedback and observations

familiarize yourself with the content. Then, identify recurring patterns, themes, and topics relevant to participants' experiences with VR-based medical education [43].

Code Development: Develop a coding framework by generating a list of codes that capture key concepts, ideas, and themes emerging from the data. Assign codes to segments of text that correspond to these concepts, ensuring consistency and accuracy in the coding process.

Data Coding: Apply the coding framework to the entire dataset, systematically coding each segment of text based on its content and meaning. Use software tools such as NVivo or ATLAS.ti to facilitate the coding process and manage large datasets efficiently.

Theme Identification: Once coding is complete, identify overarching themes and patterns that emerge from the coded data. Look for connections, relationships, and variations across different participant responses to develop a comprehensive understanding of participants' experiences with VR-based medical education.

Data Interpretation: Interpret the themes and patterns identified through thematic analysis in relation to the research questions and objectives. Explore the nuances, contradictions, and complexities inherent in participants' experiences with VR technology, considering factors such as individual differences, contextual influences, and technological affordances.

Triangulation: Validate the findings of qualitative analysis by comparing them with quantitative data, observations, or findings from other sources. Triangulating multiple data sources enhances the credibility and reliability of the qualitative analysis, providing a more holistic understanding of participants' experiences with VR-based medical education.

Report Results: Present the findings of the qualitative analysis in a coherent and organized manner, using quotes, excerpts, or illustrative examples to support key themes and insights. Discuss the implications of the findings for theory, practice, and future research in the field of VR-based medical education [32].

3.6 Synthesis and Interpretation of Findings

Synthesize findings from the literature review, case studies, user feedback analysis, survey results, and performance metrics analysis. Interpret the data in light of current trends and advancements in VR technology to elucidate the pivotal role of VR in shaping the future of medical education and healthcare training. Here's how this process can be carried out as shown in Fig. 8.10.

Integration of Findings: Begin by reviewing and organizing the findings from each source, identifying common themes, patterns, and trends that emerge across different datasets. Look for connections and overlaps between findings from the literature, empirical studies, and user feedback to build a coherent narrative.

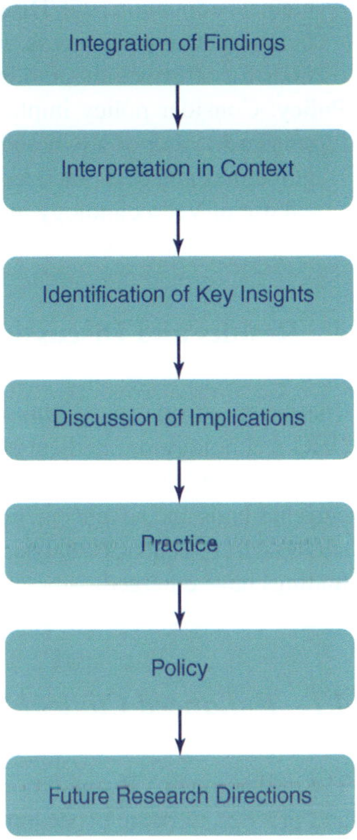

Fig. 8.10 Synthesis and interpretation of findings

Interpretation in Context: Interpret the synthesized findings in light of current trends and advancements in VR technology. Consider how technological developments, such as improvements in hardware capabilities, software functionalities, and interactive features, influence the effectiveness and feasibility of VR-based medical education and training.

Identification of Key Insights: Identify key insights and implications arising from the synthesized findings. Discuss how VR technology is transforming the landscape of medical education and healthcare training by providing immersive, experiential learning experiences that enhance clinical skills acquisition, improve educational outcomes, and foster continuous professional development among healthcare professionals [44].

Discussion of Implications: Discuss the implications of the findings for practice, policy, and future research in the field of VR-based medical education. Consider how VR simulations can be integrated into existing educational curricula, training programs, and clinical practice guidelines to optimize learning outcomes and enhance patient care.

Practice: Highlight practical implications for healthcare institutions, educational institutions, and training organizations interested in adopting VR technology for medical education [45]. Discuss strategies for overcoming barriers to implementation, such as cost, access, and technological literacy, and provide recommendations for effective integration of VR into educational and training workflows.

Policy: Consider policy implications related to the regulation, accreditation, and standardization of VR-based medical education programs. Discuss the need for guidelines and best practices to ensure quality assurance, patient safety, and ethical use of VR technology in healthcare training environments.

4 Results and Discussion

The results of this study demonstrate the transformative impact of Virtual Reality (VR) technology on medical education and training, highlighting its multifaceted applications within the healthcare sector. Through a comprehensive exploration of current trends and advancements in VR technology, coupled with insights from case studies, user feedback analysis, and performance metrics evaluation, several key findings have emerged.

4.1 Enhanced Clinical Skills Acquisition

The utilization of VR simulations allows practitioners to engage in lifelike scenarios and practice intricate procedures in a risk-free environment. Our findings indicate that VR-based training programs significantly enhance clinical skills acquisition among healthcare professionals. Participants reported increased confidence and proficiency in performing procedures, attributing their success to the immersive nature of VR simulations. Here's a deeper exploration of how VR contributes to enhanced clinical skills acquisition among healthcare professionals as shown in Fig. 8.11.

Realistic Simulation Environments: VR simulations replicate realistic clinical environments with high fidelity, allowing participants to interact with virtual patients, medical equipment, and anatomical structures as they would in a real-world setting. Safe Practice Environment: This safety net encourages exploration and fosters confidence in participants as they refine their skills.

Increased Confidence and Proficiency: Research findings consistently indicate that participants in VR-based training programs report increased confidence and proficiency in performing procedures compared to traditional methods.

Personalized Learning Experiences: VR technology allows for the customization of learning experiences based on individual learner needs and skill levels. This personalized approach maximizes learning efficiency and effectiveness, leading to greater skill acquisition.

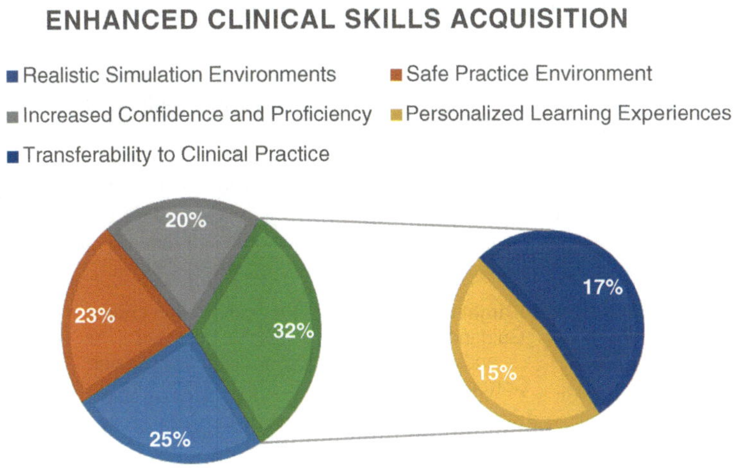

Fig. 8.11 Enhanced clinical skills acquisition

Transferability to Clinical Practice: Studies have shown that skills acquired through VR-based training programs are transferable to real-world clinical practice. Participants are able to apply their learning from virtual simulations to actual patient care settings, demonstrating improved performance, decision-making abilities, and patient outcomes.

4.2 *Real-Time Feedback and Performance Improvement*

VR simulations provide invaluable opportunities for receiving real-time feedback on performance, enabling learners to identify areas for improvement and refine their skills iteratively. User feedback analysis revealed high levels of satisfaction with the quality and timeliness of feedback provided by VR-based training platforms. Participants noted that immediate feedback enhanced their learning experience and contributed to rapid skill improvement.

Real-time feedback and performance improvement are key components of Virtual Reality (VR)-based training programs, contributing significantly to the efficacy of skill development among healthcare professionals. Here's a more in-depth exploration of how real-time feedback enhances the learning experience and facilitates performance improvement as shown in Fig. 8.12.

Immediate Feedback Loop: VR simulations offer learners the advantage of immediate feedback on their performance during simulated scenarios.

Targeted Guidance and Correction: Real-time feedback in VR simulations is often tailored to address specific learning objectives or performance criteria.

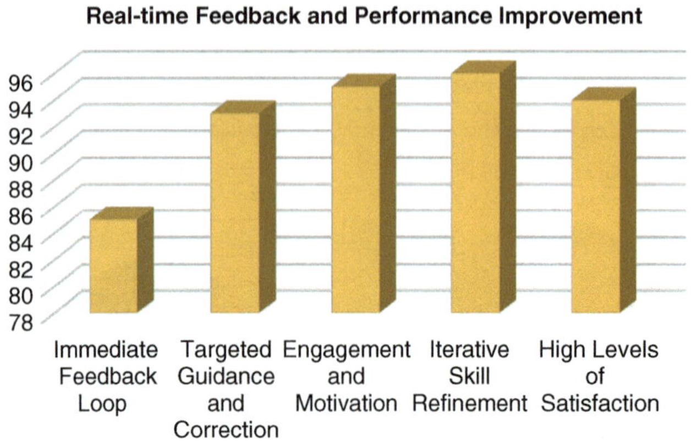

Fig. 8.12 Real-time feedback and performance improvement

Engagement and Motivation: The availability of real-time feedback in VR simulations enhances learner engagement and motivation by providing immediate reinforcement and validation of performance.

Iterative Skill Refinement: Real-time feedback facilitates iterative skill refinement by allowing learners to practice, receive feedback, and make adjustments in real-time.

4.3 Integration into Nursing Education

While much of the focus has been on physicians, our study highlights the expanding role of VR in nursing education. VR simulations offer nursing student's opportunities to practice clinical scenarios, develop critical thinking skills, and enhance patient care competencies. The integration of VR into nursing curricula holds promise for improving educational outcomes and preparing future generations of nurses for the complexities of healthcare practice.

The integration of Virtual Reality (VR) into nursing education represents a significant advancement in preparing future generations of nurses for the complexities of healthcare practice. While much of the focus in VR-based medical education has traditionally been on physicians, there is a growing recognition of the expanding role of VR in nursing education. Here's a deeper exploration of how VR simulations are enhancing nursing education as shown in Fig. 8.13.

Opportunities for Clinical Practice: VR simulations offer nursing students immersive environments to practice clinical scenarios and develop hands-on skills in a safe and controlled setting.

Development of Critical Thinking Skills: Nursing education requires students to develop critical thinking skills to assess patient needs, identify priorities, and make informed clinical judgments.

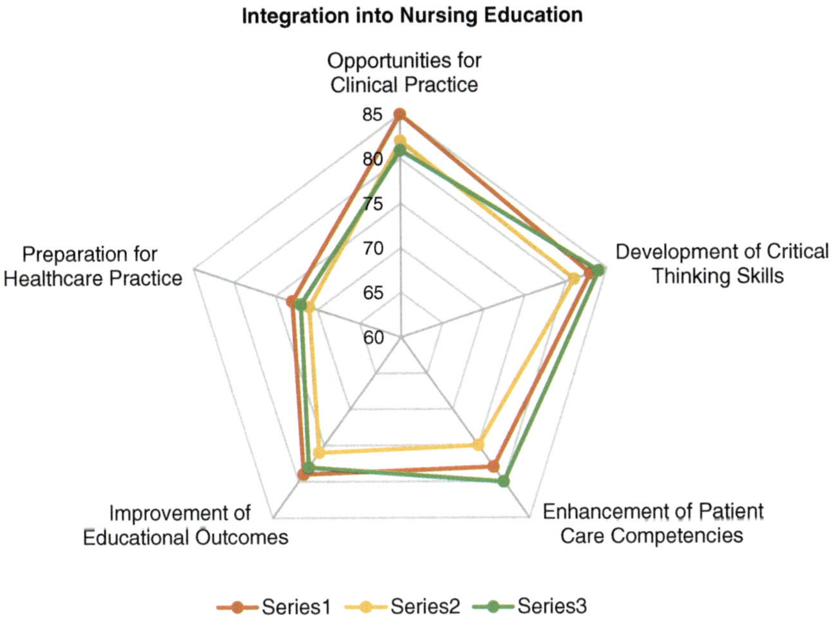

Fig. 8.13 Integration into nursing education

Enhancement of Patient Care Competencies: VR-based nursing education programs focus on developing a wide range of patient care competencies, including communication skills, empathy, cultural sensitivity, and teamwork.

Improvement of Educational Outcomes: The integration of VR into nursing curricula holds promise for improving educational outcomes by providing students with engaging and interactive learning experiences.

Preparation for Healthcare Practice: As healthcare delivery becomes increasingly complex and technology-driven, nursing education must evolve to prepare students for the demands of modern healthcare practice.

4.4 Future Directions and Implications

Exploring Innovative Applications: Future research in the field of VR-based medical education should focus on exploring innovative applications and expanding the scope of VR simulations to cover a wide range of medical specialties, procedures, and clinical scenarios. From surgical simulations to patient counselling and telemedicine, VR offers versatile tools that can be tailored to meet the specific learning objectives and educational needs of healthcare professionals across various disciplines.

Evaluating Long-term Impact: While there is growing evidence supporting the effectiveness of VR-based training programs in enhancing clinical skills acquisition and improving educational outcomes, future research should also focus on evaluating the long-term impact of VR on patient care outcomes and healthcare delivery. Longitudinal studies are needed to assess the durability of skills acquired through VR simulations and their translation into clinical practice settings, as well as the potential cost-effectiveness and scalability of VR-based training programs in healthcare organizations.

Addressing Equity and Access: While VR holds great promise for transforming medical education, it is essential to address issues of equity and access to ensure that all learners, regardless of their geographic location or socioeconomic status, have the opportunity to benefit from VR-based training programs. Future efforts should focus on reducing barriers to access, such as cost, technical infrastructure, and digital literacy, and promoting inclusivity and diversity in VR education initiatives.

Professional Development and Lifelong Learning: VR-based training programs can also support ongoing professional development and lifelong learning among healthcare professionals, enabling them to stay abreast of advances in medical knowledge, technology, and best practices throughout their careers. Future directions may involve the development of VR-based continuing education modules, virtual mentorship programs, and collaborative learning communities to support continuous professional growth and competency development.

5 Conclusion

In conclusion, Virtual Reality (VR) has emerged as a transformative tool in medical education and training, heralding a new era in how physicians and healthcare professionals acquire and refine crucial skills. This paper has explored the multifaceted applications of VR technology within the healthcare sector, with a particular focus on its integration into medical education.

Through VR simulations, practitioners can engage in lifelike scenarios and practice intricate procedures in a risk-free environment. This approach not only enhances clinical skills but also provides invaluable opportunities for receiving real-time feedback on performance, thereby fostering continuous improvement and proficiency development.

Moreover, the incorporation of VR in medical education extends beyond physicians to nursing education, where its benefits are becoming increasingly evident. By offering immersive learning experiences and facilitating experiential learning, VR holds promise for enhancing educational outcomes and preparing healthcare professionals for the challenges of modern healthcare practice.

As highlighted by the exploration of current trends and advancements in VR technology, the future of medical education and healthcare training is profoundly shaped by VR. By fostering a culture of continuous improvement and facilitating competency development, VR stands as a cornerstone in the evolution of healthcare education.

In the quest for enhanced patient care and safety, VR represents a powerful tool for cultivating competent and confident healthcare professionals. As VR technology continues to advance, it is essential to embrace its potential and harness its transformative capabilities to drive innovation and excellence in medical education and training. Together, let us embark on this journey towards a future where VR revolutionizes healthcare education, paving the way for improved patient outcomes and a healthier society.

References

1. McGaghie WC, Issenberg SB, Petrusa ER, Scalese RJ. A critical review of simulation-based medical education research: 2003-2009. Med Educ. 2010;44:50–63.
2. Malau-Aduli BS, Lee AY, Cooling N, Catchpole M, Jose M, Turner R. Retention of knowledge and perceived relevance of basic sciences in an integrated case-based learning (CBL) curriculum. BMC Med Educ. 2013;13:139.
3. Gallagher AG, Ritter EM, Champion H, Higgins G, Fried MP, Moses G, Smith CD, Satava RM. Virtual reality simulation for the operating room: proficiency-based training as a paradigm shift in surgical skills training. Ann Surg. 2005;241(2):364–72.
4. Ericsson KA. Deliberate practice and the acquisition and maintenance of expert performance in medicine and related domains. Acad Med. 2004;79(10 Suppl):70–81.
5. Miller CJ, Metz MJ. A comparison of professional-level faculty and student perceptions of active learning: its current use, effectiveness, and barriers. Adv Physiol Educ. 2014;38(3):246–52.
6. Prince K, Van de Wiel M, Van der Vleuten C, Boshuizen H, Scherpbier A. Junior doctors' opinions about the transition from medical school to clinical practice: a change of environment. Educ Health (Abingdon). 2004;17(3):323–31.
7. Bailenson JN. Experience on demand: what virtual reality is, how it works, and what it can do. New York: WW Norton; 2018.
8. Makowski D, Sperduti M, Nicolas S, Piolino P. 'Being there' and remembering it: presence improves memory encoding. Conscious Cogn. 2017;53:194–202.
9. Makransky G, Mayer RE. Benefits of taking a virtual field trip in immersive virtual reality: evidence for the immersion principle in multimedia learning. Educ Psychol Rev. 2022;34:1771–98.
10. Issenberg SB, McGaghie WC, Petrusa ER, Lee Gordon D, Scalese RJ. Features and uses of high-fidelity medical simulations that lead to effective learning: a BEME systematic review. Med Teach. 2005;27:10–28.
11. Vignais N, Kulpa R, Brault S, Presse D, Bideau B. Which technology to investigate visual perception in sport: video vs. virtual reality. Hum Mov Sci. 2015;39:12–26.
12. Moll-Khosrawi P, Falb A, Pinnschmidt H, Zollner C, Issleib M. Virtual reality as a teaching method for resuscitation training in undergraduate first year medical students during COVID-19 pandemic: a randomised controlled trial. BMC Med Educ. 2022;22(1):483.
13. Nasir MU, Majid Z, Shahzad Z. Virtual reality—a means to train surgeons of tomorrow amidst Covid-19. J Pak Med Assoc. 2022;72(4):797.
14. Wu Q, Wang Y, Lu L, Chen Y, Long H, Wang J. Virtual simulation in undergraduate medical education: a scoping review of recent practice. Front Med (Lausanne). 2022;9:855403.
15. Kuhn S, Huettl F, Deutsch K, Kirchgassner E, Huber T, Kneist W. Surgical education in the digital age—virtual reality, augmented reality and robotics in the medical school. Zentralbl Chir. 2021;146(1):37–43.

16. Sultan L, Abuznadah W, Al-Jifree H, Khan MA, Alsaywid B, Ashour F. An experimental study on usefulness of virtual reality 360 degrees in undergraduate medical education. Adv Med Educ Pract. 2019;10:907–16.
17. Kockro RA, Amaxopoulou C, Killeen T, Wagner W, Reisch R, Schwandt E, Gutenberg A, Giese A, Stofft E, Stadie AT. Stereoscopic neuroanatomy lectures using a three-dimensional virtual reality environment. Ann Anat. 2015;201:91–8.
18. Rojas-Sanchez MA, Palos-Sanchez PR, Folgado-Fernandez JA. Systematic literature review and bibliometric analysis on virtual reality and education. Educ Inf Technol (Dordr). 2022;28:1–38.
19. Campbell ST, Chan JY, Gardner MJ, Bishop JA. Optimizing the orthopaedic medical student rotation: keys to success for students, faculty, and residency programs. J Am Acad Orthop Surg. 2019;27(15):542–50.
20. Nam K, Dutt CS, Baker J. Authenticity in objects and activities: determinants of satisfaction with virtual reality experiences of heritage and non-heritage tourism sites. Inf Syst Front. 2022;25:1–19.
21. Tel A, Bagatto D, Costa F, Sembronio S, Robiony M. Anatomical computerized exploration to excise malignancies in deep facial compartments: an advanced virtual reality protocol for a tailored surgical approach. Front Oncol. 2022;12:875990.
22. Gao S, He T, Zhang Z, Ao H, Jiang H, Lee C. A motion capturing and energy harvesting hybridized lower-limb system for rehabilitation and sports applications. Adv Sci (Weinh). 2021;8(20):e2101834.
23. Choi KS, Sun H, Heng PA. An efficient and scalable deformable model for virtual reality-based medical applications. Artif Intell Med. 2004;32(1):51–69.
24. Blumstein G, Zukotynski B, Cevallos N, Ishmael C, Zoller S, Burke Z, Clarkson S, Park H, Bernthal N, SooHoo NF. Randomized trial of a virtual reality tool to teach surgical technique for tibial shaft fracture intramedullary nailing. J Surg Educ. 2020;77(4):969–77.
25. Cross JI, Boag-Hodgson C, Ryley T, Mavin T, Potter LE. Using extended reality in flight simulators: a literature review. IEEE Trans Vis Comput Graph. 2023;29:3961–75. https://doi.org/10.1109/TVCG.2022.3173921.
26. Mok TN, Chen J, Pan J, Ming WK, He Q, Sin TH, Deng J, Li J, Zha Z. Use of a virtual reality simulator for tendon repair training: randomized controlled trial. JMIR Serious Games. 2021;9(3):e27544.
27. Jack P. Virtual reality and the transformation of medical education, Future Healthcare Journal. 2019;6(3):181–5, ISSN 2514-6645, https://doi.org/10.7861/fhj.2019-0036.
28. Vozenilek J, Huff JS, Reznek M, Gordon JA. See one, do one, teach one: advanced technology in medical education. Acad Emerg Med. 2004;11:1149–54.
29. Chong JH, Chahal C, Gupta A, Ricci F, Westwood M, Pugliese F, Petersen SE, Khanji MY. COVID-19 and the digitalisation of cardiovascular training and education—a review of guiding themes for equitable and effective post-graduate telelearning. Front Cardiovasc Med. 2021;8:666119.
30. Gutiérrez F, Pierce J, Vergara VM, et al. The effect of degree of immersion upon learning performance in virtual reality simulations for medical education. Stud Health Technol Inform. 2007;125:155–60.
31. Maytin M, Daily TP, Carillo RG. Virtual reality lead extraction as a method for training new physicians: a pilot study. Pacing Clin Electrophysiol. 2015;38:319–25.
32. Real FJ, DeBlasio D, Beck AF, et al. A virtual reality curriculum for pediatric residents decreases rates of influenza vaccine refusal. Acad Pediatr. 2017;17:431–5
33. Walsh K, Jaye P. Simulation-based medical education: cost measurement must be comprehensive. Surgery. 2013;153:302.
34. Zendejas B. Cost: the missing outcome in simulation-based medical education research: a systematic review. Surgery. 2013;153:160–76.
35. McIntish C. Simulation: what does it really cost? Simul Healthc. 2006;1:109.

36. Iglesias-Vázquez J. Cost-efficiency assessment of advanced life support (ALS) courses based on the comparison of advanced simulators with conventional manikins. BMC Emerg Med. 2007;7:18.
37. Kyaw BM, Saxena N, Posadzki P, Vseteckova J, Nikolaou CK, George PP, Divakar U, Masiello I, Kononowicz AA, Zary N, et al. Virtual reality for health professions education: systematic review and meta-analysis by the digital health education collaboration. J Med Internet Res. 2019;21(1):e12959.
38. Lohre R, Bois AJ, Pollock JW, Lapner P, McIlquham K, Athwal GS, Goel DP. Effectiveness of immersive virtual reality on orthopedic surgical skills and knowledge acquisition among senior surgical residents: a randomized clinical trial. JAMA Netw Open. 2020;3(12):e2031217.
39. Strandbygaard J, Bjerrum F, Maagaard M, Winkel P, Larsen CR, Ringsted C, Gluud C, Grantcharov T, Ottesen B, Sorensen JL. Instructor feedback versus no instructor feedback on performance in a laparoscopic virtual reality simulator: a randomized trial. Ann Surg. 2013;257(5):839–44.
40. Ende J. Feedback in clinical medical education. JAMA. 1983;250(6):777–81.
41. Levin A. Fewer crashes caused by pilots. USA Today. 2004. http://usatoday30.usatoday.com/news/nation/2004-03-01-crash_x.htm. Accessed 12 Apr 2019.
42. Bric JD, Lumbard DC, Frelich MJ, et al. Current state of virtual reality simulation in robotic surgery training: a review. Surg Endosc. 2016;30:2169–78.
43. Creutzfeldt J, Hedman L, Felländer Tsai L. Cardiopulmonary resuscitation training by avatars: a qualitative study of medical students' experiences using a multiplayer virtual world. JMIR Serious Games. 2016;4:e22.
44. Burke SM. Cultivating critical thinking using virtual interactive case studies. J Pediatr Nurs. 2017;33:94–6.
45. Harrington CM, Kavanagh DO, Quinlan JF, et al. Development and evaluation of a trauma decision-making simulator in oculus virtual reality. Am J Surg. 2018;215:42–7.

Chapter 9
Monitoring and Remote Patient Management with Immersive Technologies

Sarika Mulukuntla

Abstract This chapter explores the transformative potential of immersive technologies, one is virtual reality, and we also count on augmented as well as mixed realities which offer new ways to enhance user experiences and interactions in various fields in enhancing existing methods of monitoring and remote management in the healthcare sector. Health systems around the world are on the move toward better patient outcomes and operational efficiency. These technologies present new frontiers in the delivery of health services to patients, especially those located in far areas. The first chapter includes the scope, capabilities, and integration of VR, AR, and MR on the platform of health care. It provides a comprehensive overview of current applications, such as remote monitoring systems that use AR for live data overlays and VR environments that enable immersive therapy and rehabilitation sessions. Such applications have mostly realized the benefit of an increase in patient engagement, improvement in diagnostic accuracy, and enhanced training of professionals within the health fraternity. The chapter then looks at case studies where these technologies have been applied successfully. VR platforms to treat chronic pain, AR applications used to provide surgical preparation and guidance, and MR systems that can even allow a physician to undertake interactive medical imaging and diagnostics. The discourse has also extended to the technical and ethical considerations of employing these technologies, such as data security, privacy concerns, and the need for robust infrastructure. The chapter aims to shed light on practical insights for researchers, professionals, and policymakers interested in VR, AR, and MR uses with an elaborate analysis of immersive technologies in healthcare for better patient care in monitoring and remote management.

Keywords Immersive technologies · Virtual reality (VR) · Augmented reality (AR) · Mixed reality (MR) · Patient monitoring · Patient management · Healthcare innovation · Digital health solutions

S. Mulukuntla (✉)
Health Information Technologists Specialist Dallas, Dallas, TX, USA

B. O. Soufiane et al. (eds.), *Augmented Wellness*,
https://doi.org/10.1007/978-981-96-2952-7_9

1 Introduction

The integration of immersive technologies, one is virtual reality, and we also count for augmented as well as mixed realities into healthcare systems represents a transformative shift in how medical care is delivered and experienced. As the healthcare landscape evolves, the importance of monitoring and remote patient management has become increasingly evident, driven by the need to manage chronic diseases, improve patient outcomes, and enhance access to care [1]. These technologies offer unique opportunities to visualize complex medical information, engage patients in their care, and provide remote medical services, making healthcare more efficient and accessible. Immersive technologies create engaging and interactive experiences by overlaying digital information onto the physical world or by creating entirely virtual environments. VR immerses users in a fully digital environment, often used for simulations and education. AR overlays digital content onto the real world, enhancing the user's perception and interaction with their surroundings [2]. MR combines elements of both VR and AR, allowing users to interact with digital and real-world objects simultaneously. In healthcare, these technologies are revolutionizing how medical education, diagnostics, treatment, and patient engagement are approached. For example, medical students and professionals can use VR to practice complex surgical procedures in a risk-free environment, improving their skills and confidence. AR can assist surgeons by providing real-time, 3D visualizations of a patient's anatomy during surgery, enhancing precision and reducing the risk of complications. For patients, immersive technologies can offer therapeutic experiences, such as VR-based pain management and mental health therapies, which are becoming increasingly recognized as effective treatment modalities [3].

2 Overview of the Importance of Monitoring and Remote Patient Management in Modern Healthcare

Monitoring and remote patient management have become critical components of modern healthcare, significantly enhancing the ability to deliver high-quality care. With the advent of digital health technologies, healthcare providers can now offer continuous monitoring and management of patients outside traditional clinical settings. This shift is particularly crucial for managing chronic conditions, post-operative care, and elderly patients who require consistent oversight but face challenges accessing in-person care [4].

The importance of these practices is underscored by several factors:

2.1 Chronic Disease Management

Chronic diseases, such as diabetes, hypertension, heart disease, and chronic obstructive pulmonary disease (COPD), are long-lasting conditions that require ongoing medical attention and lifestyle management. The prevalence of these diseases is increasing globally, placing a significant burden on healthcare systems. Remote monitoring and management of chronic diseases using immersive technologies can play a crucial role in improving patient outcomes and reducing healthcare costs [5].

- **Continuous Monitoring**: Immersive technologies enable continuous monitoring of vital signs and health metrics through wearable devices and sensors. For instance, a diabetic patient can use a continuous glucose monitor (CGM) integrated with a VR dashboard to track glucose levels in real time. This data can be visualized in an immersive environment, allowing both patients and healthcare providers to detect patterns, make informed decisions, and adjust treatment plans accordingly.
- **Early Intervention**: With the ability to monitor health metrics in real time, healthcare providers can detect early signs of disease exacerbation or complications. For example, VR or AR platforms can alert healthcare professionals when a patient's blood pressure exceeds a certain threshold, prompting timely intervention to prevent more severe outcomes.
- **Personalized Treatment Plans**: Immersive technologies can facilitate personalized medicine by providing detailed and individualized data. This data can be used to tailor treatment plans based on the patient's unique needs, preferences, and lifestyle. For instance, a VR-based platform can simulate different dietary or exercise regimens, helping patients understand the impact of these changes on their condition.

2.2 Access to Care

Access to healthcare remains a significant challenge, especially in rural and underserved areas. Immersive technologies can help bridge this gap by providing remote access to medical expertise and resources [6].

- **Telemedicine and Virtual Consultations**: VR and AR technologies enable virtual consultations, allowing patients to receive medical advice and treatment without the need to travel. This is particularly beneficial for patients living in remote areas or those with mobility issues. Virtual consultations can also provide access to specialists who may not be available locally.

- **Remote Diagnostics and Assessments**: AR-enabled devices can assist in remote diagnostics by overlaying medical data onto the real-world environment. For instance, a healthcare provider can use AR glasses to view a patient's medical history and current vital signs while conducting a virtual examination. This can improve diagnostic accuracy and provide a more comprehensive understanding of the patient's condition.
- **Health Education and Awareness**: Immersive technologies can be used to educate patients and communities about health conditions, preventive measures, and treatment options. For example, VR experiences can simulate the impact of smoking on lung health, helping to raise awareness and encourage healthier behaviors.

2.3 Cost-Effectiveness

The integration of immersive technologies in healthcare has the potential to significantly reduce costs for both patients and healthcare providers [7].

- **Reduced Hospital Admissions and Readmissions**: By enabling continuous monitoring and early intervention, remote patient management can prevent hospital admissions and readmissions. This not only improves patient outcomes but also reduces the financial burden on healthcare systems. For instance, timely interventions for heart failure patients can prevent costly hospitalizations.
- **Efficient Resource Allocation**: Immersive technologies can optimize the use of healthcare resources. For example, VR and AR can be used for remote training and supervision of medical staff, reducing the need for on-site trainers and minimizing travel expenses. Additionally, virtual consultations and remote diagnostics can reduce the demand for physical clinic space and medical equipment.
- **Scalability and Reach**: Immersive technologies can be scaled to reach a larger patient population without a proportional increase in costs. This scalability makes it possible to provide high-quality care to more people, including those in low-income or underserved regions.

2.4 Patient Engagement and Empowerment

Patient engagement and empowerment are critical for successful healthcare outcomes. Immersive technologies can play a key role in enhancing patient involvement in their care [8].

- **Interactive and Educational Experiences**: Immersive technologies can create engaging and interactive experiences that educate patients about their health conditions and treatment options. For example, VR simulations can demonstrate the effects of various medications or lifestyle changes, helping patients make informed decisions about their care.

- **Gamification and Motivation**: Gamification elements, such as rewards and progress tracking, can be integrated into immersive platforms to motivate patients to adhere to treatment plans. For instance, a VR-based rehabilitation program for stroke patients can include games that encourage them to complete exercises, making the process more enjoyable and motivating.
- **Personalized Patient Portals**: Immersive technologies can provide personalized patient portals that offer tailored information, resources, and support. These portals can include virtual coaches or avatars that guide patients through their treatment plans, answer questions, and provide emotional support.

2.5 Pandemic Preparedness

The COVID-19 pandemic has highlighted the importance of remote healthcare solutions and the need for preparedness in managing public health crises. Immersive technologies can play a pivotal role in enhancing pandemic preparedness and response [9].

- **Virtual Care and Social Distancing**: During pandemics, immersive technologies enable the continuation of medical care while minimizing physical contact. Virtual consultations, remote monitoring, and AR-assisted diagnostics allow healthcare providers to assess and manage patients without the need for in-person visits, reducing the risk of virus transmission.
- **Remote Training and Simulation**: VR and AR can be used for remote training and simulation exercises for healthcare professionals. This is crucial during pandemics when there is a need to quickly upskill medical staff in areas such as infectious disease management and emergency response.
- **Public Health Education and Communication**: Immersive technologies can be used to disseminate public health information and guidelines. For example, VR experiences can simulate the effects of virus spread, emphasizing the importance of preventive measures such as mask-wearing and social distancing.
- **Mental Health Support**: The psychological impact of pandemics can be significant. Immersive technologies can provide mental health support through virtual therapy sessions, relaxation exercises, and educational content on coping strategies.

By leveraging immersive technologies, healthcare systems can enhance chronic disease management, improve access to care, reduce costs, engage patients, and prepare for future pandemics. These technologies promise to transform healthcare delivery, making it more efficient, accessible, and patient-centered.

2.6 Overview of Immersive Technologies in Healthcare

Virtual reality (VR) technology has become a significant asset in the healthcare sector, offering innovative ways to enhance patient care and treatment outcomes. By immersing users in a fully digital environment, VR can simulate real-world scenarios or create new, controlled settings tailored to specific therapeutic needs. This capability makes VR uniquely suited for a variety of medical applications, from treatment to training and beyond [10].

- **Pain Management**: VR has been particularly effective in pain management, especially for burn victims and patients undergoing certain types of surgeries. By immersing patients in soothing and distracting virtual environments, VR can significantly reduce their perception of pain. For instance, VR programs can transport patients to serene landscapes or engaging interactive games, providing a psychological escape from painful treatments.
- **Surgical Simulation**: VR provides an exceptional training tool for medical students and surgeons, offering a highly realistic, risk-free environment to practice procedures. Surgical simulations can replicate the complexities of human anatomy and surgical interventions, allowing practitioners to enhance their skills without the ethical and practical concerns associated with traditional training methods.
- **Understanding Conditions and Treatments**: VR can educate patients about their health conditions and treatment plans by visually demonstrating the processes and effects on the body. For example, a patient with heart disease might explore a VR simulation of blood flowing through arteries to understand the impact of cholesterol buildup.
- **Virtual Consultations**: Through VR, patients can have immersive consultations with their healthcare providers from the comfort of their homes. This not only makes medical care more accessible but also enhances consultation experience by allowing doctors to use three-dimensional models and simulations to discuss health issues more effectively.

2.7 Advantages of Using VR in Healthcare

- **Controlled Environment**: VR offers the unique advantage of creating controlled therapeutic settings, allowing for precise management of the intensity and duration of therapeutic sessions.
- **Safety and Accessibility**: Patients can undergo therapy or training in environments that are safe and easily accessible, reducing the need for physical travel.
- **Data-Driven Insights**: VR systems can collect data on patient interactions and progress, providing valuable feedback to healthcare providers to tailor treatment plans effectively.

2.8 Example: VR for Pain Management and Rehabilitation

VR is widely used for pain management and physical rehabilitation. By immersing patients in a virtual environment, VR can distract them from pain and provide therapeutic exercises. Studies have shown that VR can significantly reduce pain perception in patients with chronic pain or undergoing painful procedures [11].

2.9 Example: VR for Medical Training

VR offers an immersive environment for medical training, allowing practitioners to simulate surgeries and other procedures without risk to patients. This hands-on experience is invaluable for training surgeons and other healthcare professionals, improving their skills and confidence.

Table 9.1 summarizes the VR Use Cases and Benefits.

An illustrative setup of VR technology in a clinical setting is represented in Fig. 9.1, showing a patient using a VR headset connected to monitoring equipment that tracks patient responses and progress [12].

2.10 Challenges and Future Directions

While VR offers substantial benefits, there are challenges to its widespread adoption:

- **Cost and Infrastructure**: The initial setup cost and the need for ongoing maintenance can be significant. The initial setup, including hardware and customized software development, can be expensive. This can limit widespread adoption, especially in under-resourced healthcare settings.

Table 9.1 VR use cases and benefits

Application area	VR use case	Benefits
Psychological therapy	PTSD treatment	Safe exposure to triggers, controlled therapy settings
Pain management	Distraction during painful procedures	Reduces perceived pain, enhances patient comfort
Physical rehabilitation	Stroke recovery exercises	Promotes neuroplasticity, tailored exercise routines

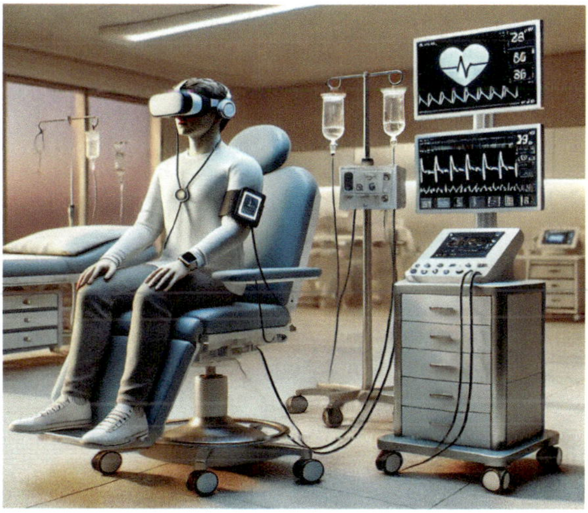

Fig. 9.1 VR technology setup in healthcare

- **Technical Training**: Healthcare providers need training to effectively implement and utilize VR technologies. Resistance from healthcare providers due to lack of training or skepticism about the effectiveness and practicality of VR solutions.
- **Patient Adaptation**: Some patients may experience discomfort or disorientation using VR headsets.

The future of VR in healthcare is promising, with ongoing advancements in technology that are likely to lower costs and enhance the effectiveness of VR applications. Continued research and development will expand its use into more areas of healthcare, providing more personalized and effective treatment options [13]. Further development of VR for medical training and simulations to provide more realistic and varied scenarios. This can improve skills and preparedness among healthcare professionals. Increasing the use of VR for therapeutic purposes, such as in mental health treatments for anxiety, PTSD, and phobias, and for physical rehabilitation. Developing more cost-effective VR solutions to ensure wider accessibility, including portable and less expensive VR systems. Creating more patient-friendly VR applications that are easy to use at home without professional supervision, increasing patient engagement and adherence to treatment. Utilizing the data collected via VR platforms for deeper insights into patient progress and outcomes, driving improvements in treatments and patient care. Leveraging VR to enhance healthcare delivery in remote or underserved regions, providing high-quality training and treatment options worldwide [14].

3 Augmented Reality (AR) in Healthcare: Enhancing Diagnostics and Education

Augmented reality (AR) in healthcare enhances the field by overlaying digital information onto the physical world, providing users with additional data or imagery to improve their perception. Integrating digital elements into the physical environment allows healthcare professionals and patients to interact with virtual components as if they were part of the physical space. AR's unique capability to provide real-time information and visual guidance makes it particularly valuable for diagnostics, surgical procedures, patient care, and medical training [15].

3.1 Diagnostic and Treatment Applications of AR

3.1.1 Enhanced Diagnostics

- **Real-Time Imaging**: AR can project imaging results (like CT scans or MRIs) directly on a patient's body, helping doctors to visualize the anatomy and functions internally without making an incision, thus improving the accuracy of diagnoses and pre-surgical planning.

- **Patient Monitoring**: Wearable AR devices can display vital health statistics in real time, enabling continuous monitoring without the need for stationary equipment.

3.1.2 Surgical Assistance

- **Surgical Navigation**: Surgeons can use AR to overlay detailed 3D models of a patient's anatomy directly onto their body during surgery, which can guide them in performing precise and minimally invasive procedures.
- **Remote Assistance**: AR can enable a surgeon in one location to see exactly what a colleague sees in another operating room and provide guidance through virtual annotations.

3.1.3 Procedural Training

- **Medical Education**: AR applications allow medical students to interact with complex anatomical models virtually. These models can be manipulated in 3D space, offering insights from different angles and fostering a deeper understanding of human anatomy.

3.2 Advantages of Using AR in Healthcare

- **Interactive Learning and Engagement**: AR enhances learning experiences by making medical education more interactive, helping students and professionals visualize and understand complex medical concepts better.
- **Improved Precision and Efficiency**: By providing real-time, hands-free access to data and patient information, AR helps clinicians perform task with greater precision and efficiency, reducing the risk of errors.
- **Cost-Effectiveness**: AR can reduce the need for physical models and other resources in medical training and patient education, lowering costs.

3.3 Example: AR for Remote Assistance and Surgery

AR technology overlays digital information onto the real world, providing healthcare professionals with real-time data and guidance. For instance, AR can be used in remote surgery where a specialist can assist a surgeon by overlaying instructions and anatomical information onto the surgical site through AR glasses. This real-time assistance improves accuracy and outcomes, especially in complex procedures [16].

3.4 Example: AR for Patient Education and Engagement

AR applications are also valuable for patient education. Patients can visualize their medical conditions and treatment plans through AR, enhancing their understanding and engagement. For example, AR apps can show patients how their medications affect their body, which can improve adherence to treatment plans [17].

Table 9.2 summarizes the AR Use Cases and Benefits.

An illustrative setup of AR technology in a clinical setting is represented in Fig. 9.2, showing a medical student using AR glasses to study a virtual anatomical model, which is overlaid onto a real-world table [18].

3.5 Challenges and Future Directions

- **Technical Limitations**: The accuracy and reliability of AR depend on the advancement of image recognition and display technologies. Improvements are necessary to enhance the stability and clarity of virtual overlays.

Table 9.2 AR use cases in healthcare

Application area	AR use case	Benefits
Diagnostics	Overlaying imaging data on patient	Enhanced precision in diagnosis and planning
Surgical assistance	Real-time surgical navigation	Increased accuracy, reduced invasiveness

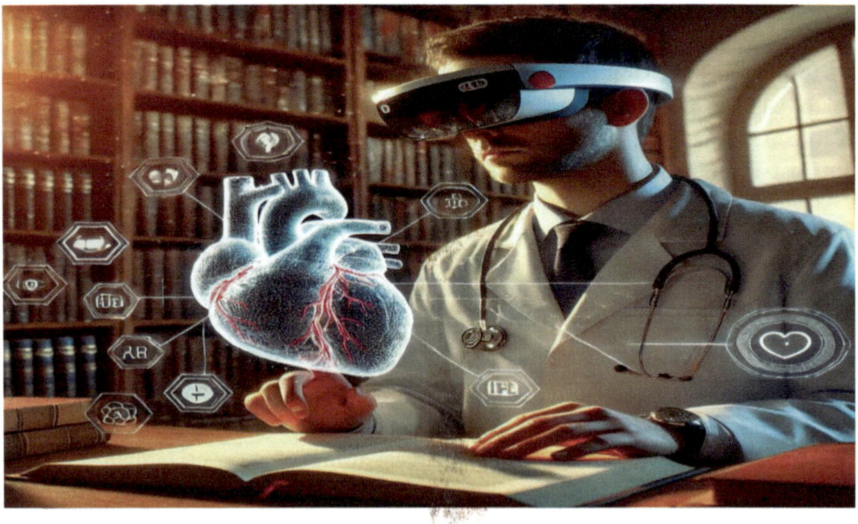

Fig. 9.2 AR in medical training

- **Integration with Existing Systems**: Integrating AR with current medical systems and workflows can be complex and requires careful planning and execution.
- **User Experience Issues**: Ensuring comfort and usability for long periods, particularly in surgical settings, remains a challenge.

The ongoing development and refinement of AR technology will likely continue to expand its applications within healthcare. Future advancements could lead to more widespread adoption of AR in routine medical procedures and training, making it an indispensable tool in modern healthcare. This expanded explanation of AR in healthcare illustrates its potential to significantly impact diagnostics, patient care, and medical education, offering a blend of real-time data integration and interactive learning opportunities. Enhancing real-time information during surgeries and procedures [19]. AR can project vital data directly into the surgeon's field of view, improving precision and outcomes. AR can facilitate remote guidance for surgeries and consultations, where specialists can "see" what the onsite medical professional sees and provide immediate advice. Expanding the use of AR for educational purposes, allowing medical students and professionals to practice skills on virtual patients in a controlled, realistic environment. Using AR to provide patients with visual information about their health conditions or treatments, thereby enhancing understanding and engagement. Integrating AR with imaging technologies to enhance the visualization of complex anatomical structures, aiding in diagnostics and treatment planning. Developing lightweight, wearable AR devices tailored for continuous use in clinical environments without hindering user mobility. Focusing on creating interoperable systems that can function seamlessly across different healthcare environments and platforms. Making AR tools more accessible and affordable to healthcare providers globally, particularly in low-resource settings [20].

4 Mixed Reality (MR) in Healthcare: Integrating the Physical and Digital Worlds to Improve Medical Practice

Mixed Reality (MR) merges physical and digital environments, allowing real and virtual objects to coexist and interact in real time. This technology integrates the immersive aspects of Virtual Reality (VR) with the augmentative visualizations of Augmented Reality (AR), resulting in a powerful tool for visualizing and manipulating complex medical data. MR's ability to blend the real world with virtual elements provides a unique platform for medical training, procedural planning, and patient care.

4.1 Applications of MR in Medical Practice

4.1.1 Complex Surgical Planning

- **Pre-operative Procedures**: Surgeons use MR to visualize complex surgeries by overlaying three-dimensional holographic data onto the physical surgical field. This preparation allows surgeons to examine multiple layers of an organ, foresee potential challenges, and plan the surgical approach with higher precision.
- **Intra-operative Navigation**: During surgery, MR provides real-time, holographic guidance, helping surgeons navigate complex anatomical structures without looking away from the surgical site.

4.1.2 Medical Training and Education

- **Interactive Learning**: MR creates dynamic, interactive training environments where medical students can engage with realistic 3D models of human anatomy, performing virtual dissections and procedures that feel tangible.
- **Collaborative Learning**: MR allows for collaborative sessions where students and instructors can interact with the same virtual objects simultaneously, despite being in different locations.

4.1.3 Patient Care and Rehabilitation

- **Physical Rehabilitation**: MR systems support the rehabilitation of patients by creating interactive environments that adapt to their physical capabilities, motivating them through gamified exercises that track and respond to their progress in real time.
- **Patient Education**: MR can enhance the understanding and engagement of patients by visualizing their medical conditions or the effects of proposed treatments, leading to better-informed health decisions.

4.2 Advantages of Using MR in Healthcare

- **Precision and Efficiency**: The precision of MR applications in surgical settings enhances the efficiency of procedures, potentially reducing operation times and improving outcomes.
- **Engagement and Motivation**: For patients, especially in rehabilitation, MR's immersive and interactive nature can lead to greater motivation and adherence to treatment protocols.

4.3 Example: MR for Remote Consultations

MR technology facilitates remote consultations by allowing doctors to interact with 3D representations of patients' medical data, such as CT scans or MRI results. This interactive approach enables more accurate diagnoses and better patient management, even when the doctor and patient are not in the same location.

4.4 Example: MR for Surgical Planning and Simulation

Mixed Reality combines elements of both AR and VR, creating a hybrid environment where digital and physical objects coexist and interact. In surgical planning, MR can be used to create detailed 3D models of a patient's anatomy. Surgeons can interact with these models to plan and practice complex surgeries, improving precision and outcomes.

Table 9.3 summarizes the MR Use Cases and Benefits.

An illustrative setup of MR technology in a clinical setting is represented in Fig. 9.3, an image depicting a surgeon using MR goggles to view holographic anatomical structures during a surgical procedure, enhancing the precision and safety of the operation [21].

4.5 Challenges and Future Directions

- **Hardware and Cost**: The development and deployment of MR technology involve significant costs, primarily due to the sophisticated hardware required.
- **Technical Complexity**: The integration of MR into existing medical systems poses technical challenges, requiring robust software capable of handling complex data in real time.
- **User Adaptation**: There is a learning curve associated with using MR technology effectively, and continuous training is necessary to maximize its benefits.

The ongoing advancements in MR technology promise to further enhance its applications in healthcare. As MR devices become more accessible and their

Table 9.3 MR use cases in healthcare

Application area	MR use case	Benefits
Surgical planning	Pre-operative visualization	Enhances surgical precision and planning
Medical education	Interactive anatomical models	Improves depth of understanding, interactive learning
Patient rehabilitation	Gamified therapy sessions	Increase patient engagement and therapy adherence

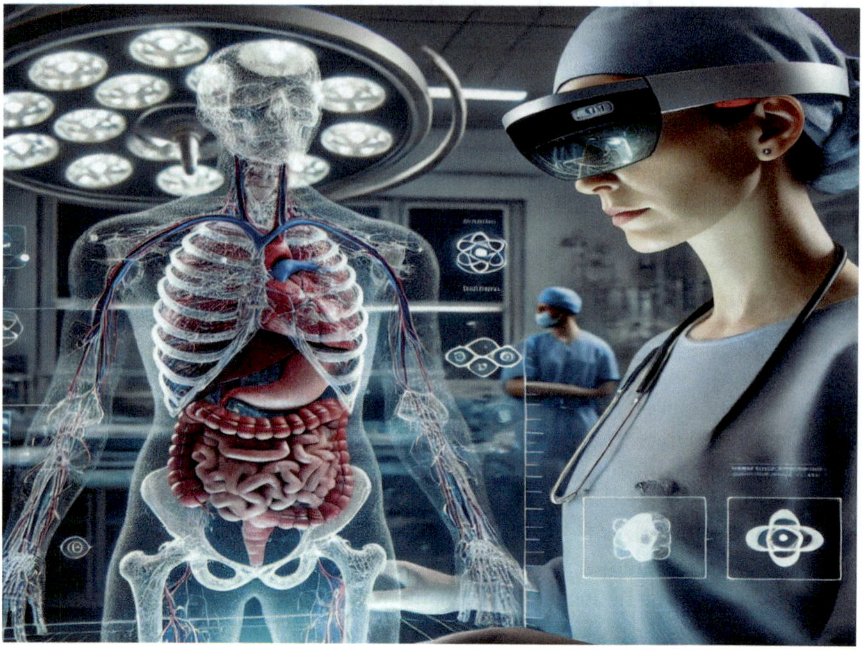

Fig. 9.3 MR for surgical visualization

interfaces more user-friendly, their integration into daily medical practice is likely to grow, making them an integral part of modern healthcare delivery [22]. This chapter provides a detailed exploration of how MR is transforming healthcare by enabling more precise medical procedures, enriching education, and improving patient care through advanced, interactive visualizations. Leveraging MR for remote assistance in surgeries and diagnostics can enable specialists worldwide to collaborate in real-time, offering their expertise without the need for physical presence. Expanding the use of MR in medical training environments to provide highly interactive and immersive learning experiences, mimicking real-life medical scenarios without any associated risks. Utilizing MR to explain complex medical conditions or procedures to patients in a more comprehensible and interactive manner, potentially improves understanding and compliance with treatment plans. Developing MR applications for rehabilitation and therapy that can adapt to the progress and needs of patients, offering a more personalized and engaging treatment approach. Innovation in wearable MR technology that is more compact, comfortable, and capable of being used in various medical settings, from operating rooms to outpatient clinics. Establishing standards for interoperability among different MR systems and healthcare databases to ensure seamless integration and communication across platforms. Making MR technology more accessible and affordable for a wider range of healthcare providers, including those in resource-limited settings [23].

5 Current Trends and Developments in Immersive Technologies

Immersive technologies, including Virtual Reality (VR), Augmented Reality (AR), and Mixed Reality (MR), are rapidly evolving, driven by advancements in hardware, software, and user interface design. These technologies are increasingly being integrated into various sectors, including healthcare, where they offer new ways to visualize, interact with, and understand medical data and environments [24].

5.1 Enhanced Hardware Capabilities

- **Increased Resolution and Field of View**: Modern VR headsets offer higher resolution displays and a wider field of view, providing more realistic and immersive experiences. This advancement is crucial in medical simulations and training, where detailed visualization can significantly enhance learning outcomes.
- **Improved Tracking and Sensing**: Enhanced motion tracking and sensor technologies allow for more precise tracking of movements and interactions within immersive environments. This is particularly beneficial for applications like surgical training and rehabilitation, where accurate tracking of hand and body movements is essential.
- **Wearable and Portable Devices**: The development of lighter, more comfortable, and portable devices, such as standalone VR headsets and AR glasses, is making immersive experiences more accessible and user-friendly. These devices can be used in a variety of healthcare settings, from clinics to patients' homes.

5.2 Software and Content Development

- **Realistic Simulations and Virtual Environments**: Advances in graphics rendering and simulation algorithms are enabling the creation of highly realistic virtual environments. These environments are used for medical training, therapeutic interventions, and patient education.
- **AI Integration**: Artificial Intelligence (AI) is being integrated into immersive technologies to enhance user experiences. For example, AI algorithms can create adaptive learning experiences in medical training simulations or provide personalized patient education in virtual environments.
- **Cross-Platform Compatibility**: Efforts to make immersive content compatible across different platforms (e.g., VR, AR, MR) and devices are increasing. This compatibility ensures that content can be accessed and utilized in various contexts, making it more versatile and widely applicable.

5.3 Expanding Applications in Healthcare

- **Telemedicine and Remote Consultations**: The use of immersive technologies for remote consultations is growing, enabling healthcare providers to conduct virtual visits and examinations. These technologies enhance the telemedicine experience by providing 3D visualizations and interactive tools [25].
- **Surgical Planning and Simulation**: Surgeons are using VR and AR for preoperative planning and simulations, allowing them to practice complex procedures and visualize patient anatomy in 3D. This practice can improve surgical accuracy and reduce the risk of complications.
- **Mental Health and Pain Management**: Immersive experiences are being used to treat mental health conditions such as anxiety, PTSD, and phobias. VR-based therapy can provide controlled environments for exposure therapy. Additionally, VR is being used as a tool for pain management, distracting patients and reducing the perception of pain during medical procedures.

6 The Role of Immersive Technologies in Healthcare Settings

Immersive technologies are playing an increasingly important role in transforming various aspects of healthcare delivery, from diagnosis and treatment to education and patient engagement.

6.1 Medical Training and Education

- **Simulation-Based Learning**: Immersive technologies provide medical students and professionals with realistic simulations for training purposes. These simulations can replicate surgical procedures, emergency scenarios, and patient interactions, allowing learners to practice and refine their skills in a safe environment.
- **Interactive Anatomy and Physiology**: AR and VR can be used to create interactive 3D models of human anatomy and physiology. These models enhance the learning experience by allowing users to explore and interact with different body systems, providing a deeper understanding of complex structures and functions.
- **Continuing Medical Education (CME)**: Immersive technologies enable health care professionals to engage in CME activities remotely. VR-based courses and workshops can simulate real-world clinical scenarios, helping practitioners stay updated on the latest medical advancements and techniques.

6.2 Patient Care and Treatment

- **Remote Monitoring and Management**: Immersive technologies enable the remote monitoring of patients' health conditions, providing real-time data to healthcare providers. For example, AR-enabled wearable devices can display vital signs and other health metrics, allowing for continuous monitoring and timely interventions.
- **Therapeutic Interventions**: VR is being used as a therapeutic tool for various conditions, including chronic pain, PTSD, and anxiety disorders. Virtual environments can be tailored to provide exposure therapy, relaxation exercises, and cognitive behavioral therapy (CBT) sessions.
- **Enhanced Patient Engagement**: Immersive technologies can improve patient engagement by providing interactive and educational experiences. For instance, VR experience can help patients understand surgical procedures, treatment plans, and potential outcomes, making them more informed and involved in their care.

6.3 Diagnosis and Surgical Assistance

- **Preoperative Planning**: Surgeons can use VR and AR to visualize and plan complex surgeries. These technologies provide detailed 3D representations of patient anatomy, allowing surgeons to practice procedures and identify potential challenges before entering the operating room.
- **Intraoperative Assistance**: AR can assist surgeons during procedures by overlaying critical information, such as anatomical structures and surgical guides, onto the patient's body. This real-time data can improve precision and reduce the likelihood of errors.
- **Postoperative Care**: Immersive technologies can also aid postoperative care by providing virtual follow-up visits and monitoring recovery progress. Patients can use VR platforms for guided rehabilitation exercises, ensuring they adhere to prescribed protocols.

6.4 Public Health and Education

- **Health Awareness Campaigns**: Immersive technologies can be used to create impactful public health campaigns. For example, VR experiences can simulate the effects of smoking or poor diet, raising awareness and promoting healthy behaviors.
- **Disaster Preparedness and Response**: Immersive simulations can prepare healthcare professionals and the public for disaster scenarios, such as pandemics

or natural disasters. These simulations can train responders, test emergency plans, and educate the public on safety measures.

In summary, immersive technologies are revolutionizing healthcare by enhancing medical training, improving patient care, assisting in diagnosis and surgery, and supporting public health initiatives. Their continued development and integration into healthcare settings promise to improve outcomes, increase access to care, and create more engaging and effective healthcare experiences.

7 Case Studies and Examples

- **Real-World Examples of Healthcare Providers Using Immersive Technologies**: Immersive technologies have been adopted by various healthcare providers worldwide to enhance patient care, training, and operational efficiency. Here are some notable examples:
- **Cleveland Clinic and Virtual Reality for Pain Management**: The Cleveland Clinic has implemented virtual reality (VR) as a tool for pain management in patients undergoing painful procedures or those with chronic pain conditions. By immersing patients in calming VR environments, they can reduce their perception of pain and anxiety, leading to a more comfortable experience.
- **Stanford Children's Health and Augmented Reality in Surgery**: Surgeons at Stanford Children's Health have utilized augmented reality (AR) to perform complex surgeries. By overlaying 3D models of patients' anatomy onto their bodies during surgery, surgeons can better visualize internal structures and navigate during procedures. This has been particularly useful in pediatric surgeries, where precise and minimally invasive approaches are crucial.
- **Cedars-Sinai Medical Center and VR for Medical Training**: Cedars-Sinai has integrated VR into its medical training programs, providing medical students and residents with realistic simulations of surgical procedures and emergency scenarios. This technology allows trainees to practice and refine their skills in a controlled, risk-free environment, improving their readiness for real-world situations.
- **Mayo Clinic and AR for Patient Education**: The Mayo Clinic uses AR to enhance patient education. By providing patients with AR visualizations of their conditions and potential treatments, the clinic helps patients better understand their health issues and the procedures they will undergo. This approach has been shown to improve patient satisfaction and engagement.
- **University of Oxford and VR for Mental Health Therapy**: Researchers at the University of Oxford have developed VR-based therapy programs for treating mental health conditions such as social anxiety and PTSD. These programs create controlled virtual environments where patients can confront and overcome their fears under the guidance of a therapist, demonstrating promising results in clinical trials.

8 Case Studies Demonstrating the Impact on Patient Outcomes and Healthcare Delivery

8.1 Case Study 1: VR in Pain Management for Burn Patients

- **Background**: Burn injuries are associated with intense pain, especially during wound care and dressing changes. Traditional pain management techniques may not be sufficient, leading to high levels of distress and discomfort for patients.
- **Intervention**: A study conducted at the University of Washington's Harborview Burn Center explored the use of VR as a complementary pain management tool. Patients were immersed in a VR environment known as "Snow World," designed to be engaging and distracting.
- **Results**: The study found that patients who used VR during wound care reported significantly lower pain levels compared to those who did not use VR. The immersive nature of the VR experience helped shift patients' focus away from their pain, reducing their reliance on pain medication and improving their overall experience.

8.2 Case Study 2: AR in Orthopedic Surgery

- **Background**: Accurate placement of screws and implants is critical in orthopedic surgeries, such as spine surgeries. Traditional imaging techniques provide limited real-time guidance, potentially leading to complications or the need for revision surgeries.
- **Intervention**: At the Johns Hopkins Hospital, AR technology was integrated into the operating room, allowing surgeons to overlay 3D images of the patient's spine onto their view. This AR system provided real-time guidance, highlighting the optimal paths for surgical instruments.
- **Results**: The use of AR significantly improved the accuracy of screw placement, reducing the risk of complications and enhancing patient outcomes. Surgeons also reported increased confidence and efficiency, leading to shorter surgery times and faster recovery for patients.

8.3 Case Study 3: VR in Stroke Rehabilitation

- **Background**: Rehabilitation is crucial for stroke patients to regain motor function and independence. Traditional rehabilitation exercises can be repetitive and monotonous, leading to low patient engagement and adherence.
- **Intervention**: The Stroke Center at the University of California, Los Angeles (UCLA), implemented a VR-based rehabilitation program called "VIRTUES"

(Virtual Reality Training and Upper Extremity Stroke Rehabilitation). The program included a series of VR exercises designed to improve motor skills and coordination.

- **Results**: Patients participating in the VR program showed significant improvements in upper extremity function compared to those receiving standard therapy. The interactive and gamified nature of the VR exercises increased patient engagement and motivation, leading to better adherence and more effective rehabilitation.

8.4 Case Study 4: AR for Remote Surgical Assistance

- **Background**: In remote or resource-limited settings, access to specialized surgical expertise can be limited. This can impact on the quality of care for complex cases that require specialized skills.
- **Intervention**: In a collaborative effort between Microsoft and the University of Alberta, AR technology was used to provide remote surgical assistance. Using AR headsets, surgeons in remote locations could receive real-time guidance from specialists located elsewhere. The specialists could see the surgical field and provide step-by-step instructions overlaid on the AR display.
- **Results**: This approach enabled complex surgeries to be performed in remote settings with the support of expert guidance. It also facilitated knowledge transfer and training for local surgeons, improving the overall quality of care in underserved areas.

8.5 Case Study 5: VR for Cognitive Behavioral Therapy (CBT) in PTSD

- **Background**: Traditional CBT for PTSD involves exposing patients to memories or situations that trigger their symptoms in a controlled setting. However, it can be challenging to recreate these triggers accurately and safely.
- **Intervention**: The University of Southern California Institute for Creative Technologies developed a VR-based CBT program called "Bravemind" for veterans with PTSD. The program created realistic virtual environments that allowed patients to confront and process their traumatic experiences under the guidance of a therapist.
- **Results**: The VR-based CBT program showed significant reductions in PTSD symptoms among participants. The immersive nature of VR allowed for more controlled and effective exposure therapy, helping patients process their trauma and develop coping mechanisms.

9 Conclusion

The chapter has explored the transformative potential of immersive technologies in healthcare, focusing on their applications in monitoring, remote patient management, training, and patient care. The growing importance of monitoring and remote patient management, particularly in chronic disease management, access to care, cost-effectiveness, patient engagement, and pandemic preparedness. The types of immersive technologies and their current trends, such as enhanced hardware capabilities, AI integration, and the expansion of telehealth services. The significant roles these technologies play in medical training, patient care, diagnostics, and public health, as well as emerging trends and potential future applications. The critical role of AI and machine learning in enhancing immersive experiences, providing personalized, predictive, and efficient healthcare solutions. In summary, incorporating immersive technologies into patient monitoring and management marks a notable progression in healthcare. These technologies offer unprecedented opportunities for enhancing patient engagement, improving the precision of monitoring, and delivering personalized care in ways that were previously unattainable.

The adoption of VR, AR, and MR in healthcare settings enables a more intuitive and immersive approach to patient management, allowing for real-time visualization of medical data and simulations of complex medical conditions. This can lead to better understanding and faster decision-making by healthcare professionals. Additionally, immersive technologies can significantly elevate the quality of patient education and compliance, providing interactive and engaging methods to explain diagnoses, treatments, and procedures.

References

1. Freeman D, Reeve S, Robinson A, Ehlers A, Clark D, Spanlang B, Slater M. Virtual reality in the assessment, understanding, and treatment of mental health disorders. Psychol Med. 2017;47(14):2393–400.
2. Valmaggia LR, Latif L, Kempton MJ, Rus-Calafell M. Virtual reality in the psychological treatment for mental health problems: a systematic review of recent evidence. Psychiatry Res. 2016;236:189–95.
3. Maples-Keller JL, Bunnell BE, Kim SJ, Rothbaum BO. The use of virtual reality technology in the treatment of anxiety and other psychiatric disorders. Harv Rev Psychiatry. 2017;25(3):103–13.
4. Rizzo AS, Koenig ST. Is clinical virtual reality ready for primetime? Neuropsychology. 2017;31(8):877–99.
5. Barsom EZ, Graafland M, Schijven MP. Systematic review on the effectiveness of augmented reality applications in medical training. Surg Endosc. 2016;30(10):4174–83.
6. Bork F, Stratmann L, Enssle S, Eck U. The benefits of augmented reality in educational environments—a systematic literature review. J Univ Comput Sci. 2016;22(12):1548–68.
7. Chen PJ, Huang CD, Yeh SL. Augmented reality-assisted pedicle screw installation surgical navigation system. J Med Biol Eng. 2017;37(1):27–36.

8. Khor WS, Baker B, Amin K, Chan A, Patel K, Wong J. Augmented and virtual reality in surgery—the digital surgical environment: applications, limitations and legal pitfalls. Ann Transl Med. 2016;4(23):454.
9. Pucher PH, Johnston MJ, Dalal M, Rajaretnam N, Arora S. The future of the digital surgical environment in the context of mixed reality: opportunities and threats. Ann Surg. 2016;264(6):933–8.
10. Pratt P, Ives M, Lawton G, Simmons J, Radev N, Spyropoulou L, Yang GZ. Through the HoloLens™ looking glass: augmented reality for extremity reconstruction surgery using 3D vascular models with perforating vessels. Eur Radiol Exp. 2018;2(1):2.
11. Ferrari V, Megali G, Troia E, Pietrabissa A, Mosca F. A mixed-reality surgical trainer with haptic feedback for robotic surgery. Biomed Eng Online. 2017;16(1):1–18.
12. Hooper J, Tsiridis E, Feng J, Schwarzkopf R, Waren D, Long WJ, Einhorn TA. MR-guided robotic arm for orthopedic surgery: technological advancements and clinical applications. J Robot Surg. 2018;12(3):345–53.
13. Baran S. Augmented reality in medical education and surgery. J Healthc Technol. 2020;12(3):45–58.
14. Jones T, Smith M, Brown R. Virtual reality for chronic pain management. Pain Manag J. 2016;9(2):85–93.
15. Liu Y, Ma X, Chen Z. Mixed reality for surgical planning and simulation. Int J Med Inform. 2018;112:67–76.
16. Moorhouse N, Glover S, Khan A. Augmented reality in patient education. Healthc Commun Rev. 2019;14(1):22–9.
17. Peterson E, Roberts H, Li Y. Mixed reality for remote medical consultations. Telemed e-Health. 2021;27(6):583–9.
18. Smith A, Hamilton R. Virtual reality in medical training: a systematic review. Med Educ Today. 2017;37(4):33–41.
19. Rizzo AS, Shilling R. Virtual reality therapy for PTSD: a clinical trial. J Anxiety Disord. 2020;64:102–10.
20. Liu D, Martin R. Remote surgical assistance using augmented reality. J Telemed Telecare. 2020;26(6):356–63.
21. Roberts S, Clarke J. Augmented reality in orthopedic surgery. Orthop Surg J. 2022;14(4):453–62.
22. Hoffman HG, Chambers GT. VR analgesia: reducing pain in burn patients. J Pain Res. 2020;13:123–30.
23. Zhang X, Lee H. AI-driven advancements in VR healthcare applications. Artif Intell Med. 2021;70(2):130–45.
24. Evans L, Norton J. Next-generation applications of VR and AR in healthcare. J Innov Health Sci. 2023;15(3):210–22.
25. Chen Y, Wang M. Virtual reality for stroke rehabilitation: a clinical trial. Neurorehabil Neural Repair. 2021;35(5):434–42.

Chapter 10
Current State of AR and VR in Healthcare

K. M. Kirthika, S. Sangeetha, R. Immanual, and N. Sanjana

Abstract Augmented Reality (AR) and Virtual Reality (VR) technologies rapidly transform the healthcare landscape, offering innovative solutions to enhance patient care, medical education, and therapeutic interventions. This chapter delves into the current state of AR and VR in healthcare, exploring their diverse applications and their profound impact on reshaping various aspects of the industry. The chapter examines the multifaceted applications of AR in healthcare settings, including patient education and engagement through interactive visualizations, enhancing diagnostic procedures through real-time guidance, gamified rehabilitation programs, and empowering patients with chronic conditions through real-time feedback and monitoring. Additionally, it explores the transformative potential of VR in exposure therapy for anxiety and phobias, pain management and distraction, surgical training and simulation, and mental health assessments and interventions. While highlighting the numerous benefits and applications, the chapter also addresses the challenges and considerations associated with implementing AR and VR in healthcare, such as accessibility and cost, data privacy and security concerns, and user experience and design considerations. This chapter paves the way for discussions on the future of these immersive technologies in healthcare by providing a comprehensive overview of the current landscape and exploring the opportunities and challenges. It underscores the immense potential of AR and VR to lead to a more patient-centered, efficient, and accessible healthcare system that prioritizes personalized care and improved outcomes.

K. M. Kirthika (✉) · N. Sanjana
Department of Computer Science and Engineering, Sri Ramakrishna Institute of Technology, Coimbatore, India

S. Sangeetha
Department of Electrical and Electronics Engineering, Sri Ramakrishna Institute of Technology, Coimbatore, India

R. Immanual
Department of Mechanical Engineering, Sri Ramakrishna Institute of Technology, Coimbatore, India

B. O. Soufiane et al. (eds.), *Augmented Wellness*,
https://doi.org/10.1007/978-981-96-2952-7_10

Keywords Augmented reality healthcare · Virtual reality healthcare · Medical education and training VR · AR surgical guidance · VR exposure therapy · AR/VR patient engagement

1 Introduction to AR and VR in Healthcare

The healthcare industry has consistently been at the vanguard of technological innovation, perpetually seeking novel methods to enhance treatment outcomes, optimize medical processes, and deliver superior patient care. Immersive technologies, such as virtual reality (VR) and augmented reality (AR), have experienced rapid advancements in recent years, unlocking groundbreaking opportunities within the healthcare sector. These innovative tools possess the capacity to revolutionize various aspects of medical practice, including therapeutic interventions, patient experiences, medical education, and remote consultations. AR technology augments or enhances a user's perception of reality by overlaying digital content, such as text, animations, and images, onto their real-world view [1]. In healthcare, AR can profoundly transform how medical professionals interact with and convey complex information by enabling them to visualize and manipulate data in a more immersive and intuitive manner. On the other hand, VR immerses users in a completely simulated, computer-generated environment, transcending the limitations of the physical world and providing a sense of presence and immersion [2]. This technology allows individuals to be transported into virtual environments, opening up numerous possibilities for therapeutic interventions, medical training, and even remote consultations.

The integration of AR and VR technologies into healthcare practices presents significant opportunities to elevate patient experiences, improve treatment outcomes, and streamline medical procedures. These immersive solutions offer a wide range of applications across various domains within the healthcare landscape, spanning from patient education and engagement to diagnostic processes, rehabilitation programs, and chronic disease management. One of the primary advantages of AR in healthcare is its potential to enhance patient education and engagement through interactive visualizations. By merging digital and physical information, AR enables patients to gain a deeper understanding of their medical conditions, treatment plans, and procedures [3]. This interactive approach can lead to improved patient adherence, informed decision-making, and ultimately better health outcomes.

Moreover, AR holds transformative potential in revolutionizing diagnostic procedures by providing medical professionals with valuable guidance during surgical interventions or other complex medical procedures. By superimposing digital imagery, annotations, and real-time data onto the patient's anatomy, AR can equip healthcare professionals with invaluable insights, potentially reducing the risk of errors and enhancing the accuracy of medical interventions [4]. Additionally, gamified experiences facilitated by AR offer exciting possibilities in the domain of rehabilitation programs, making the process more engaging and motivating for patients.

Through interactive AR applications, patients can participate in personalized virtual exercises and activities while receiving instant feedback and guidance [5]. This approach can improve patient compliance, accelerate recovery, and enhance treatment outcomes.

In the context of chronic disease management, AR technology empowers patients by providing real-time feedback and monitoring capabilities, enabling them to better understand and manage their conditions [6, 7]. For example, AR applications can seamlessly overlay digital information onto the user's environment, providing reminders for medication adherence, tracking vital signs, and offering personalized recommendations for lifestyle modifications.

While AR and VR technologies present immense potential in healthcare, it is crucial to address the challenges and considerations associated with their implementation. Accessibility and affordability remain significant concerns, as AR and VR experiences often require costly technology and software, which may be prohibitive for individuals or healthcare institutions with limited resources [8]. Furthermore, data privacy and security are critical considerations, particularly when integrating AR and VR in healthcare, as it often involves the collection and processing of sensitive patient data [9]. Robust data protection measures and strict compliance with privacy regulations are essential to ensure the ethical and responsible use of these technologies. Figure 10.1 shows that Applications of AR and VR in Healthcare.

Virtual reality (VR) transforms healthcare by providing a safe space for medical professionals to learn new skills. Doctors and nurses can train in anatomy, practice surgery, and learn infection control using VR simulations. This builds their confidence and skills without harming patients. VR also aids in diagnostics and surgery planning, allowing surgeons to virtually see inside a patient's body and prepare for complex procedure. Figure 10.2 shows that Virtual reality benefits in the medical field.

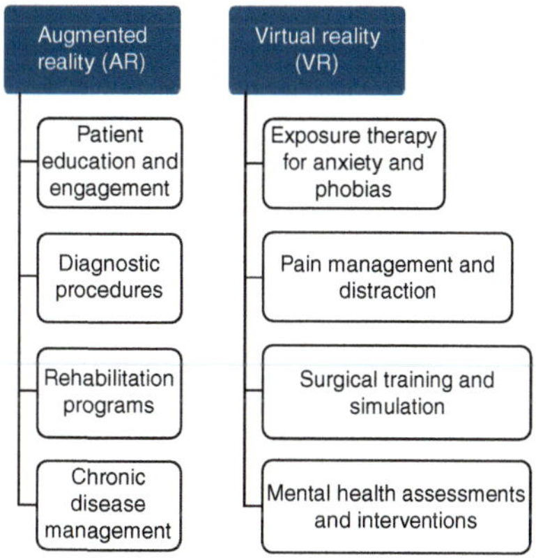

Fig. 10.1 Applications of AR and VR in healthcare

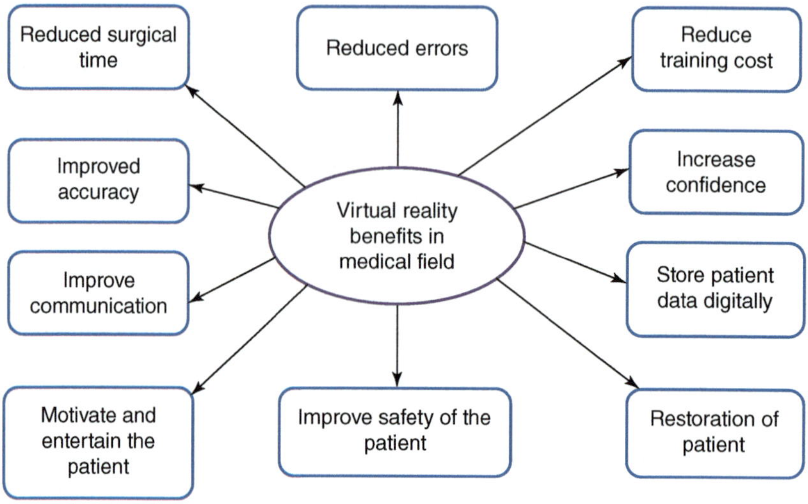

Fig. 10.2 Virtual reality benefits in the medical field

2 AR Applications in Healthcare

Augmented Reality (AR) applications in healthcare are a domain ripe with potential for revolutionizing patient education and engagement. AR is a promising tool that bridges the gap between patients and complex medical information in the ever-evolving healthcare technology landscape.

- Patient Education and Engagement
- Diagnostic Procedures
- Rehabilitation Programs
- Chronic Disease Management

2.1 *Patient Education and Engagement*

Patient Education and Engagement through AR entail the utilization of interactive digital overlays to enhance the understanding of medical conditions, treatments, and procedures. By superimposing virtual elements onto the real world, AR provides patients with immersive experiences that facilitate learning in a visually engaging manner. For instance, patients can use a smartphone or tablet to visualize 3D models of anatomical structures or medical devices overlaid onto their own bodies, allowing for a personalized and interactive learning experience.

Simplifying complicated medical concepts and processes to make them more understandable and accessible to people with different levels of health literacy is one of AR's main advantages in patient education. In order to obtain a better understanding of their health and available treatments, patients might engage with virtual representations of their anatomy or medical disorders. As a result of this participatory approach, patients feel more empowered and included in the healthcare decision-making process, which eventually improves patient outcomes and treatment plan adherence.

It has been demonstrated that using AR applications in patient education increases information retention and engagement. AR draws patients in and promotes active engagement in their healthcare journey by offering immersive and interactive learning experiences. Research has shown that patients interacting with augmented reality (AR) educational resources retain information better and are more likely to follow doctor's orders. Consequently, by providing interactive, tailored, and immersive learning experiences, augmented reality apps in healthcare have the potential to transform patient education and engagement completely. Through augmented reality (AR) technology, healthcare professionals may give patients the information and resources they need to engage in their care actively, improving patient outcomes and increasing patient literacy.

2.2 *Diagnostic Procedures*

Augmented reality (AR) applications have become a game-changer for improving diagnostic processes, providing healthcare practitioners with cutting-edge methods for precise and effective evaluations. AR technology can completely change how diagnostic tests are performed and understood by fusing digital and physical information.

AR-enabled diagnostic procedures provide healthcare providers with immersive, real-time visualizations of patient data in various applications, from surgical navigation to medical imaging. One noteworthy instance is the application of augmented reality (AR) in medical imaging. It allows radiologists to superimpose 3D reconstructions of anatomical structures on real-time scans, improving their ability to precisely and perceive intricate interior structures. This enhanced view improves the diagnostic process by enabling more precise abnormality diagnosis and better treatment planning. Moreover, AR technology has played a significant role in surgical navigation by providing physicians with a dynamic and user-friendly way of visualizing patient anatomy during procedures. AR helps surgeons navigate complex anatomical features more confidently and accurately by superimposing preoperative imaging data onto the surgery area in real time. Patient safety is increased, surgical accuracy is increased, and the chance of complications is decreased thanks to this improved visualization. Enhancing medical education and training has also demonstrated potential when incorporating AR into diagnostic procedures. AR technology allows medical practitioners to practice diagnostic skills in a realistic virtual environment by offering immersive and engaging learning experiences. By using a hands-on approach, healthcare practitioners become more proficient in evaluating complex medical data and improving their diagnoses' accuracy. By giving medical practitioners cutting-edge resources for precise and effective evaluations, augmented reality applications in healthcare have the potential to transform diagnostic processes completely. Healthcare professionals may raise the bar for patient care by using AR technology to improve treatment outcomes, increase diagnostic accuracy, and more.

2.3 Rehabilitation Programs

Augmented reality (AR) applications in healthcare have become a viable means of improving rehabilitation programs and providing creative ways to raise patient outcomes and care quality. AR technology can completely change how rehabilitative therapies are administered and experienced by fusing digital and physical aspects.

AR-enabled rehabilitation programs use various applications, from cognitive to physical therapy, giving patients fascinating and interactive experiences that aid in recuperation and enhance their functional abilities. One noteworthy instance is the application of augmented reality (AR) in physical therapy, where patients can participate in interactive games and exercises superimposed on their environment to improve motor skills and encourage movement. In addition to raising patient motivation, this gamified approach to rehabilitation makes it easier to implement focused, individualized therapy sessions. Furthermore, AR technology has proven to be a valuable resource for cognitive rehabilitation, providing patients with neurological disorders with cutting-edge methods to improve cognitive function and stimulate brain plasticity. Patients can practice their mental skills in memory, attention, and problem-solving by using AR applications to provide interactive cognitive challenges and simulations. Thanks to this individualized approach to cognitive rehabilitation, patients can take an active role in their treatment and see significant gains in their cognitive abilities. Enhancing patient participation and adherence to treatment plans has been demonstrated to be another benefit of integrating AR into rehabilitation programs. AR technology engages patients and motivates them to actively participate in their recovery by providing immersive and interactive rehabilitation experiences. Research has indicated that patients participating in AR-based rehabilitation programs show increased motivation and adherence, improving rehabilitation results and quality of life.

2.4 Chronic Disease Management

Augmented Reality (AR) apps have become a promising area in managing chronic diseases in the healthcare environment. They provide creative ways to improve patient outcomes and raise the standard of care for patients with long-term medical disorders. AR technology can completely change how chronic diseases are tracked, treated, and managed by fusing digital overlays into the real world.

Chronic Disease Management through AR encompasses various applications, from remote monitoring to patient education and self-management tools, providing individuals with chronic conditions with personalized and interactive resources to support their health journey. One notable example is the use of AR in remote monitoring, where patients can access real-time health data and receive personalized feedback and guidance through AR-enabled devices, empowering them to actively participate in their care and make informed decisions about their health. Furthermore, AR technology has been instrumental in patient education and self-management for chronic diseases, offering individuals interactive tools and resources to enhance their understanding of their condition and treatment plan. By superimposing virtual information onto the physical environment, AR applications can provide patients with visualizations of disease processes, medication instructions, and lifestyle recommendations, fostering greater health literacy and adherence to treatment regimens. This personalized approach to patient education promotes self-efficacy and empowers individuals to take control of their health.

Integrating AR into chronic disease management has also shown promise in improving patient engagement and health outcomes. By offering interactive and immersive experiences, AR technology captivates patients' attention and encourages active participation in their care, leading to better treatment adherence and health outcomes. Studies have demonstrated that patients who engage with AR-based chronic disease management tools exhibit improved self-care behaviors and enhanced quality of life, highlighting the potential of AR technology to transform the management of chronic conditions.AR applications in healthcare promise to revolutionize chronic disease management by providing individuals with innovative tools and resources to support their health journey. By leveraging AR technology, healthcare providers can enhance patient education, promote self-management, and optimize care delivery for individuals with chronic diseases, ultimately improving health outcomes and quality of life. Figure 10.3 shows that Augmented Reality in Healthcare Platform.

3 VR Applications in Healthcare

3.1 Exposure Therapy for Anxiety and Phobias

Virtual reality-based exposure therapy (VRET) is an innovative approach used for treating anxiety and phobias, including dental phobia [12]. This therapy method gradually exposes individuals to virtual environments that simulate their feared situations or stimuli in a controlled setting. The procedure typically begins with an assessment to identify the specific fears or phobias, followed by creating customized virtual environments replicating the feared scenarios.

During the exposure sessions, individuals interact with these virtual environments, starting from less anxiety-provoking situations and progressively moving to

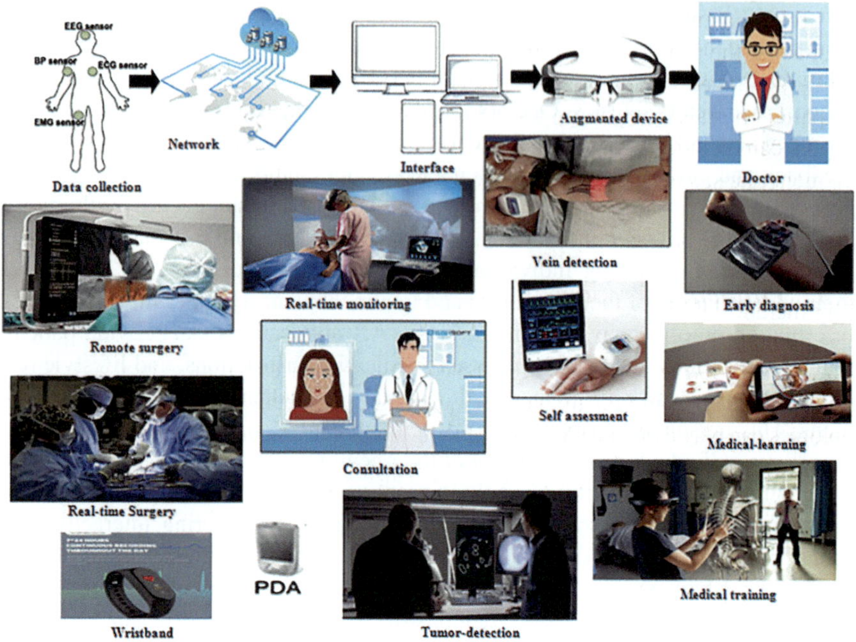

Fig. 10.3 Augmented reality in healthcare platform [10, 11]

more challenging ones. Throughout the sessions, therapists provide cognitive restructuring to help individuals challenge and reframe negative thoughts associated with their fears. Additionally, individuals learn coping strategies, such as deep breathing exercises and relaxation techniques, to manage anxiety during exposure [13].

In the context of dental phobia, VRET offers a novel approach by immersing patients in a virtual dental operatory, where they encounter virtual representations of a dentist and dental tools. The exposure hierarchy is systematically introduced, starting from passive sitting in the chair to more anxiety-provoking scenarios like oral inspection and drilling sounds. Real-time monitoring of physiological responses, such as heart rate, ensures safety during the single-session therapy. Overall, VRET provides a safe and immersive environment for individuals to confront their fears and gradually reduce anxiety, leading to improved treatment outcomes and quality of care.

3.2 Pain Management and Distraction

Virtual reality (VR) technology has emerged as a promising tool for pain management and distraction during medical procedures [14]. The therapy procedure typically begins with assessing the patient's pain levels, medical history, and suitability for VR therapy. Patients are then prepared by being introduced to the VR equipment, addressing any concerns, and selecting a suitable virtual environment based on their preferences and the procedure performed [14]. During immersion, patients wear a VR headset or use a mobile device to enter the chosen virtual world, which may include calming scenes, interactive activities, or engaging games. The immersive nature of VR helps distract patients from the medical procedure, shifting their focus to the virtual environment and potentially reducing their perception of pain by altering the body's pain modulation system. Healthcare providers monitor the patient's response throughout the procedure and adjust the VR experience as needed. After the procedure, patients' pain levels and overall experience with VR therapy are evaluated, and follow-up sessions may be scheduled for continued pain management. Detailed documentation of the VR therapy session, including the virtual environment used, the patient's response, and any adjustments made is recorded in the patient's medical records.

By integrating VR into pain management strategies, healthcare providers can offer patients a non-pharmacological and immersive approach to alleviate pain and anxiety during medical procedures. Further research and advancements in VR applications can potentially enhance patient comfort and treatment outcomes in healthcare settings.

3.3 Surgical Training and Simulation

Virtual reality (VR) based surgical training and simulation have revolutionized surgical education and practice, providing a realistic and immersive environment for training purposes [15–17]. The procedure typically begins with a needs assessment to identify the specific training objectives, surgical procedures, and skills to be developed. A suitable VR platform or simulator, such as those based on Unreal or Unity engines, is then selected to align with the training goals. A comprehensive training curriculum is developed, outlining the sequence of training modules, specific skills to be practiced, and assessment criteria for evaluating proficiency [15, 17]. Trainees engage in hands-on training using the VR simulation, which replicates real-life surgical scenarios, allowing them to practice techniques, improve hand-eye coordination, and familiarize themselves with surgical tools and equipment [15–17]. The VR system provides real-time feedback on the trainee's performance, enabling immediate correction of errors and continuous improvement. Progress tracking through data analytics and performance metrics allows instructors to identify areas for improvement and tailor training programs to individual learning needs.

VR-based surgical training is often integrated with traditional surgical education methods, such as lectures and workshops, to provide a comprehensive learning experience [15, 17]. Regular evaluation and improvement of the VR training program are undertaken based on feedback from trainees, instructors, and performance metrics. By utilizing VR technology in surgical training, medical professionals can enhance their surgical skills, improve patient outcomes, and advance the healthcare field in a safe and controlled environment without risk of harm. This innovative approach offers a transformative solution for medical education and training.

3.4 Mental Health Assessments and Interventions

Virtual reality (VR) technology has been increasingly utilized in mental health assessments and interventions, particularly in the treatment of conditions such as post-traumatic stress disorder (PTSD) in military personnel [16]. Researchers conducted initial validations of VR environments to assess memory functioning and attentional processing. These assessments provide a novel and immersive way to evaluate cognitive abilities in a controlled virtual setting. They additionally explored the effectiveness of virtual reality exposure therapy (VRET) for anxiety disorders and PTSD in active duty soldiers, demonstrating positive outcomes. The procedure typically involves the presentation of emotionally evocative VR combat-related scenarios within the context of professional care, allowing for treatment options previously not feasible with traditional methods. By leveraging the advancements in VR technology and the positive outcomes reported in the literature, VR-based mental health assessments and interventions offer a promising avenue for enhancing clinical practice and research in military behavioral healthcare. Virtual reality (VR) technology has been increasingly utilized in mental health assessments and interventions, offering innovative treatment approaches. Researchers highlighted the importance of considering basic theoretical and pragmatic issues when developing VR applications for clinical psychology and neuropsychology [18]. The process typically involves.

- selecting a suitable VR approach based on existing methods,
- assessing the compatibility of the VR environment with the target clinical population,
- evaluating the level of presence experienced by users,
- addressing navigation factors,
- considering potential side effects,
- ensuring generalization of outcomes, and
- Incorporating sound methodological and data analysis practices.

VR applications have shown promise in fear reduction, pain management, stress reduction, cognitive rehabilitation, and assessment of various mental health conditions. By integrating VR technology into mental health practices, professionals can enhance treatment outcomes and provide more personalized and effective

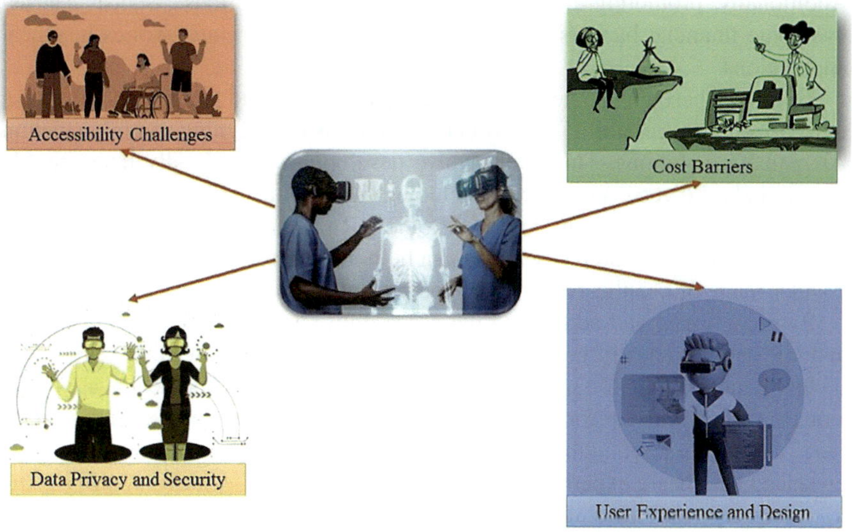

Fig. 10.4 Challenges and considerations in AR and VR healthcare applications

interventions for individuals experiencing psychological challenges. Figure 10.4 shows that Challenges and Considerations in AR and VR healthcare applications.

4 Challenges and Considerations

4.1 Accessibility and Cost

Integrating Augmented Reality (AR) and Virtual Reality (VR) technologies in healthcare present's opportunities and challenges. One significant hurdle is ensuring accessibility for individuals with disabilities. The lack of built-in accessibility features tailored to accommodate various impairments, such as visual or cognitive challenges, hinders the effective utilization of AR/VR solutions by users with disabilities in healthcare settings. Furthermore, the cost associated with AR/VR devices and accessibility tools poses a considerable barrier to adoption. Affordability is a key concern, as impaired users should not face additional expenses for necessary accessibility features. The financial constraints related to acquiring AR/VR devices, software, and training for healthcare professionals can be substantial, potentially limiting the widespread implementation of these technologies in healthcare environments and impeding their potential benefits for patient care and medical training.

Addressing these accessibility and cost challenges requires collaborative efforts from industry manufacturers, researchers, and healthcare providers. Developing built-in accessibility features without additional costs for users with impairments is crucial to enhancing the inclusivity of AR/VR technologies in healthcare.

Additionally, promoting affordability and equal access to these devices can help overcome financial barriers and facilitate their integration into various healthcare applications.

To tackle these challenges, healthcare organizations and technology developers must prioritize accessibility and cost considerations in designing and implementing AR and VR solutions. This may involve developing user-friendly interfaces that accommodate diverse user needs, including those with disabilities. Furthermore, efforts to reduce the overall cost of AR and VR technologies through innovation, scalability, and strategic partnerships can help make these tools more accessible to a wider range of healthcare providers and patients.

By recognizing and addressing the accessibility and cost considerations of AR and VR technologies in healthcare, stakeholders can work towards creating more inclusive and sustainable solutions that benefit both patients and healthcare professionals. Collaborative efforts to improve accessibility features, reduce financial barriers, and promote equitable access to AR/VR innovations can pave the way for enhanced healthcare experiences and outcomes.

4.2 Data Privacy and Security

The adoption of augmented reality (AR) and virtual reality (VR) technologies in healthcare raises critical data privacy and security concerns, necessitating robust measures to safeguard sensitive patient information [10, 11]. While AR and VR offer potential benefits for patient care and medical training, they also introduce vulnerabilities that require the implementation of stringent security protocols.

Protecting patients' confidential information from unauthorized access or data breaches is a significant challenge when integrating AR and VR technologies in healthcare settings. As these technologies involve collecting and transmitting personal health data, implementing strong security measures such as encryption, authentication mechanisms, and secure network connections is essential to prevent data leaks, cyber-attacks, and maintain data confidentiality, integrity, and availability [19].

Compliance with regulatory standards like HIPAA adds complexity to data privacy and security considerations in AR and VR healthcare applications. Implementing robust data encryption protocols, secure authentication mechanisms, and regular security audits are crucial to address these challenges. Educating healthcare professionals and technology users about data privacy best practices and potential security threats can help mitigate risks associated with AR and VR applications in healthcare, ensuring patient confidentiality and data protection.

4.3 User Experience and Design

Integrating Augmented Reality (AR) and Virtual Reality (VR) technologies in healthcare presents several user experience and design challenges that require careful consideration to ensure successful implementation and optimal outcomes. These challenges encompass technical limitations, ethical concerns, user acceptance, and the need for design approaches tailored to the unique requirements of healthcare settings.

Addressing technical limitations, such as device weight, battery life, and processing power, is crucial to ensure seamless user experiences without compromising functionality. Ethical principles, including patient privacy, data security, and informed consent, must be prioritized in the design process to safeguard sensitive information and maintain compliance with regulatory standards.

User acceptance is another key consideration, as healthcare professionals and patients may have varying familiarity and comfort with AR and VR technologies [20]. Creating user-friendly interfaces, providing adequate training and support, and addressing user concerns are essential to promote widespread adoption in healthcare environments. Furthermore, designing AR and VR applications for healthcare necessitates a tailored approach that considers users' specific needs and preferences in clinical settings. Customization, personalization, and user-centered design principles are vital to creating intuitive, effective, and engaging applications for healthcare professionals and patients [21]. Additional challenges and considerations in user experience and design for AR and VR in healthcare include:

- Complexity of healthcare data: Presenting complex and vast healthcare data clearly and understandably through user interfaces.
- User interaction: Designing intuitive and user-friendly interfaces for efficient AR and VR systems interaction during critical tasks.
- Integration with existing systems: Ensuring seamless integration of AR and VR applications with existing healthcare systems and workflows.
- Privacy and security: Implementing robust measures to maintain patient confidentiality and data security.
- Training and education: Providing comprehensive educational resources for healthcare professionals to use AR and VR technologies effectively.
- Regulatory compliance: Adhering to healthcare regulations and standards to ensure patient safety and data security.
- Technical challenges: Addressing latency, resolution, and device compatibility issues to deliver a seamless and immersive user experience.

By articulating these challenges and considerations clearly and concisely, healthcare professionals can collaborate with designers and developers to create innovative solutions that enhance patient care, improve workflows, and optimize healthcare delivery through the effective utilization of AR and VR technologies.

4.4 Challenges

The development of affordable and easily integrable AR and VR technology into current healthcare systems is crucial in addressing the accessibility dilemma. To develop scalable and reasonably priced solutions, government organizations, healthcare providers, and technology companies may work together [8]. Adoption hurdles can also be addressed by giving healthcare workers the guidance and assistance they need to use and incorporate these technologies into their work [22].

When integrating AR and VR technologies in healthcare, it is imperative to create strong data protection mechanisms and follow pertinent laws, such as HIPAA and GDPR, in order to address concerns about data privacy [9]. This entails putting safe data transfer and storage procedures into place, getting patients' informed consent, and routinely reviewing and updating security measures to take into account new threats. Furthermore, trust may be built and responsible use of these technologies encouraged by teaching patients and healthcare professionals about their rights and best practices for data protection [23].

Connectivity with IoMT: By fostering a more connected, individualized, and data-driven approach to patient care, the Internet of Medical Things (IoMT) in conjunction with AR and VR technologies holds the potential to completely transform the delivery of healthcare. Real-time patient data can be gathered by IoMT devices, including wearable sensors and smart medical equipment. AR and VR interfaces can then be used to see and evaluate the data [24]. Improved patient outcomes and lower healthcare costs can result from this integration's ability to provide predictive analytics, individualized treatment regimens, and remote monitoring.

For instance, real-time patient vital signs and alarms from IoMT sensors might be shown on an AR-enabled smart glasses device, enabling medical practitioners to quickly monitor and respond to changes in a patient's condition [6, 7]. Similar to this, IoMT device data might be incorporated into VR simulations to generate dynamic, individualized training situations that improve medical professionals' and students' skills and ability to make decisions [25].

4.5 Comparative Analysis

The ways in which AR and VR technologies are used in healthcare settings might differ greatly between industrialized and developing nations, as well as between urban and rural areas. The adoption of AR and VR solutions is aided by the more easy access to advanced technology and infrastructure found in industrialized nations and urban areas. In these situations, healthcare facilities might have the funds and know-how to purchase state-of-the-art AR and VR systems for patient care, surgical guidance, and medical education.

However, due to a lack of funding, inadequate infrastructure, and gaps in digital literacy, the adoption of AR and VR technology may present difficulties in rural areas and underdeveloped nations. In these situations, healthcare facilities might not have the resources—financial, technological, and human—to implement and manage AR and VR systems efficiently. The benefits that these technologies bring may not be distributed equally as a result of the differences in access and implementation.

Initiatives centered on technology transfer, capacity building, and partnerships between developed and developing nations can be extremely important in closing this gap. Collaborations across healthcare facilities, tech firms, and global organizations can help exchange best practices, resources, and information, which makes it possible to customize AR and VR solutions to suit regional requirements. Furthermore, the creation of inexpensive, portable, and easy-to-use AR and VR solutions can aid in removing adoption hurdles in environments with limited resources.

5 The Future of AR and VR in Healthcare

Virtual Reality (VR) and Augmented Reality (AR) hold great promise for revolutionizing various aspects of the healthcare sector, including patient care, medical education, and treatment delivery. As these technologies advance and gain traction, they have the potential to drastically alter the healthcare industry. It is projected that these technologies will play a major role in enhancing medical education and training. Virtual reality therapy is growing quickly, with promising applications in pain management, rehabilitation, and mental health care. Virtual Reality-Based Treatment (VRBT) has shown promising results in treating conditions such as anxiety disorders, phobias, and Post-Traumatic Stress Disorder (PTSD) through the use of controlled exposure and desensitization processes. The healthcare sector may open up new avenues for boosting patient outcomes, raising the standard of treatment, and advancing scientific research by embracing these cutting-edge technologies.

5.1 Ongoing Research and Development

AR and VR can facilitate collaborative research by enabling researchers from different locations to visualize and interact with complex data sets in a shared virtual environment. This can accelerate the discovery process, foster interdisciplinary collaboration, and lead to drug development, medical imaging, and disease modeling breakthroughs. To expand the implementation of new technology, which is the need of the hour to promote reliable health care to people, challenges related to technology integration, user adoption, data privacy, and regulatory frameworks have to be overcome. Collaboration among healthcare providers, technology companies, researchers, and policymakers will be essential in ensuring these technologies' safe, ethical, and effective implementation.

After the release of GG (Google Glass), a head-mounted display (HMD) featuring produced items placed into real-time photos, there was a resurgence of interest in accessible augmented reality (AR). This enthusiasm persisted until January 2015, when production of the HMD was halted [26]. This technology enabled a high-resolution video camera and a lightweight wearable with an overlaid viewing screen. It possesses capabilities like wireless and cloud accessibility that are found on smart phones. Numerous comparable devices designed specifically for surgical enhancement have since been developed due to the device's open-source development platform, encouraging innovative medical and surgical applications [27].

Numerous research studies have shown how AR can help surgeons navigate complex anatomy during minimally invasive procedures. Robotic laparoscopic restricted partial nephrectomy guided by preoperative imaging and intra-operative 3D overlay was demonstrated by Su et al. [28]. AR-guided surgery has shown interest in minimal access to limited partial nephrectomy. With AR guiding, surgical incisions can be marked within the laparoscopic view by projecting 3D images onto the picture.

Applications for live stream and recorded data include trainee-trainer exchanges within the framework of work-based assessments. Junior team members might provide an off-site senior member with a real-time perspective of challenging cases using telementoring, allowing them to get quick assistance. This would benefit specialties like trauma management, patient monitoring, and injury assessment where visual examination is necessary. A safe way to provide professional diagnosis and advice is through telementoring, which can lower the risk of errors in management and needless patient transfers [29].

5.2 *Expanding Applications*

Implementing Augmented Reality in the medical field can revolutionize surgical planning and guidance with the help of three-dimensional visualizations of patient anatomy and pathologies. AR can overlay critical information onto the surgical field, enhancing the surgeon's situational awareness and precision. VR can facilitate preoperative planning, enabling surgeons to rehearse complex procedures and anticipate potential challenges before entering the operating room. Without endangering patient safety, immersive virtual environments may replicate intricate surgical processes, anatomical models, and patient scenarios to give medical professionals and students a realistic hands-on experience. This can enhance one's capacity for learning new skills, making sound decisions, and being ready for high-stakes scenarios.

During consultations or emergencies, distant medical providers can receive real-time visual information and guidance from AR-enabled devices. VR can enable immersive virtual consultations, allowing patients to interact with healthcare providers in a simulated environment, potentially improving access to care in remote or underserved areas. Thus, telemedicine and remote care capabilities can transform healthcare into a new regime. Interacting with interactive virtual environments allows Patients to learn more about their medical issues, available treatments, and self-care guidelines. Improved treatment plan adherence, better health outcomes, and greater patient empowerment are all possible benefits of this deeper understanding.

The AccuVein AV400 device is one particular instance of AR being used in a real-world healthcare setting. It uses AR technology to project a map of the patient's veins onto their skin, making it easier for medical personnel to locate veins for IV insertions and blood draws [30]. According to Kaddoura et al. [31], the use of augmented reality has been demonstrated to increase the precision and effectiveness of these operations while also lowering patient discomfort and raising patient satisfaction.

One particular application of virtual reality (VR) in medicine is the use of VR exposure therapy to treat anxiety disorders like phobias and post-traumatic stress disorder (PTSD). Virtual reality exposure therapy has been shown to be beneficial in treating certain phobias, including fear of heights and flying, according to a 2017 study by Botella et al. Through a safe and regulated virtual environment, patients were progressively exposed to the stimuli that they dreaded, enabling them to face and overcome their concerns.

5.3 *Transforming Healthcare Delivery*

When used as a safe and entertaining diversion during medical procedures or treating chronic pain disorders, Virtual Reality (VR) can be a very effective tool for pain management. AR, on the other hand, can improve patient education and engagement through the overlay of pertinent medical data over a real-world setting, leading to improved comprehension and treatment plan adherence. Much technological advancement has been identified as potentially disruptive to the surgical profession as the surgical environment continues to advance in the digital era. Because Augmented Reality (AR) and Virtual Reality (VR) are quickly becoming more widely available, easily accessible, and—most importantly—affordable, it seems inevitable that they will be used in healthcare to increase data usage for medical purposes. Research is currently underway to find useful applications for a surgeon's toolkit concerning anatomy, intraoperative surgery, and post-operative rehabilitation.

The majority of virtual reality trainers, especially those for robotic and laparoscopic surgery, teach fundamental motor skills that are connected to key surgical skills. These include grasping, dexterity, grasping instruments, controlling cameras, manipulating endodontist tools, and three-dimensional vision. The majority of VR modules feature abstract tasks, such moving objects to learn specific abilities or hanging hoops from hooks. The effectiveness of VR simulation in teaching basic surgical skills has been recognized [32]; nonetheless, additional training is required to close the gap between this kind of discrete skill training and carrying out entire surgical procedures in the operating theater.

5.4 Technological Insights

In order to produce immersive and engaging experiences, AR and VR systems require sophisticated hardware and software components. AR devices overlay digital content over the physical world using a combination of cameras, sensors, and displays. Examples of these devices are the Microsoft HoloLens and Magic Leap. By tracking the user's position and orientation using computer vision algorithms, these gadgets allow virtual items to be precisely aligned with their physical surroundings. With the help of AR software development kits (SDKs) like Google's ARCore and Apple's ARKit, developers can create AR apps for a variety of platforms.

Conversely, Virtual Reality (VR) systems immerse users entirely in a computer-generated environment through the use of specialist headsets like the Oculus Rift or HTC Vive. To create a realistic and engaging experience, these headsets include haptic feedback, motion tracking sensors, and high-resolution displays. While game engines like Unity and Unreal Engine make it easier to construct immersive VR environments, VR platforms like SteamVR and Oculus SDK enable developers to create VR applications and games.

Advanced haptic feedback systems, hand tracking, and eye tracking are examples of recent technology advancements in AR and VR. In addition to facilitating more natural user interactions, eye-tracking technology can yield useful information for diagnostics and medical research. Hand tracking improves the immersive experience by enabling users to interact with virtual items using natural hand gestures. Realistic tactile sensations are provided by sophisticated haptic feedback devices, like the HaptX Gloves, and are especially helpful for surgical training and simulation.

5.5 Patient and Practitioner Perspectives

As a patient who underwent surgery with the assistance of an AR system, I was amazed by the technology. The surgeon explained how the AR overlays helped guide the procedure, and I felt more confident knowing that they had this advanced tool to ensure precision and safety. Patient testimonial [4].

Virtual reality has been a game-changer in our pain management practice. We've seen patients who were previously struggling with chronic pain find relief and improved quality of life through VR interventions. The immersive nature of VR helps patients focus on something other than their pain, and the results have been impressive. Healthcare practitioner [14].

5.6 *Case Study*

Case Study 1: Surgical Navigation Using Augmented Reality Sola-no-Rojas et al.'s work is a noteworthy case study that illustrates the usefulness of AR in surgical navigation (2022). Using the Micro-soft HoloLens, the researchers created an augmented reality system to help surgeons perform laparoscopic cholecystectomy procedures. The device gave surgeons improved visualization and guidance by superimposing 3D anatomical models and important components onto the patient's body in real-time. In comparison to conventional laparoscopic procedures, the AR system reduced the risk of complications, enhanced surgical precision, and shortened the total operation time, according to the study.

Case Study 2: Pain Management with Virtual Reality The case study by Mosso Vázquez et al. [14] provides a strong illustration of VR's influence on pain management. In order to reduce patients' pain and anxiety during ambulatory surgery, the researchers employed virtual reality distraction therapy. Using a VR headgear, patients were submerged in a relaxing virtual world while having surgery. In comparison to the control group, the study found that patients who had VR distraction again had much lower pain and anxiety scores. This case study demonstrates how virtual reality (VR) has the potential to enhance patient experiences during medical operations and offer non-pharmacological pain alleviation.

5.7 *Analysis of AR and VR Applications in Healthcare*

A synoptic view of applications of Augmented Reality and Virtual Reality in healthcare.

The use of Augmented Reality (AR) and Virtual Reality (VR) systems in the healthcare domain has promising implications in changing patients' care, medical training, and therapeutic procedures. The above technologies are creative ways through which different facets of delivery and practice in the healthcare sector can be improved.

1. Patient Education and Engagement: Research shows that using AR, patients' experiences can be enriched by application-based experiences, including interactive visualizations [3]. Through the digital and non-digital data integration, AR helps patients to gain more profound understandings of their health status, and thus enhance the treatment compliance and decision-making.
2. Diagnostic Procedures and Surgical Guidance: Using AR technology in health care has demonstrated that it has the potential to help in improving the diagnostic processes through facilitating real time assistance during difficult operations [4]. Thus, applying digital images and annotations on the patient's body, AR can provide crucial information to healthcare professionals which may minimize and enhance the efficiency of their operations.

3. Rehabilitation and Chronic Disease Management: The general use of AR integrated games has been introduced in the rehabilitation programs wherein patients look forward to the process [5]. In the care of chronic diseases, augmented reality technology is used in the sense that all patient inputs are real-time and can be monitored leading to better management of the diseases [6, 7].
4. Virtual Reality in Mental Health: VR proved useful in the treatment of mental health disorders with focus on the application of exposure therapy for anxiety disorders and phobia. Since it is possible to provide a moderated exposure to the feared stimuli, VRET has beneficial effects in treating conditions like PTSD [13].
5. Pain Management and Distraction: VR has been described as a useful technique in the field of pain control and reducing distress in medical procedures [33]. Research has described that VR changes the perceived pain and anxiety levels by lessening the scores among patients who undergo ambulatory surgery [14].
6. Medical Education and Surgical Training: The applications of Virtual Reality and Augmented Reality have included redefining medical education and surgical skill training through the possibilities of having effective and realistic models depicting tough procedures [15, 17]. These technologies enable the learners and the medical practitioners to practice in safe endowments with nadir risks hence enabling the learners and the practitioners to nexus ample procedure on human subjects.

5.8 *Key Results*

1. Mobile AR applications have demonstrated benefits in enhancing patients' knowledge about diseases and possibilities of the further therapy, which presumably may promote better medication compliance and more rational choices [3].
2. AR augments the diagnostic processes because it offers real-time information when conducting intricate medical procedures, which may help decrease the standard of mistakes added to the precision [4].
3. Studies of using VRET have shown promising results in the treatment of mental health disorders, but mainly ranged from anxiety disorders, phobias, and PTSD [13].
4. Concerning the impact on pain and anxiety, the authors found that using VR distraction therapy it is possible to enhance the reduction of scores in patients who undergo ambulatory surgery [14].
5. Some of the applications of VR and AR are in the training of medical students and surgeons, where VK simulate real life surgical procedures [15, 17].
6. Some of the difficulties and limitations of using AR and VR in the healthcare context are financial costs of integrating AR and VR [8], the problem of data protection [9], and UX design that should be adapted for the clinical environment [21].

6 Conclusion and Future Scope

Integrating Augmented Reality (AR) and Virtual Reality (VR) technologies into healthcare practices holds immense promise for transforming care delivery and advancing medical education. As these immersive technologies continue to evolve and gain traction, their impact on the healthcare industry is poised to be profound and far-reaching. While the current state of AR and VR in healthcare demonstrates diverse applications, from enhancing patient education and engagement to revolutionizing surgical training and therapeutic interventions, ongoing research and development efforts are crucial to expand their capabilities and address existing challenges.

Addressing issues related to accessibility, cost, data privacy, and security will be paramount to ensure the equitable and responsible implementation of these technologies. Collaboration among healthcare providers, technology companies, researchers, and policymakers is essential to develop comprehensive frameworks and guidelines prioritizing patient safety, ethical considerations, and regulatory compliance.

Moreover, the user experience and design of AR and VR applications in healthcare settings must be carefully considered, focusing on user-friendliness, intuitive interfaces, and seamless integration into existing workflows. Tailoring these technologies to healthcare professionals' and patients' specific needs and preferences is crucial for successful adoption and optimal utilization.

As the future of AR and VR in healthcare unfolds, their applications are expected to expand further, revolutionizing surgical planning and guidance, facilitating remote medical consultations, and enhancing patient education and engagement. The potential for these technologies to transform healthcare delivery is immense, offering the promise of a more patient-centered, efficient, and accessible system that prioritizes personalized care and improved outcomes. By embracing the potential of AR and VR in healthcare while addressing the associated challenges, stakeholders can unlock new opportunities for innovation, collaboration, and advancements in medical practices, ultimately leading to enhanced patient experiences, improved healthcare outcomes, and a more sustainable and equitable healthcare system.

Virtual reality (VR) and augmented reality (AR) technologies will play a crucial role in the provision of healthcare. Patient care will be revolutionized by the smooth integration of AR, VR, and the Internet of Medical Things (IoMT), enabling individualized, data-driven, and immersive experiences. Technological developments in AR and VR hardware, software, and analytics will tackle issues with accessibility, affordability, and data privacy, resulting in a wider use and lower cost of these technologies. In order to guarantee the ethical and responsible use of AR and VR in healthcare, legislators, technology companies, and healthcare providers must work together to set detailed norms and standards.

References

1. Munoz-Saavedra L, Villalta MF, SarabiaCachadiña E, Pérez JL. Augmented reality in healthcare: applications and future trends. Electronics. 2021;10(20):2501. https://doi.org/10.3390/electronics10202501.
2. Pan J, Liu J, Xing N, Hu F, Avram C. Virtual reality assisted therapy for anxiety: a narrative review. Front Psychol. 2021;12:908. https://doi.org/10.3389/fpsyg.2021.628412.
3. Brigham TJ. Reality check: basics of augmented, virtual and mixed reality. Med Ref Serv Q. 2017;36(2):171–8. https://doi.org/10.1080/02763869.2017.1293987.
4. Solano-Rojas E, Bobadilla-Rivas F, Sesma-Vazquez S, Melesio-Portillo R, García-Peña Á, Palafox-Avalos B. Integrating augmented reality into healthcare: a systematic review of applications, benefits, and challenges. Appl Sci. 2022;12(4):1792. https://doi.org/10.3390/app12041792.
5. Aravind G, Darekar A, Fung J, Lamontagne A. Virtual reality-based navigation task to assess cognitive function in individuals with stroke: a pilot study. Int J Virtual Reality. 2020;20(1):2–15.
6. Syed S, Rahman M, Shad MM, Al-Ramahah E, Ahmad S, Ali M, Zafar A. Augmented reality and its pragmatic application in chronic diseases: appraisal and systemic review. In: 2020 IEEE international IOT, electronics and mechatronics conference (IEMTRONICS). Piscataway: IEEE; 2020. p. 1–6. https://doi.org/10.1109/IEMTRONICS51293.2020.9216499.
7. Syed S, Rahman M, Shad MM, Al-Ramahah E, Ahmad S, Ali M, Zafar A. Augmented reality and its pragmatic application in chronic diseases: appraisal and systemic review. In: 2020 IEEE international IOT, electronics and mechatronics conference (IEMTRONICS). Piscataway: IEEE; 2020. p. 1–6. https://doi.org/10.1109/IEMTRONICS51293.2020.92164994.
8. Spicer A, Brivaki E, Yunker AG, Mulrooney A. Augmented reality for extending healthcare across geographies. J IES. 2021;64(1):70–83. https://doi.org/10.1108/JIES-10-2019-0058.
9. Fernandez-Vigas C, Diaz-Redondo RP, Lok S, Yu S, Borozdina Y, Goorahoo D. Privacy and security considerations in the use of virtual and augmented reality in health care. Virtual Reality. 2021;25(4):739–58. https://doi.org/10.1007/s10055-021-00550-x.
10. Ara J, Bhuiyan H, Bhuiyan YA, Bhyan SB, Bhuiyan MI. Ar-based modern healthcare: a review. arXiv preprint arXiv:2101.06364; 2021.
11. Ara J, Karim FB, Alsubaie MSA, Bhuiyan YA, Bhuiyan MI, Bhyan SB, Bhuiyan H. Comprehensive analysis of augmented reality technology in modern healthcare system. Int J Adv Comput Sci Appl. 2021;12(6):840–9.
12. Gujjar KR, Van Wijk A, Sharma R, De Jongh A. Virtual reality exposure therapy for the treatment of dental phobia: a controlled feasibility study. Behav Cogn Psychother. 2018;46(3):367–73.
13. Zhang W, Paudel D, Shi R, Liang J, Liu J, Zeng X, et al. Virtual reality exposure therapy (VRET) for anxiety due to fear of COVID-19 infection: a case series. Neuropsychiatr Dis Treat. 2020;16:2669–75.
14. Mosso Vázquez JL, Mosso Lara D, Mosso Lara JL, Miller I, Wiederhold MD, Wiederhold BK. Pain distraction during ambulatory surgery: virtual reality and mobile devices. Cyberpsychol Behav Soc Netw. 2019;22(1):15–21.
15. Sheik-Ali S, Edgcombe H, Paton C. Next-generation virtual and augmented reality in surgical education: a narrative review. Surg Technol Int. 2019;35:27–35.
16. Al-Hiyari N, Jusoh S. The current trends of virtual reality applications in medical education. In: 2020 12th international conference on electronics, computers and artificial intelligence (ECAI). Piscataway: IEEE; 2020. p. 1–6.
17. Sheik-Ali S, Edgcombe H, Paton C. Next-generation virtual and augmented reality in surgical education: a narrative review. Surg Technol Int. 2019;33:27–35.
18. Rizzo AA, Wiederhold MD, Buckwalter JG. Basic issues in the use of virtual environments for mental health applications. Stud Health Technol Inform. 1998;58:21–42.
19. De Guzman JA, Thilakarathna K, Seneviratne A. Security and privacy approaches in mixed reality: a literature survey. ACM Comput Surv. 2019;52(6):1–37.

20. Feng B, Ng LL. Facilitating writing performance of EFL learners via virtual reality: immersion, presence, embodiment. Front Psychol. 2023;14:1134242.
21. Gall D, Roth D, Stauffert JP, Zarges J, Latoschik ME. Embodiment in virtual reality intensifies emotional responses to virtual stimuli. Front Psychol. 2021;12:674179.
22. Kouri P, Pasonen-Seppänen S, Rantanen M, Laitinen A. Augmented reality in surgical education: a systematic review. Appl Sci. 2021;11(9):3952. https://doi.org/10.3390/app11093952.
23. Takyi H. Security, privacy, confidentiality and integrity of emerging healthcare technologies: a framework for quality of life technologies to be HIPAA/HITECH compliant, with emphasis on health kiosk design [Doctoral dissertation]. University of Pittsburgh; 2019.
24. Ahmadpour N, Pedell S, Mayasari A, Beh J. Co-creating and assessing future wellbeing technology using design fiction. She Ji: J Des Econ Innov. 2019;5(3):209–30. https://doi.org/10.1016/j.sheji.2019.08.003.
25. Barsom EZ, Graafland M, Schijven MP. Systematic review on the effectiveness of augmented reality applications in medical training. Surg Endosc. 2020;34(10):4174–83. https://doi.org/10.1007/s00464-020-07498-4.
26. Yu J, Ferniany W, Guthrie B, et al. Lessons learned from Google glass: telemedical spark or unfulfilled promise? Surg Innov. 2016;23:156–65.
27. Mitrasinovic S, Camacho E, Trivedi N, et al. Clinical and surgical applications of smart glasses. Technol Health Care. 2015;23:381–401.
28. Su LM, Vagvolgyi BP, Agarwal R, et al. Augmented reality during robot assisted laparoscopic partial nephrectomy: toward real-time 3D-CT to stereoscopic video registration. Urology. 2009;73:896–900.
29. Gardiner S, Hartzell TL. Telemedicine and plastic surgery: a review of its applications, limitations and legal pitfalls. J Plast Reconstr Aesthet Surg. 2012;65:e47–53.
30. AccuVein Inc. AccuVein AV400 Vein Viewing System. 2021. https://www.accuvein.com/products/catalog/av400-vein-viewing-system/.
31. Kaddoura M, Vandyke O, Smallwood C, Gonzalez KM. Perceived benefits and challenges of repeated exposure to high fidelity simulation experiences of first degree accelerated bachelor nursing students. Nurse Educ Today. 2020;90:104228. https://doi.org/10.1016/j.nedt.2020.104228.
32. Raison N, Harrison P, Abe T, Aydin A, Ahmed K, Dasgupta P. Procedural virtual reality simulation training for robotic surgery: a randomised controlled trial. Surg Endosc. 2021;35:6897–902.
33. Mahrer NE, Gold JI. The use of virtual reality for pain control: a review. Curr Pain Headache Rep. 2009;13:100–9.

Chapter 11
Paradise of Virtual Reality (VR) and Augmented Reality (AR) in Mental Health Treatment: Breaking Down Barriers Advancing Supremely Smart Healthcare for Futuristic Growth

Bhupinder Singh and **Christian Kaunert**

Abstract Technological advancements continue to reshape the way it approaches mental health treatment in the rapidly evolving landscape of healthcare. Among the most promising innovations are Virtual Reality (VR) and Augmented Reality (AR), technologies that hold the potential to revolutionize therapeutic interventions for mental health conditions. Traditional mental health treatments often face challenges such as accessibility, stigma, and limited engagement. VR and AR break down these barriers by providing immersive and interactive experiences that transcend the limitations of traditional therapeutic methods. VR environments are designed to simulate calming and serene landscapes, providing an immersive escape for individuals seeking relaxation and stress reduction. Virtual Reality and Augmented Reality are transforming the landscape of healthcare. These immersive technologies offer not only novel therapeutic interventions but also the prospect of making mental health care more accessible and effective. As it embarks on this journey into the future, the smart integration of VR and AR stands as a beacon of hope for a world where mental health treatment is personalized, engaging and profoundly impactful. The intersection of technology and well-being is creating a future where smart healthcare is synonymous with compassionate and innovative mental health care. This chapter explores the therapeutic applications of VR and AR in mental health, examining their potential to revolutionize traditional treatment methods and enhance overall patient outcomes. By providing immersive and personalized interventions, these technologies represent a significant stride toward supremely smart healthcare in the realm of mental well-being. It also reconnoiters the applications of VR and AR in

B. Singh (✉)
Sharda University, Greater Noida, India

C. Kaunert
Dublin City University, Dublin, Ireland

University of South Wales, Newport, UK

B. O. Soufiane et al. (eds.), *Augmented Wellness*,
https://doi.org/10.1007/978-981-96-2952-7_11

mental health treatment, showcasing their transformative impact on smart healthcare for futuristic advances.

Keywords VR · AR · Mental health · Treatment · Patient outcome

1 Introduction

Virtual Reality has found a profound application in mental health treatment and in the context of therapy, VR creates simulated environments that allow individuals to confront and navigate various scenarios in a controlled and supportive setting [1]. This is particularly beneficial for exposure therapy, a proven method for treating anxiety disorders, phobias, and post-traumatic stress disorder [2]. A patient suffering from social anxiety can engage in virtual scenarios that mimic real-world social situations. This exposure, carefully tailored to the individual's needs, enables gradual desensitization, empowering the patient to confront and manage anxiety in a safe and controlled environment [3].

Augmented Reality enhances our perception of the real world by overlaying digital information onto our physical surroundings [4]. In mental health treatment, AR provides real-time support and information, offering a unique approach to managing stress, anxiety and mood disorders [5]. Through the use of AR glasses or smartphone apps, users can access guided meditation sessions or receive real-time prompts for stress-relieving exercises [6]. This integration of digital elements into the user's immediate environment facilitates on-the-spot interventions, promoting mental well-being in everyday life [7]. VR and AR are increasingly being used in innovative ways to support mental health beyond traditional therapeutic approaches [8]. VR-based biofeedback applications enable individuals to visualize and manage physiological responses to stress, promoting self-regulation. The most important strategies for maximizing patient-specific healthcare decisions are personalized medicine [9]. The use of customized medicine has been made easier by artificial intelligence in a number of medical fields, including genomics and imaging diagnostics [8]. As a result, the necessity for tailored therapy in mental health applications is increasing [9]. From a psychological perspective, every mental health illness requires a different strategy and course of therapy [10]. As a result, technical solutions designed for one set of circumstances are usually not transferable to another [11]. Besides, different people with the same condition may present with different triggers, experiences, and symptoms, making technology solutions that are rigid and non-adaptive insufficient [12].

1.1 *Background: Exploring Historical Context of Technology in Mental Health Treatment*

Emotional, psychological and social well-being are all included in mental health which affects our thoughts, feelings and actions in addition to our capacity to handle stress, communicate with others, and make wise decisions [13]. It is important at all stages of life, from infancy to maturity [14]. Both physical and mental well-being is essential to total wellness [15]. Depression, for example, increases one's vulnerability to a number of physical health problems, especially long-term ailments like diabetes, heart disease and stroke. On the other hand, persistent physical illnesses may increase the risk of mental health issues [16]. Figure 11.1 resume the historic context of technological Development in Mental Health Treatment.

- **Ancient Faiths**—Abnormal conduct was frequently linked to supernatural entities such as gods, demons, or evil spirits in prehistoric societies. As a result, trephination and exorcism were common forms of therapy [16]. Greek philosophers, such as Hippocrates and Plato, brought a more naturalistic perspective to medicine by equating mental illnesses with bodily conditions that had inherited and biological causes. Roman philosophers such as Galen rejected the notion of demonic possession, emphasizing instead physical and psychological factors. But during the Middle Ages, there was a return to supernatural explanations, and severe therapies and exorcism were commonplace [17].
- **The Medieval Era**—In the Middle Ages, the Church's hegemony brought back paranormal justifications for mental illness along with punitive measures like imprisonment and exorcism. The comprehension of mental illnesses [18]. However, events such as the Black Death and an increasing demand for governmental control over nonreligious concerns led to a progressive return towards scientific and medical explanations towards the end of this century [19].
- **The Age of Renaissance**—The Renaissance saw a shift away from supernatural theories of mental disease and toward humanistic viewpoints [20]. The practice of witch-hunting claims that many people who were charged with witchcraft

Fig. 11.1 Travels the historical context of technological development in mental health treatment

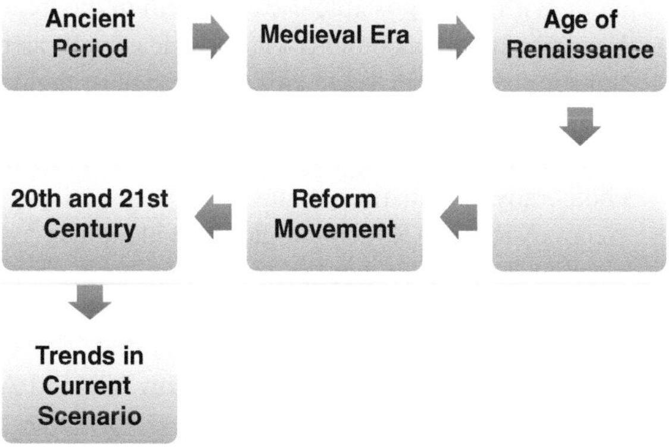

were genuinely mentally ill. Asylums were also established during this time, yet their patient care frequently fell short of humanitarian standards [21].

- **Reform Movement**—The moral treatment movement, which promoted considerate and compassionate care for mentally ill individuals, gained traction in the late eighteenth and early nineteenth centuries [22]. Compassionate techniques to treating mental diseases were pioneered by individuals which resulted in notable improvements in patient outcomes. But issues with asylum overpopulation and insufficient funding prevented all patients from receiving compassionate care [23].
- **The twentieth and twenty-first Century**—Throughout the twentieth century, conflicting theories on mental disease came to light, with the biological and psychological theories gaining ground. Significant changes in mental health care were brought about by the deinstitutionalization and increase of psychotropic medications however worries about overprescription and subpar care remained [24]. Contemporary approaches to mental health, which prioritize comprehensive and culturally sensitive care have also been influenced by managed health care, multicultural psychology, psychologists' prescription rights, and preventative research [25].
- **Views and Trends in Current Scenario**—There is still a high prevalence of mental illness, despite tremendous efforts to lower stigma and increase access to high-quality care. More specialized therapies have been made possible by our growing understanding of the biology and psychological foundations of mental illnesses [26]. But issues like insufficient care and inequalities in access to care continue to exist, underscoring the continuous requirement for all-encompassing methods of mental health promotion and management [27].

1.2 Objectives of the Chapter

This chapter has the following objectives to:

- Summarize of augmented reality (AR) and virtual reality (VR) technologies, along with some possible uses for them in the treatment of mental illness.
- Fundamental ideas behind VR and AR as well as their distinctions and applications in therapeutic treatments for a range of mental health issues.
- Explore the approaches like cognitive-behavioral therapy (CBT) and mindfulness-based interventions in addition to more established ones like medication, psychotherapy, and institutionalization.
- Access VR and AR technologies provide in the context of mental health therapy covered in detail to conventional techniques, these immersive technologies can offer a therapeutic setting that is more engaging, adaptable, and controllable.
- Thorough rundown of the many ways that VR and AR are being used to treat different mental health issues and how VR and AR are being utilized to treat

illnesses including PTSD, depression, anxiety disorders, phobias, autistic spectrum disorder and schizophrenia.
- Possible future paths and areas for growth such as technological developments, multidisciplinary cooperation, and legislative efforts to include VR and AR into standard mental health procedures.

1.3 Structure of the Chapter

This chapter comprehensively explores the various dimensions of Virtual Reality (VR) and Augmented Reality (AR) in Mental Health Treatment: Breaking down Barriers Advancing Supremely Smart Healthcare for Futuristic Growth. Section 2 elaborates the VR and AR as Emerging Technologies with Potential to Reshape Mental Health Treatment. Section 3 expresses the Therapeutic Applications of Virtual Reality. Section 4 lays down the Augmented Reality in Mental Health Treatment. Section 5 discusses the Innovations beyond Traditional Therapies. Finally, Section 6 Conclude the Chapter with Future Scope.

2 VR and AR as Emerging Technologies with Potential to Reshape Mental Health Treatment

The emerging technologies such as virtual reality (VR) and augmented reality (AR) have the potential to improve current approaches to mental health diagnosis and treatment [28]. Virtual reality (VR) provides an immersive experience that separates users from the real world; this is usually accomplished using a 360° head-mounted display. Augmented reality (AR) superimposes virtual aspects on the physical world [28]. This is demonstrated by features included in well-known social media platforms [29]. Patients can have their world probed, broadened and challenged via the use of AR and VR [30]. These technologies have the potential to improve overall wellbeing, support learning, entertain, and treat maladaptive behaviors or cognitive patterns by utilizing visual, aural, haptic, somatosensory and olfactory inputs. The first commercial VR products hit the market in the 1980s, while the technology itself dates back to the 1960s. Virtual reality (VR) was first widely used in creative industries like gaming, entertainment and shopping. However, it has gradually made its way into the healthcare space, where it is being used for medical imaging, surgical teamwork, and medical education [30].

Virtual reality (VR) has been used as a patient care tool in a variety of medical specialties. It has helped with stroke rehabilitation, supported balance in adults with Parkinson's disease, trained children with cerebral palsy to use their upper extremities and managed pain during childhood vaccinations. Real-world video feeds are integrated into AR which first appeared in the 1990s as an extension of the fully synthetic VR environment, in order to improve the virtual experience. By superimposing real-time, three-dimensional information from noninvasive imaging modalities like MRI, CT or ultrasound on the patient on the operating table, augmented reality (AR) principally aims to improve surgical visibility in the medical field. When combined, AR and VR form the "metaverse" and represent digital medicines which are evidence-based interventions driven by software [31].

2.1 *Evolution of VR and AR and Their Entrance into Healthcare*

Virtual reality (VR) and augmented reality (AR) technologies have surfaced as revolutionary instruments within medical contexts, fundamentally altering healthcare methodologies and reshaping interactions between medical practitioners and patients within healthcare settings. Mental illness is serious in character has the power to destroy joy from all aspect of a person's life. Its enormous effects on relationships, productivity, longevity and health necessitate acknowledgment and swift action. However, for a variety of reasons, mental health is frequently neglected. But things don't have to stay this way. Modern virtual reality therapy for mental health is now available, and it is provided with discretion and perfection. Advanced degree-holding trained therapists at XR-Health provide customized, drug-free therapies that have the capacity to change lives [32]. Virtual reality (VR) is a sensation of presence that users get from computer-generated environments that can mimic or deviate from the actual world. VR was first conceptualized in 1956 when filmmaker Morton Heilig created the Sensorama device, which provided a multi-sensory experience similar to being within a movie, like riding a motorcycle through a city [33].

The first VR/AR head-mounted display linked to a computer was made in 1968 by Ivan Sutherland and Bob Sproull. It was a prototype for subsequent VR headsets used in classrooms. In 1990, NASA unveiled the Virtual Interface Environment Workstation (VIEW), which allowed astronomers to conduct virtual space exploration and set up VR teaching facilities. Later developments included the Matrix, Landmark VR PTSD therapy, and SEGA initiatives. Large corporations such as Google, Sony, and Facebook drove the rapid advancement of virtual reality technology in 2014 [34]. While Sony revealed their VR project for the PlayStation and Facebook purchased Oculus, Google Cardboard introduced 360-degree photo VR to the general public. With the debut of devices like the Microsoft Hololens, Playstation VR, HTC VIVE, and Oculus Quest, this trend persisted [35].

In October 2020, Facebook released Oculus Quest 2 which at the time was the most potent and reasonably priced wireless VR headset available. Virtual reality (VR) is being used in education in the twenty-first century to allow for the study of historical places, art galleries, and isolated areas. The incorporation of virtual reality (VR) into gaming platforms and media depictions highlights its broad popularity and illustrates its progression from theoretical concepts to tangible technology [36].

2.2 *Importance of Innovative Approaches to Address the Growing Mental Health Challenge*

Augmented and virtual reality is at the center of a fascinating change that is being sparked by the integration of technology into healthcare. Not only are these immersive technologies changing the way that healthcare is delivered, but they are also showing signs of substantial market growth. In a time when technology is always pushing the boundaries of innovation, few industries have seen a transformation as significant as the healthcare industry [37]. Virtual reality (VR) and augmented reality (AR) are two notable examples of cutting-edge technology changing the face of healthcare. These immersive digital experiences are becoming a necessary part of our current reality, no longer exclusive to science fiction. They are radically changing the way we approach healthcare and how we traverse the fields of diagnosis, therapy, and other related areas [38].

3 Therapeutic Applications of Virtual Reality

Virtual reality (VR) technology has advanced quickly from its beginnings in entertainment and gaming to finding intriguing uses in a variety of industries, including healthcare. VR therapy is one such application that has gained a lot of attention recently as an inventive method of treating mental health issues. In virtual reality (VR) treatment, patients are submerged in 3D computer-generated landscapes that mimic actual situations or fictional worlds designed to treat a variety of psychological and emotional issues. These interactive virtual worlds let patients securely move, engage with items, and face difficult circumstances. To address certain phobias, anxiety, PTSD and other mental health concerns, therapists might create customized situations that support a more targeted and successful therapy strategy [39].

VR therapy helps patients face their fears and anxieties in a safe and controlled setting by exposing them to real-life circumstances or treating them in controlled scenarios. Because VR therapy is immersive, it enables therapists to oversee and guide patients through the exposure process, resulting in a customized and progressive approach to treatment. Virtual reality (VR) has several uses in psychiatry. It may be used for interactive diagnostics with patients by simulating events and analyzing patient reactions, or it can be used as immersive exposure treatment for disorders including PTSD, social anxiety and phobias. Virtual reality (VR) is also useful in cognitive-behavioral therapy, therapist training, remote care consultations, and therapeutic technique research [40]. Panelists advise converting 3D films into virtual reality (VR) experiences using readily available commercial products like MetaQuest or Google headgear, or even less expensive solutions like Google Cardboard [41]. There are multifarious principal advantages of VR Therapy:

- **Enhanced Immersion**: Virtual reality (VR) therapy offers a degree of immersion that is sometimes absent from traditional therapy techniques, enabling patients to experience scenarios more authentically and improving therapeutic results [42].
- **Personalized Care**: VR settings may be tailored by therapists to each patient's unique requirements, imitating anxiety-inducing situations like tight quarters, heights or public speaking [43].
- **Safe Environments**: Virtual reality treatment provides a safe haven where patients may face their concerns without the dangers of going out into the real world. As patients advance, therapists may control the degree of exposure and progressively increase obstacles [44].
- **Overcoming Stigma**: By offering a covert and private treatment option, virtual reality therapy helps to lessen the stigma attached to seeking assistance for mental health issues, hence motivating more people to do so [45].

3.1 Use of VR in Exposure Therapy for Anxiety Disorders and Phobias: VR Therapy Applications

- **Anxiety and Phobias**: With enabling patients to face their fears in a secure setting, virtual reality therapy has shown promise in the treatment of anxiety disorders, PTSD, and specific phobias such as—a fear of flying, heights or spiders [46].
- **Addiction Treatment**: With mimicking circumstances that result in drug cravings, virtual reality therapy is being investigated as a helpful tool for addiction treatment. This allows patients to practice coping mechanisms in a risk-free setting [47].

3.2 VR Applications in Cognitive Behavioral Therapy and Mindfulness Training

Virtual reality (VR) is a rapidly developing technology that provides an entirely immersive experience by submerging users in a three-dimensional computer-generated world. Users create a virtual environment by integrating visual and aural inputs through a head-mounted display. It has developed into a useful tool for clinical training over time, especially in areas like neurosurgery and orthopedic procedural teaching. Also, its therapeutic uses are still growing because of its proven efficacy in treating mental disorders, treating pain and facilitating physical rehabilitation. The use of virtual reality (VR) for cancer patients' pain, both acute and chronic and during traumatic medical procedures is examples of recent occurrences [48].

3.3 Exploring the Potential of VR for Biofeedback and Physiological Regulation in Stress Management

A cutting-edge tool in the realm of digital immersion, virtual reality is frequently used for both leisure and teaching. Its promise, however, also includes therapeutic uses meant to promote attention and relaxation. Researchers are delving deeper into its applicability and efficacy in assessing, controlling and mitigating stress. Historically, the two main methods of treating pain have been medicine and intrusive procedures. But with the current opioid epidemic and advances in our understanding of chronic pain, there is a rising awareness of the need for non-invasive, non-pharmacological therapeutic options. The majority of well-known guidelines now support treating chronic pain with a multimodal, multidisciplinary approach that gives non-pharmacological therapies priority. These might include a variety of strategies like physical treatment, exercise plans, dietary changes, behavioral therapy and other types of psychological counseling [49].

4 Augmented Reality in Mental Health Treatment: Concept Relevance and Applications

Augmented Reality (AR) is a cutting-edge technology that modifies one's experience of reality by fusing virtual and physical components. Augmented Reality (AR) produces an immersive experience where virtual and real items coexist and interact in real time by superimposing computer-generated visuals over the actual world. The use of augmented reality (AR) and other aided technologies in the medical industry has increased significantly from training to clinical practice. The ability to significantly alter the physical world gives AR treatments a competitive advantage over conventional techniques [50]. There are applications of AR in Mental Health Treatment as:

- **Education and Training in Medicine**—Medical education is being revitalized by the use of AR and VR in healthcare. Through realistic simulations, students may practice identifying rare medical illnesses or performing complex surgical operations. Medical practitioners may practice and improve their decision-making skills in a virtual environment that is very realistic without putting real patients in peril. AR glasses represent yet another use of augmented reality in healthcare. Augmented Reality (AR) is a cutting-edge technology that modifies one's experience of reality by fusing virtual and physical components [51]. Augmented Reality (AR) produces an immersive experience where virtual and real items coexist and interact in real time by superimposing computer-generated visuals over the actual world. The use of augmented reality (AR) and other aided technologies in the medical industry has increased significantly, from training to clinical practice. The ability to significantly alter the physical world gives AR treatments a competitive advantage over conventional techniques. There are applications of augmented reality (AR) in order to obtain a better understanding of its possible effects on people who have mental health issues, especially when combined with gamification [52].

- **Therapy and Pain Management**—Virtual reality has become an important technique for pain treatment. Patients might distract themselves from the agony of difficult operations by submerging themselves in relaxing virtual reality surroundings. The FDA has approved VR headsets that use cognitive behavioral therapy methods to help patients relax and control their discomfort. VR therapy has demonstrated potential in the treatment of anxiety disorders, PTSD and phobias [53].
- **Planning and Training for Surgery**—VR technology is being used by surgeons to practice complex procedures. Surgeons are able to practice skills and customize surgical plans by creating 3D models of a patient's anatomy and modeling surgical operations in a virtual setting. Through realistic surgical training modules from companies like Osso VR, doctors may become acquainted with novel techniques and medical equipment. This lowers problems, shortens the duration of the operation, and improves surgical precision [54].
- **Physical Therapy**—Physical treatment techniques are being revolutionized by virtual and augmented reality. Businesses use virtual reality technology to provide customized workouts for physical rehabilitation. In virtual worlds, patients recuperating from operations or injuries can participate in interactive activities that enhance therapy sessions and allow therapists to properly evaluate progress and modify treatments [55].
- **Remote Diagnoses and Consultations**—The need for telemedicine has significantly increased, especially during times of global health catastrophes. There are recent developments in AR and VR are improving remote consultations by giving medical professionals access to real-time patient data and vital signs, providing a more thorough and engaging experience for patients as well as healthcare providers. The accuracy of remote diagnosis and training sessions is improved by these immersive technologies [56].

5 Innovations Beyond Traditional Therapies

Robotic-Assisted Surgery (RAS) systems have garnered increasing attention in recent times because of their potential to improve surgical accuracy, dexterity, and accessibility to minimally invasive treatments. Augmented reality (AR) combined with robotic-assisted surgery offers a sophisticated interface that improves user perception. The power of AR to increase situational awareness will greatly improve surgeons' ability to make judgments in the moment. In the upcoming years, it is anticipated that the integration of AR will transform a number of RAS fields, including preoperative imaging and 3D rendering, intraoperative robotic ultrasound application, robotic liver resections, intraoperative reconstructions and tracking systems and robotic hepatic surgery [57]. If these general tech solutions do not adjust to the variety of circumstances that therapists and patients meet on a regular basis, they may be perceived as insufficient by both parties. Customization should be an option for technological solutions meant to assist patients and mental health practitioners. In order to create and build appropriate and adaptable technological solutions, this approach should entail collaboration between technology designers, researchers, developers, mental health specialists and end users. These customizable elements should also be easy to use so that patients won't reject them or interrupt providers' regular practices. When it comes to XR settings, one of the biggest challenges is creating experiences that are adaptable and adjustable enough to change their interfaces and digital 3D material to fit the demands of different patients in different circumstances [58].

For a long time, mental health illnesses have been diagnosed and treated with the use of virtual and augmented reality. Technological developments have increased consumer and practitioner accessibility to commercial solutions. Still, a number of obstacles and restrictions prevent these technologies from being widely used in regular professional practice. Depression and obsessive-compulsive disorder are two disorders that have previously gotten less attention but are currently receiving greater attention [59]. Hardware and equipment advancements like Mixed Reality Head Mounted Displays might lead to new opportunities in the mental health sector. Virtual reality (VR), mixed reality (MR), and augmented reality (AR) are all included in extended reality (XR), which has several uses. Emerging literature investigates multimodal interactions, such as auditory, olfactory, and tactile input, whereas XR predominantly focuses on visual experiences. The possibility for improving treatment results by including such cues into XR experiences connected to mental health issues is yet somewhat unexplored. Virtual reality technology has drawbacks as notable obstacles include the variety of VR technology, regulatory procedures that cannot keep up with technological improvements, and skepticism among potential consumers. There are also worries about the absence of regulatory channels, which might cause moral dilemmas [60].

6 Conclusion and Future Scope

Millions of people worldwide are severely impacted by mental health illnesses, which also places a heavy financial and social cost on those affected. The most neglected global health concerns are mental health. The significant effects that mental health illnesses have on people, families and society as a whole are not well understood, which contributes significantly to this neglect. It is a common misperception that there is little that can be done to treat these disorders, even though there are many useful pharmaceutical, psychological, and social therapies available. The action is further hampered by the widespread stigma associated with mental health problems, which occasionally results in abuses of human rights. Because of this, most people who suffer from mental health illnesses do not get the attention and treatment they require which might significantly enhance their quality of life. With further research proving its efficacy and potential to become a crucial component of mental health treatment procedures, VR and AR has a bright future ahead of it. VR treatment is positioned as a useful tool in improving mental healthcare results and boosting patients' quality of life because to its unique methodology, safe exposure, enhanced engagement and individualized experiences.

References

1. Singha R, Singha S. Mental health treatment: exploring the potential of augmented reality and virtual reality. In: Applications of virtual and augmented reality for health and wellbeing. Hershey: IGI Global; 2024. p. 91–110.
2. Shekar V, Kachhi Z. Technology applications in virtual and augmented reality for human welfare: the new future. In: Entrepreneurship and creativity in the metaverse. Hershey: IGI Global; 2024. p. 179–97.
3. Khan T, Civas M, Cetinkaya O, Abbasi NA, Akan OB. Nanosensor networks for smart health care. In: Nanosensors for smart cities. Amsterdam: Elsevier; 2020. p. 387–403.
4. De Clara L. The neuropsychological impact of immersive experiences in the metaverse and virtual and augmented reality. In: Applications of virtual and augmented reality for health and wellbeing. Hershey: IGI Global; 2024. p. 148–66.
5. Agoulmine N, Kim K, Kim S, Rim T, Lee JS, Meyyappan M. Enabling communication and cooperation in bio-nanosensor networks: toward innovative healthcare solutions. IEEE Wirel Commun. 2012;19(5):42–51.
6. Dorj UO, Lee M, Choi JY, Lee YK, Jeong G. The intelligent healthcare data management system using nanosensors. J Sens. 2017;2017:1–9.
7. Rani P. Nanosensors and their potential role in internet of medical things. In: Nanosensors for futuristic smart and intelligent healthcare systems. Boca Raton: CRC Press; 2022. p. 293–317.
8. Ahirwar R, Khan N. Smart wireless nanosensor systems for human healthcare. Boca Raton: CRC Press; 2022. p. 265–92.
9. Han B, Tomer V, Nguyen TA, Farmani A, Singh PK, editors. Nanosensors for smart cities. Amsterdam: Elsevier; 2020.
10. Sarkar S, Misra S. From micro to nano: the evolution of wireless sensor-based health care. IEEE Pulse. 2016;7(1):21–5.
11. Rizwan A, Zoha A, Zhang R, Ahmad W, Arshad K, Ali NA, et al. A review on the role of nano-communication in future healthcare systems: a big data analytics perspective. IEEE Access. 2018;6:41903–20.
12. Singh B, Kaunert C. Augmented reality and virtual reality modules for mindfulness: boosting emotional intelligence and mental wellness. In: Applications of virtual and augmented reality for health and wellbeing. Hershey: IGI Global; 2024. p. 111–28.
13. Datta SPA. Future healthcare: bioinformatics, nano-sensors, and emerging innovations. In: Nanosensors: theory and applications in industry, healthcare and defense. London: Routledge; 2016. p. 247.
14. Pramanik PKD, Solanki A, Debnath A, Nayyar A, El-Sappagh S, Kwak KS. Advancing modern healthcare with nanotechnology, nanobiosensors, and internet of nano things: taxonomies, applications, architecture, and challenges. IEEE Access. 2020;8:65230–66.
15. Nazari A. Nanosensors for smart cities: an introduction. In: Nanosensors for smart cities. Amsterdam: Elsevier; 2020. p. 3–8.

16. Sahi KSS, Kaushik S. Smart nanosensors for healthcare monitoring and disease detection using AIoT framework. In: Nanosensors for futuristic smart and intelligent healthcare systems. Boca Raton: CRC Press; 2022. p. 387–400.
17. Oukhatar A, Bakhouya M, El Ouadghiri D. Electromagnetic-based wireless nano-sensors network: architectures and applications. J Commun. 2021;16(1):8–19.
18. Ramasamy M, Kumar PS, Varadan VK. Wearable nanosensor systems and their applications in healthcare. In: Nanosensors, biosensors, info-tech sensors and 3D systems 2017, vol. 10167. Bellingham: SPIE; 2017. p. 1016703.
19. Paliwal G, Bunglowala A. Nanosensor and actuator technologies for wearable mobile patient monitoring systems: a review. In: Workshop on computational models of natural argument. Singapore: Springer Singapore; 2018. p. 83–95.
20. He L, Eastburn M. Smart nanosensor networks for body injury detection. In: 2022 IEEE international conference on smart internet of things (SmartIoT). Piscataway: IEEE; 2022. p. 15–9.
21. Ali NA, Abu-Elkheir M. Internet of nano-things healthcare applications: requirements, opportunities, and challenges. In: 2015 IEEE 11th international conference on wireless and mobile computing, networking and communications (WiMob). Piscataway: IEEE; 2015. p. 9–14.
22. Pirmagomedov R, Hudoev I, Shangina D. Simulation of medical sensor nanonetwork applications traffic. In: Distributed computer and communication networks: 19th international conference, DCCN 2016, Moscow, Russia, November 21-25, 2016, revised selected papers 19. Berlin: Springer International Publishing; 2016. p. 430–41.
23. Mujawar MA, Gohel H, Bhardwaj SK, Srinivasan S, Hickman N, Kaushik A. Nano-enabled biosensing systems for intelligent healthcare: towards COVID-19 management. Mater Today Chem. 2020;17:100306.
24. Omanović-Mikličanin E, Maksimović M, Vujović V. The future of healthcare: nanomedicine and internet of nano things. Folia Medica Facultatis Medicinae Universitatis Saraeviensis. 2015;50(1):23–8.
25. Udoh EE, Hermel M, Bharmal MI, Nayak A, Patel S, Butlin M, Bhavnani SP. Nanosensor technologies and the digital transformation of healthcare. Pers Med. 2023;20(3):251–69.
26. Chelliah R, Wei S, Daliri EBM, Rubab M, Elahi F, Yeon SJ, et al. Development of nanosensors based intelligent packaging systems: food quality and medicine. Nanomaterials. 2021;11(6):1515.
27. Topel SD, Al-Turjman F. Nanosensors for the internet of nano-things (IoNT): an overview. In: Internet of nano-things and wireless body area networks (WBAN). London: Routledge; 2019. p. 21–44.
28. Dinesh Kumar K. Human health monitoring mobile phone application by using the wireless nanosensor based embedded system. In: 2013 international conference on information communication and embedded systems (ICICES). Piscataway: IEEE; 2013. p. 889–92.
29. Materon EM, Gómez FR, Joshi N, Dalmaschio CJ, Carrilho E, Oliveira ON Jr. Smart materials for electrochemical flexible nanosensors: advances and applications. In: Nanosensors for smart manufacturing. Amsterdam: Elsevier; 2021. p. 347–71.
30. Kumar S, Kumari P. Flexible nano smart sensors. In: Nanosensors for smart manufacturing, vol. 199. Amsterdam: Elsevier; 2021.
31. Gaobotse G, Mbunge E, Batani J, Muchemwa B. Non-invasive smart implants in healthcare: redefining healthcare services delivery through sensors and emerging digital health technologies. Sens Int. 2022;3:100156.
32. Vivekananthan V, Khandelwal G, Alluri NR, Kim SJ. E-skin for futuristic nanosensor technology for the healthcare system. In: Nanosensors for futuristic smart and intelligent healthcare systems. Boca Raton: CRC Press; 2022. p. 133–57.
33. Ramasamy M, Varadan VK. Wireless nanosensor system for diagnosis of sleep disorders. In: Nanosensors, biosensors, and info-tech sensors and systems 2016, vol. 9802. Bellingham: SPIE; 2016. p. 200–5.
34. Pradhan MR, Mago B, Ateeq K. A classification-based sensor data processing method for the internet of things assimilated wearable sensor technology. Clust Comput. 2023;26(1):807–22.

35. Tang WH, Ho WH, Chen YJ. Data assimilation and multisource decision-making in systems biology based on unobtrusive internet-of-things devices. Biomed Eng Online. 2018;17(2):1–13.
36. Manogaran G, Alazab M, Song H, Kumar N. CDP-UA: cognitive data processing method wearable sensor data uncertainty analysis in the internet of things assisted smart medical healthcare systems. IEEE J Biomed Health Inform. 2021;25(10):3691–9.
37. Hassanalieragh M, Page A, Soyata T, Sharma G, Aktas M, Mateos G, et al. Health monitoring and management using internet-of-things (IoT) sensing with cloud-based processing: opportunities and challenges. In: 2015 IEEE international conference on services computing. Piscataway: IEEE; 2015. p. 285–92.
38. Hiran KK, Doshi R, Patel M, editors. Applications of virtual and augmented reality for health and wellbeing. Hershey: IGI Global; 2024.
39. Singh B, Kaunert C, Vig K. Reinventing influence of artificial intelligence (AI) on digital consumer lensing transforming consumer recommendation model: exploring stimulus artificial intelligence on consumer shopping decisions. In: Musiolik T, Rodriguez R, Kannan H, editors. AI impacts in digital consumer behavior. Hershey: IGI Global; 2024. p. 141–69. https://doi.org/10.4018/979-8-3693-1918-5.ch006.
40. Singh B. Featuring consumer choices of consumable products for health benefits: evolving issues from tort and product liabilities. J Law Torts Consumer Protect Law. 2024;7(1).
41. Singh B. Unleashing alternative dispute resolution (ADR) in resolving complex legal-technical issues arising in cyberspace lensing E-commerce and intellectual property: proliferation of E-commerce digital economy. Revista Brasileira de Alternative Dispute Resolution-Brazilian Journal of Alternative Dispute Resolution-RBADR. 2023;5(10):81–105.
42. Singh B, Kaunert C. Integration of cutting-edge technologies such as internet of things (IoT) and 5G in health monitoring systems: a comprehensive legal analysis and futuristic outcomes. GLS Law J. 2024;6(1):13–20.
43. Singh B. Green infrastructure in real estate landscapes: pillars of sustainable development and vision for tomorrow. Natl J Real Estate Law. 2024;7(1):4–8.
44. Suma DV. Wearable IoT based distributed framework for ubiquitous computing. J Ubiquitous Comput Commun Technol. 2021;3(1):23–32.
45. Jenkins T. Wearable medical sensor devices, machine and deep learning algorithms, and internet of things-based healthcare systems in COVID-19 patient screening, diagnosis, monitoring, and treatment. Am J Med Res. 2022;9(1):49–64.
46. Malik A, Kumar A. Assimilation of blockchain with internet of things (IoT) with possible issues and solutions for better connectivity and proper security. In: New trends and applications in internet of things (IoT) and big data analytics. Cham: Springer International Publishing; 2022. p. 187–207.
47. Byerly K, Vagner L, Grecu I, Grecu G, Lăzăroiu G. Real-time big data processing and wearable internet of medical things sensor devices for health monitoring. Am J Med Res. 2019;6(2):67–72.
48. Garg D, Dubey N, Goel P, Ramoliya D, Ganatra A, Kotecha K. Improvisation in spinal surgery using AR (augmented reality), MR (mixed reality), and VR (virtual reality). Eng Proc. 2024;59(1):186.
49. Kumar PM, Khan LU, Hong CS. Affirmative fusion process for improving wearable sensor data availability in artificial intelligence of medical things. IEEE Sensors J. 2022.
50. Han X. A novel assimilated navigation model based on advanced optical systems (AOS), internet of things (IoT) and artificial intelligence (AI). Opt Quant Electron. 2023;55(7):655.
51. Dash D, Farooq R, Panda JS, Sandhyavani KV. Internet of things (IoT): the new paradigm of HRM and skill development in the fourth industrial revolution (industry 4.0). IUP J Inf Technol. 2019;15(4):7–30.
52. Goethals I. Real-time and remote health monitoring internet of things-based systems: digital therapeutics, wearable and implantable medical devices, and body sensor networks. Am J Med Res. 2019;6(2):43–8.

53. Banerjee A, Chakraborty C, Rathi M Sr. Medical imaging, artificial intelligence, internet of things, wearable devices in terahertz healthcare technologies. In: Terahertz biomedical and healthcare technologies. Amsterdam: Elsevier; 2020. p. 145–65.
54. Chen ET. The internet of things: opportunities, issues, and challenges. In: The internet of things in the modern business environment. Hershey: IGI Global; 2017. p. 167–87.
55. Barnes R, Zvarikova K. Artificial intelligence-enabled wearable medical devices, clinical and diagnostic decision support systems, and internet of things-based healthcare applications in COVID-19 prevention, screening, and treatment. Am J Med Res. 2021;8(2):9–22.
56. Desai D, Shende P. Integration of internet of things with quantum dots: a state-of-the-art of medicine. Curr Pharm Des. 2021;27(17):2068–75.
57. Albalawi U, Joshi S. Secure and trusted telemedicine in internet of things IoT. In: 2018 IEEE 4th world forum on internet of things (WF-IoT). Piscataway: IEEE; 2018. p. 30–4.
58. Hurley D, Popescu GH. Medical big data and wearable internet of things healthcare systems in remotely monitoring and caring for confirmed or suspected COVID-19 patients. Am J Med Res. 2021;8(2):78–90.
59. Rubi J, Vijayalakshmi A, Venkatesan S. Integration of biomedical engineering in augmented reality and virtual reality applications. In: Applications of virtual and augmented reality for health and wellbeing. Hershey: IGI Global; 2024. p. 41–54.
60. Okoro YO, Ayo-Farai O, Maduka CP, Okongwu CC, Sodamade OT. The role of technology in enhancing mental health advocacy: a systematic review. Int J Appl Res Soc Sci. 2024;6(1):37–50.

Chapter 12
Embedding Virtual Reality Technologies as an Early Career Neuroscience Research Training Methodology Advancing Medical Research

Lorenz S. Neuwirth and Maxime Ros

Abstract Over the past two decades, advancements in neuroscience have outpaced the availability of resources and funding for undergraduate and graduate programs, particularly at Primarily Undergraduate Institutions (PUIs) and Minority Serving Institutions (MSIs). This disparity has widened the gap in early-career skills development, limiting opportunities for students and faculty at under-resourced institutions. Additionally, selective funding favouring established Research 1 (R1) universities exacerbates inequalities, forcing new researchers into power-imbalanced collaborations with senior faculty. Despite these challenges, innovative pedagogical strategies such as applied research mentoring, case studies, and computational neuroscience courses have emerged at PUIs to bridge the gap. Cost-effective technologies, particularly open-source software and virtual reality (VR), offer promising avenues for enhancing neuroscience education and research, providing immersive learning and procedural skill development. Thus, VR in particular, has demonstrated therapeutic applications and the increased potential for transforming neuroscience education. The future of neuroscience education lies in leveraging these VR tools, fostering interdisciplinary research, and addressing funding inequities to create a more inclusive and equitable field for next generation neuroscientists.

Keywords Neuroscience education · Funding disparities · Primarily undergraduate institutions (PUIs) · Virtual reality (VR) · Innovative pedagogy

---

L. S. Neuwirth (✉)
Department of Psychology, SUNY Old Westbury, Old Westbury, NY, USA

SUNY Neuroscience Research Institute, Old Westbury, NY, USA
e-mail: neuwirthl@oldwestbury.edu

M. Ros
Education Sciences School—LIRDEF, Montpellier University, Montpellier, Occitanie, France

B. O. Soufiane et al. (eds.), *Augmented Wellness*,
https://doi.org/10.1007/978-981-96-2952-7_12

1 Introduction

Research advancements in the neurosciences have evolved significantly over the last 20 years, but resources and funding for undergraduate and graduate neuroscience programs have failed to keep up with this pace. Arguably, this situation has inadvertently provoked a dual problem: (1) creating wider academic achievements gap in preparing early career skills for students interested in neuroscience careers in both education and research; and (2) forcing students and faculty in the neurosciences to become more resourceful within their means to remain relevant in a technology and big data dominated era of education. Parallel to these real problems are the government's selective funding directed towards specific neurological and neurodevelopmental disorders more so than creative scientific explorations, which has created a very niche grant "cash cow" mechanism for well-funded and established Research 1 universities (R1s). What can be argued is that this situation creates a disadvantage for new faculty and researchers entering the field at the early career level to be forced into collaborations with more senior neuroscience researchers possessing real stature and power dynamics over them as a new model for future funding between primarily undergraduate institutions (PUIs; which are often Minority Serving Institutions [MSIs]) and R1s [1–3]. Publishing trends from 2006 to 2015 have shown that neuroscience research is being directed towards autism, meta-analyses, functional connectivity, default mode networks and modeling, and neuroimaging; thereby, changing the landscape of neuroscience research [4]. To this point, many colleges and universities that have these technologies and instruments are typically R1s, containing or directly linked with a hospital system, and do not reflect the larger number of colleges/universities in the United States (US) representing PUIs that lack these resources. This grant funding selectivity further creates an inequity system in neuroscience education, that as a byproduct, makes PUI funding more rare or to some degree almost unattainable and subsequently reduces the neuroscience education and research advancements of a large majority of students in their early careers that has persisted and grown worse over the last two decades.

However, in surveilling the undergraduate population within the US it was reported that faculty have attempted to creatively circumvent these problems through innovative learning spaces that were found to be critically in need and were especially lacking in under resourced and underfunded colleges/universities [5–7]. Despite these resource and funding setbacks, many faculty have tried using a range of creative pedagogical interventions to compensate by: utilizing applied research mentoring programs ([8, 9]), creative writing in neuroscience [10], creating nested and scaffolding neuroscience educational programs as a blueprint to learning [11, 12], integrating recent advances in the field into their curriculum [13], teaching computational neuroscience at PUIs [14], the use of case studies to guide explorations [15], teaching courses on publishing in neuroscience [16], increasing brain literacy [17], and using a historical context [18], to name a few. Yet, given these creative approaches, undergraduate- and graduate-level neuroscience students are still lacking in their early career prerequisite training; especially, procedural skills from PUIs. Today, these challenges are being approached through two distinct cost-effective means for neuroscience educational pedagogy and training: (1) open software and hardware utilization [19]; and (2) virtual reality (VR) integration into the curriculum with procedural applied learning [20].

Additionally, others have proposed seven distinct challenges that are beginning to and will continue to shape twenty-first century neuroscience through: (1) becoming big science; (2) developing data ladders between single areas of the brain with different levels of organizations that cross-link and cross-reference between humans and other pre-clinical animal models; (3) developing novel hardware and software to simulate brain activity and computational models that best simulate human properties; (4) improved classification and simulation of brain diseases to better inform diagnosis and effective drug discovery; (5) exploit current knowledge to build new brain-inspired and brain-based technologies benefiting both society and industry; and (6) facilitating neuroscience research to deliver huge societal-linked goals to address the widespread social concern regarding brain disorders and diseases, respectfully [21]. Moreover, in the last two decades, the attentional systems and its underlying neural and nodal networks have become both more integrated and more promising in elucidating the spectrum of attentional disorders coupled with comorbid diagnoses for next generation neuroscientists to further resolve [22].

Reinforcing the position that Altimus et al. [23] and Martin [24] similarly reported across a near 20 year time span, that as the field of neuroscience moves forward it needs to break down the silos of laboratory investigation and move towards more integrative neuroscientific approaches to best advance the field (i.e., cautions are raised here between PUIs and R1s as stated above); while also reducing and working towards eliminating sex biases and omission in neuroscience publication and communication [25–27]. Consistent with these ideas, Vázquez-Guardado et al. [28] noted the advancement in neurotechnologies with broad ranging potential and applications for neuroscience research through the modern use of electrical, optical, and microfluidic methodologies to study brain function. However, considering the aforementioned factors and concerns, it begs the question is society "ready for real-world neuroscience" as a whole, consistent with the position that cognitive neuroscientists have already taken [29]?

Here we propose an underutilized avenue for the future of neuroscience through utilizing VR in both education as a good example of viable application [30] as well as in neuroscience research and therapy [31]. Regarding the latter, VR has been used therapeutically to display simulations of the embodied experience for a range of behavioral health conditions (e.g., anxiety; [32]) and studies have suggested that as early as 2017, that it has been ready for clinical usage to treat patients [33]. Therefore, in this context we consider how VR has evolved over the last two decades and propose how it can be used to further support current and future advancements of medical research and education through complementary learning approaches for next-generation neuroscience students coming from mainly PUIs.

2 The Last 20 Years: Examining the Integration of Neuroscience and VR

It is interesting to see that the first medical articles related to VR (i.e., non-immersive VR) covered content within the neurosciences, where they studied autism [34]. The initial goal of this work was to understand the interaction of a person with autism with their environment, but it also provided insight into perceptual visual, psycho-social, and emotional processing of VR effects on users with autism [34]. Continuing from this original paper and covering the last two decades of advancement, VR technologies have not only become more integrated, but the technology itself has rapidly upgraded with new capabilities (e.g., immersion, augmented virtual reality, first-person point-of-view 180° stereoscopic vision, etc.; [35–38]). These fundamental changes in VR from gaming and entertainment, to psychotherapy, supplemental, and complementary medical student training and pedagogy, have in turn, played critical roles in advancing surgical neuroscience methods ([39–43]). One case in point, is the ability for more resourced universities and hospitals to develop curriculum within the available VR platforms [44] and to then deploy it to other universities and hospitals that may not have neuroscientists, surgeons, or other authorities skilled in the technique to deliver the same procedural skillsets to students enrolled in under resourced institutions of higher learning (i.e., PUIs; [45–51]). Deploying VR in neuroscience education and research has further value between international collaborations, educational exchange, and developing premier training and learning opportunities between developed and developing countries (e.g., Collaborative Online International Learning [COIL] programs that are implemented as Course-based Undergraduate Research Experiences [CURE] within the curriculum; [52–60]; Coumans and Winter 2024). However, addressing these neuroscience infrastructural needs between the under resourced colleges and universities and well-resourced college and universities within the US (i.e., the PUI and R1 conundrum) remain problematic, but given the evolution of VR [61], VR-based pedagogical methods might be a practical under-utilized solution to address this critical issue.

3 The Unprecedented Coronavirus-19: Increased Educational and Research Value of VR

When the Coronavirus-19 (COVID-19) occurred, New York became the first "hot spot of contagion" in the US in which all curriculum shifted drastically from in-person learning to virtual/remote learning educational formats [62]. This situation made science procedural laboratory learning and other hands-on training/applied curricular challenges more difficult to overcome, especially for under resourced colleges and universities. This situation forced faculty to reimagine the curriculum in different ways, with different resources or with little to no resources given the additional constraint of the paused and then slow reinstatement of the technology supply chain. One such example was to record in real-time and deploy a timely COVID-19 healthcare guideline through First-Person Point-of-View (FP-POV) to nurses treating patients that were COVID-19 positive during the pandemic [63]. These concurrent and unprecedented challenges became the catalyst by which VR as illustrated in the prior example began to evolve ever so quickly in both educational and research aspects; especially, for neuroscience [64–70]. For example, as a by-product of COVID-19, many medical school and similar allied healthcare related graduate programs had to shift their curriculum to VR-based platforms for dentistry [71], nursing as apart of neuroeducation [64], neuroscience education and outreach from high school students [65] to undergraduate-level students [72], and graduate-level students in neuroanatomy [73–75], practicing surgical procedures on cadavers and models [50, 76], and neurosurgery [77] to name a few. However, within that context, validation of effective VR vs. anything done in VR as a rapid response to address the COVID-19 challenges became more obscure. Then researchers began to systematically look into VR with greater scrutiny and in many cases, it was validated when deployed in the neuroscience curriculum [78].

4 Undergraduate Neuroscience VR Education: A Changing Climate for Developing Early Career Procedural Skills

Considering the evolution of VR in the context of higher education, it is not surprising that more undergraduate-level colleges and universities are attempting to integrate more VR into their neuroscience curriculum. The use of VR has vast impacts on students' visuospatial abilities whether learning through desktop and immersive VR environments [79, 80]. The VR experience exploits the use of the student's mirror neuron system to envision their perceived VR experience as their own (i.e., an extended proprioceptive experience with a high degree of procedural learning integrative experiences) that serves to reinforce and guide the learner in real-time [81–83]. The neural correlates of these VR-experiences directly connect with the behavioral gesturing of acting out the movements in VR while in an immersive experience anywhere the student may be [84]; thereby facilitating a contrived, yet effective, applied learning experience from afar ([85–87]; *for review on the instructional design see* [87]). However, to overcome cost limitations companies like Revinax®, Google, and others have used cardboards or other cheap modifiable glasses that can be attached to students' cellular phones to use the VR experience from downloaded Apps ([88, 89]). These cost-effective solutions for undergraduate students, especially for PUIs, can truly leverage their educational gains and begin to reduce the education gaps that have become wider over the last 20-years in the neurosciences. Thus, the climate for developing early career procedural skills in the neurosciences may be well positioned to utilize VR in a range of curricular contexts (*For Review, see* [90]). As VR has become more mainstream over the last 20-years, it appears that the literature has not fully recognized nor conceptualized its value in the field of neuroscience along with its many subdisciplines of specialty.

5 Results

In reviewing the literature on VR and neuroscience education over the last 20-years, a set of three article database searches (i.e., *Google Scholar*, *PubMed*, and *Science Direct*) were conducted from the years covering 2004–2024 using the following *key terms*: *"VR AND Neuroscience Education"*; *"Virtual Reality AND Neuroscience Education"*; *"AR AND Neuroscience Education"*; *"Augmented Reality AND Neuroscience Education"*; *"FP-POV AND Neuroscience Education"*; and *"First-Person POV AND Neuroscience Education"*, respectively. The 2024 year was not included as we are currently in this year and the data would be truncated and underestimating what the year total might be.

For the *"VR AND Neuroscience"* and the *"Virtual Reality AND Neuroscience"* key words, they did not differ in the number of articles returned; thus, the *"VR AND Neuroscience Education" key terms* were used. Figure 12.1 illustrates an increasing trend line in all three article databases with *Google Scholar* (i.e., an increase from 859 articles from 2004 to 14,100 articles in 2023 [a 1,541.44% increase in publications]), then *PubMed* (i.e., an increase from 0 articles from 2004 to 53 articles in 2023 [i.e., a 53,000% increase in publications]), and finally *Science Direct* (i.e., an increase from 16 articles from 2004 to 317 in 2023 [i.e., a 1,882.25% increase in publications]). Thus, the best place to identify current and more broad ranging research articles on VR and neuroscience education would be either *Google Scholar* or *Science Direct*.

Next, for the *"AR AND Neuroscience"* and the *"Augmented Reality AND Neuroscience"* key words, they did not differ in the number of articles returned; thus, the *"AR AND Neuroscience Education" key terms* were used. Figure 12.2 illustrates an increasing trend line in all three article databases with *Google Scholar* (i.e., an increase from 114 articles from 2004 to 5230 articles in 2023 [a 4,487.72% increase in publications]), then *PubMed* (i.e., an increase from 0 articles from 2004 to 15 articles in 2023 [i.e., a 14,900% increase in publications]), and finally *Science Direct* (i.e., an increase from 18 articles from 2004 to 165 in 2023 [i.e., a 816.67% increase in publications]). Thus, the best place to identify current and more broad ranging research articles on AR and neuroscience education would be either *Google Scholar* or *Science Direct*.

Lastly, for the *"FP-POV AND Neuroscience"* and the *"First-Person POV AND Neuroscience" key words*, the *"FP-POV AND Neuroscience"* key words did not return any articles; thus, the *"First-Person POV AND Neuroscience Education" key terms* were used. Figure 12.3 illustrates an increasing trend line in all three article databases with *Google Scholar* (i.e., an increase from 220 articles from 2004 to 1900 articles in 2023 [a 763.64% increase in publications]), then *PubMed* (i.e., an increase from 0 articles from 2004 to 0 articles in 2023 [i.e., no publications retrieved

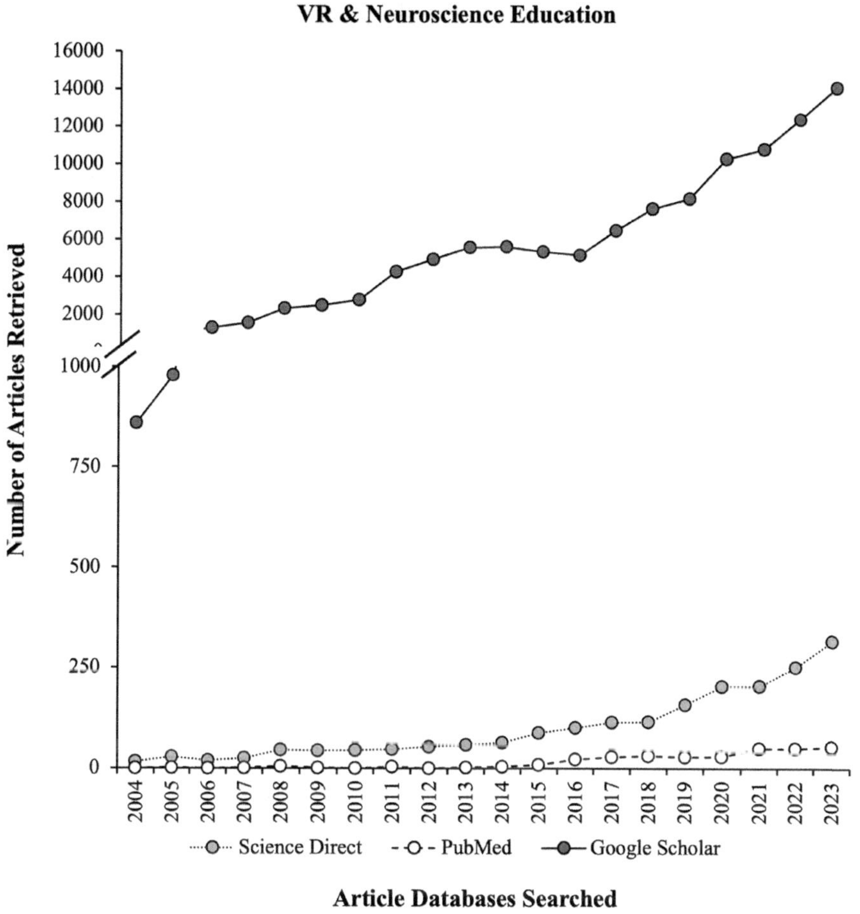

Fig. 12.1 Comparisons between Google Scholar, PubMed, and Science Direct publications searched with the key words "VR AND Neuroscience Education" from 2004 to 2023. The results showed an increasing trend in research published in these areas across all three databases with the following growth rates: Google Scholar (1,541.44%), PubMed (53,000%), and Science Direct (1,882.25%), respectively. (Source © 2024, Neuwirth et al., Used with permission)

within that database covering 20 years]), and finally *Science Direct* (i.e., an increase from 4 articles from 2004 to 110 in 2023 [i.e., a 2,650% increase in publications]). Thus, the best place to identify current and more broad ranging research articles on FP-POV and neuroscience education would be either *Google Scholar* or *Science Direct.*

Although it is important to maintain scientific control and reliability with conducted literature review searches of this kind, one should equally be both methodological and cautious that over the last 20-years new combinations of search terms based on Artificial Intelligent (AI) programming may opt for different proxies for gathering more information in databases like *Google Scholar* than what might be the

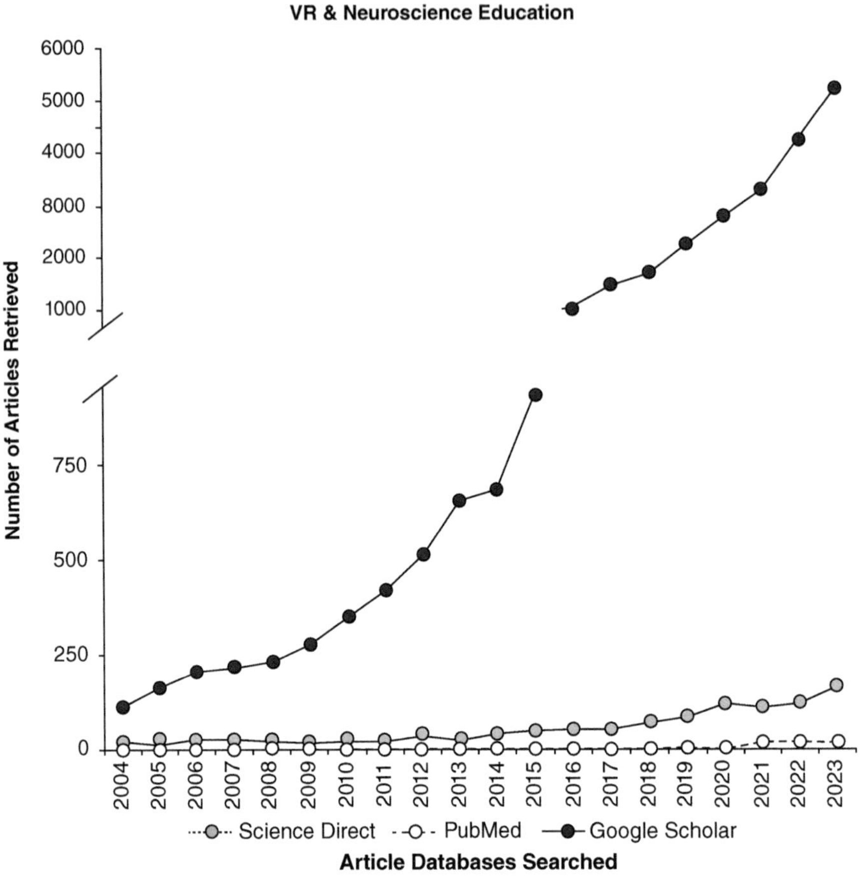

Fig. 12.2 Comparisons between Google Scholar, PubMed, and Science Direct publications searched with the key words "AR AND Neuroscience Education" from 2004 to 2023. The results showed an increasing trend in research published in these areas across all three databases with the following growth rates: Google Scholar (4,487.72%), PubMed (14,900%), and Science Direct (816.67%), respectively. (Source © 2024, Neuwirth et al., Used with permission)

typically utilized in databases like *PubMed* and *Science Direct*. Additionally, PubMed and Science Direct have numerous journals where articles that are submitted for publication and accepted are subjected to either 3–5 year embargo agreements or are limited to pay subscription accessibility (i.e., pay wall barriers to more immediate public access and education). Conversely, these journals can also be on the top-tier side of publications with high impact factors in any neuroscience sub-field that requires substantial money from researchers and educators to pay Article Process Charges (APCs) that can range from a few hundred dollars to ten thousand dollars. Additionally, most researchers and educators at PUIs and MSIs are unable to have their administration cover APCs, unlike Private Universities and Colleges within the US. The latter point is critical as most researchers and educators in the modern world

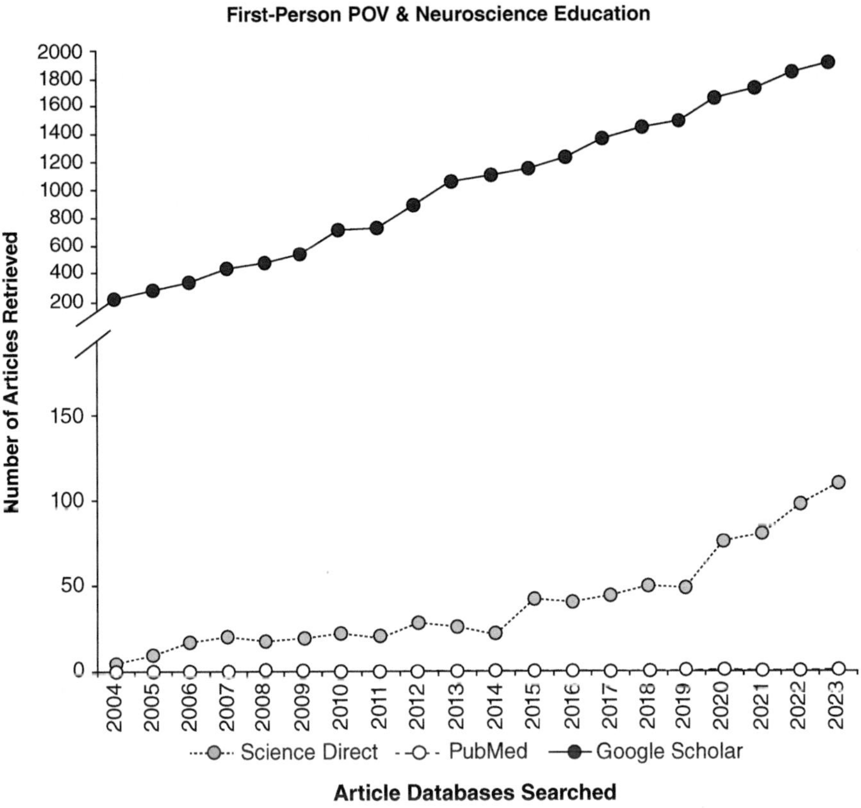

Fig. 12.3 Comparisons between Google Scholar, PubMed, and Science Direct publications searched with the key words "First-Person POV AND Neuroscience Education" from 2004 to 2023. The results showed an increasing trend in research published in these areas across only two databases with the following growth rates: Google Scholar (763.64%) and Science Direct (2650%),whereas PubMed (0%) never had a paper retrieved in that database, respectively. (Source © 2024, Neuwirth et al., Used with permission)

are coming from PUIs/MSIs and are struggling to obtain federal and state grant funding within the US at funding rates between approximately 1.6–3.6% in the neurosciences. Thus, researchers and educators may opt to intentionally seek out lower impact factor journals to publish their work for either free/no APC costs or this may appear as the work is less impactful to other professionals in the field, but may also contribute, in part, as to why Google Scholar may have more articles available within its search capabilities Additionally, from a principled and ethical position, researchers and educators may opt to only publish in journals without pay wall barriers to ensure the public has immediate access to new knowledge generated in their given fields.

Lastly, PUI and MSI colleges and universities may not be able to afford these subscriptions for these high impact factor journals and as a result may redirect themselves and their students to use other no-cost and more accessible (i.e., no paywall to access) publications that may begin to leverage the gap in these initialized practices from high impact journals. However, what is gleaned from these analyses on these three database search engines are that Google Scholars' use of AI can return many more hits with the key words chosen, but more work might need to be done to filter out irrelevant articles which may be cumbersome until *Google* produces better selection filters like *PubMed* and *Science Direct*. It is not unreasonable to think that Google will swiftly address this issue and make its *Google Scholar* database equal if not better than *PubMed* and *Science Direct* for these research-based queries for conducting literature reviews and background reading for experimental studies across any scientific discipline. Here, we just provide one example of this case-in-point for using VR in neuroscience education.

6 Discussion

The range of student's perquisite skills for any curriculum can vary considerably. Implementing VR pedagogy into the curriculum can have substantial advantages besides cost savings to the student and the higher educational industry in supplying students with such VR classrooms or laboratories as well as portable cardboard VR options to partake in similar experiences anywhere in the world. The latter allows for more international student enrollment and collaboration that can also promote more diverse perspectives in the classroom emanating from many differing life experiences. Moreover, the VR programs can serve as intentional pedagogy directed at upskilling, reskilling, and training early career undergraduate students with competitive workforce ready skills in a unique procedural way. This situation enables under resourced PUIs to competitively teach their students with applied learning experiences in which the student's behavioral demonstration of their learned gestures from the VR experience can be assessed as a practicum competency for their overall learning. Students can also, through these early career experiences, learn to make less errors in the contrived VR experience when they begin to be tested in real practicum settings [76]. This is important as undergraduate students have less hands-on training, opportunities to practice skills, less attempts to demonstrate proficiency, and full comprehension of the ability to ethically and safely learn from their errors when demonstrating these learned behaviors in laboratory classes or practicum-based courses. Finally, VR can also serve as a pretext to understand neuroscientific principles underlying its efficiency. Considering the aforementioned factors, VR pedagogical instruction for undergraduate students in the neuroscience can serve to reduce training times, costly resources, and may make students more prepared with job- and crisis-ready skills unlike prior generations of students in the neurosciences.

7 Conclusion

Thus, VR has shown much promise over the last 20-years as a timely and unique, yet underutilized, neuroscience pedagogical tool and research method in which undergraduate students could greatly benefit from if engaged earlier in their careers. This encouraging direction to integrate VR early in the undergraduate student curriculum for the neurosciences is further motivated by its attractiveness and ease of delivering the curriculum across different teaching modalities (i.e., in-person, online, and remote learning). Such curricular flexibility offers more equitable solutions to help students from PUIs to also overcome several educational barriers to resources, access, and other inequities. The growing recognition over the last two decades that VR can be creatively and effectively used within the neurosciences to overcome these educational barriers is becoming more clear and less disputed, but not implemented as frequently as we would have thought by now. If PUIs want to level the playing field for providing their undergraduate students with earlier job market ready procedural skills, then they should consider ways to adopt and integrate cost effective VR solutions for their neuroscience programs.

References

1. Hardwick JC, Smith JS. Undergraduate neuroscience faculty: results from a survey of faculty for undergraduate neuroscience members. J Undergrad Neurosci Educ. 2010;8(2):A101–7.
2. Kreitzer MA, Malchow RP. Engaging undergraduates in a unique neuroscience research opportunity: a collaborative research experience between a primarily undergraduate institution (PUI) and a major research institution. J Undergrad Neurosci Educ. 2013;12(1):A85–92.
3. Rovnyak D, Shields GC. A roadmap to successful collaborations between primarily undergraduate institutions and research institutions. In: Credit where credit is due: respecting authorship and intellectual property. Washington, DC: American Chemical Society; 2018. p. 105–27. https://doi.org/10.1021/bk-2018-1291.ch011.
4. Yeung AWK, Goto TK, Leung WK. The changing landscape of neuroscience research, 2006–2015: a bibliometric study. Front Neurosci. 2017;11:120. https://doi.org/10.3389/fnins.2017.00120.
5. Ramos RL, Esposito AW, O'Malley S, Smith PT, Grisham W. Undergraduate neuroscience education in the US: quantitative comparisons of programs and graduates in the broader context of undergraduate life sciences education. J Undergrad Neurosci Educ. 2016;15(1):A1–4.
6. Ramos RL, Fokas GJ, Bhambri A, Smith PT, Hallas BH, Brumberg JC. Undergraduate neuroscience education in the US: an analysis using data from the National Center for Education Statistics. J Undergrad Neurosci Educ. 2011;9(2):A66–70.
7. Ramirez JJ. Undergraduate neuroscience education: meeting the challenges of the 21st century. Neurosci Lett. 2020;739:135418. https://doi.org/10.1016/j.neulet.2020.135418.
8. Ramos RL. Virtual was the reality in neuroscience education during the COVID-19 pandemic. J Undergrad Res. 2020;19(1):E1–E2.
9. Mickley GA, Kenmuir C, Remmers-Roeber D. Mentoring undergraduate students in neuroscience research: a model system at Baldwin-Wallace College. J Undergrad Neurosci Educ. 2003;1(2):A28–35.
10. Adams J. "Writing in neuroscience": a course designed for neuroscience undergraduate students. J Undergrad Neurosci Educ. 2011;10(1):A50–7.

11. Wiertalak EP, Hardwick J, Kerchner M, Parfitt K, Ramirez JJ. The new blueprints: undergraduate neuroscience education in the twenty-first century. J Undergrad Neurosci Educ. 2018;16(3):A244–51.
12. Wiertelak EP, Ramirez JJ. Undergraduate neuroscience education: blueprints for the 21st century. J Undergrad Neurosci Educ. 2008;6(2):A34–9.
13. Cleland CL. Integrating recent advances in neuroscience into undergraduate neuroscience and physiology courses. Adv Physiol Educ. 2002;26(1–4):271–277. https://doi.org/10.1152/advan.00044.2002.
14. Oprisan SA. Interdisciplinary curriculum for computational neuroscience at primarily undergraduate institutions. J Comput Sci. 2022;61:101642. https://doi.org/10.1016/j.jocs.2022.101642.
15. Meil WM. The use of case studies in teaching undergraduate neuroscience. J Undergrad Neurosci Educ. 2007;5(2):A53–62.
16. Jones LS, Black LC, Bright L, Meekins C, Thakur V, Warren C. An undergraduate course on publishing neuroscience. J Undergrad Neurosci Educ. 2006;4(2):A60–7.
17. Salomon D, Martin-Harris L, Mullen B, Odegaard B, Zvinyatskovskiy A, Chandler SH. Brain literate: making neuroscience accessible to a wider audience of undergraduates. J Undergrad Neurosci Educ. 2015;13(2):A64–73.
18. Neuwirth LS, Dacius TF Jr, Mukherji BR. Teaching neuroanatomy through a historical context. J Undergrad Neurosci Educ. 2018;16(2):E26–31.
19. Chagas AM, Canli T, Ziadlou D, Forlano PM, Smaddar S, Chua E, Baskerville KA, Poon K, Neuwirth LS. Editorial: Using open neuroscience to advance equity in the pedagogy and research infrastructure in colleges/universities still financially impacted by COVID-19: the emergence of a global resource network aimed at integrating neuroscience and society. J Undergrad Neurosci Educ. 2023;21(2):E2–7. https://pubmed.ncbi.nlm.nih.gov/37588641
20. Ros M, Weaver L, Neuwirth LS. Virtual reality stereoscopic 180-degree video-based immersive environments: Applications for training surgeons and other medical professionals. In JE Stefaniak (Ed.) Cases on Instructional Design and Performance Outcomes in Medical Education. 2020 pp. 92–110. Hershey, PA: IGI Global. https://www.igi-global.com/book/casesinstructional-design-performance-outcomes/244674.
21. Markram H. Seven challenges for neuroscience. Funct Neurol. 2013;28(3):145–51.
22. Petersen SE, Posner MI. The attention system of the human brain: 20 years after. Annu Rev Neurosci. 2012;35:73–89. https://doi.org/10.1146/annurev-neuro-062111-150525.
23. Altimus CM, Marlin BJ, Charalambakis NE, Colón-Rodrígues A, Glover EJ, Izbicki P, Johnson A, Lourenco MV, Makinson RA, McQuail J, Obeso I, Padilla-Coreano N, Wells MF. The next 50 years of neuroscience. J Neurosci. 2020;40(1):101–6. https://doi.org/10.1523/JNEUROSCI.0744-19.2019.
24. Martin JB. The integration of neurology, psychiatry, and neuroscience in the 21st century. Am J Psychiatry. 2002;159(5):695–704. https://doi.org/10.1176/appi.ajp.159.5.695.
25. Mamlouk GM, Dorris DM, Barrett LR, Meitzen J. Sex bias and omission in neuroscience research is influenced by research model and journal, but not reported NIH funding. Front Neuroendocrinol. 2020;57:100835. https://doi.org/10.1016/j.yfrne.2020.100835.
26. Shansky RM, Woolley CS. Considering sex as a biological variable will be valuable for neuroscience research. J Neurosci. 2016;36(47):11817–22. https://doi.org/10.1523/JNEUROSCI.1390-16.2016.
27. Will TR, Proaño SB, Thomas AM, Kunz LM, Thompson KC, Ginnari LA, Jones CH, Lucas SC, Reavis EM, Dorris DM, Meitzen J. Problems and progress regarding sex bias and omission in neuroscience research. eNeuro. 2017;4(6):ENEURO.0278-17.2017. https://doi.org/10.1523/ENEURO.0278-17.2017.
28. Vázquez-Guardado A, Yang Y, Bandodkar AJ, Rogers JA. Recent advances in neurotechnologies with broad potential for neuroscience research. Nat Neurosci. 2020;23:1522–36. https://doi.org/10.1038/s41593-020-00739-8.

29. Matusz PJ, Dikker S, Huth AG, Perrodin C. Are we ready for real-world neuroscience? J Cogn Neurosci. 2019;31(3):327–38. https://doi.org/10.1162/jocn_e_01276.
30. Devonshire IM, Dommett EJ. Neuroscience: viable applications in education? Neuroscientist. 2010;16(4):349–56. https://doi.org/10.1177/1073858410370900.
31. Bohil CJ, Alicea B, Biocca FA. Virtual reality in neuroscience research and therapy. Nat Rev Neurosci. 2011;12(12):752–62. https://doi.org/10.1038/nrn322.
32. Riva G, Wiederhold BK, Mantovani F. Neuroscience of virtual reality: from virtual exposure to embodied medicine. Cyberpsychol Behav Soc Netw. 2019;22(1):82–96. https://doi.org/10.1089/cyber.2017.29099.gri.
33. Rizzo AS, Koenig ST. Is clinical virtual reality ready for primetime? Neuropsychology. 2017;31(8):877–99. https://doi.org/10.1037/neu0000405.
34. Max ML, Burke JC. Virtual reality for autism communication and education, with lessons for medical training simulators. Stud Health Technol Inform. 1997;39:46–53. https://pubmed.ncbi.nlm.nih.gov/10184722.
35. Cipresso P, Giglioli IAC, Raya MA, Riva G. The past, present, and future of virtual and augmented reality research: a network and cluster analysis of the literature. Front Psychol. 2018;9:309500. https://doi.org/10.3389/fpsyg.2018.02086.
36. Hamad A, Jia B. How virtual reality technology has changed our lives: an overview of the current and potential applications and limitations. Int J Environ Res Public Health. 2022;19(18):11278. https://doi.org/10.3390/ijerph191811278.
37. Liu D, Bhagat KK, Goa Y, Chang TW, Huang R. The potentials and trends of virtual reality in education: a bibliometric analysis on top research studies in the last two decades. In: Virtual, augmented, and mixed realities in education. Singapore: Springer; 2017. p. 105–30. https://doi.org/10.1007/978-981-10-5490-7_7.
38. Wohlgenannt I, Simons A, Stieglitz S. Virtual reality. Bus Inf Syst Eng. 2020;62:455–61. https://doi.org/10.1007/s12599-020-00658-9.
39. Meier AH, Rawn CL, Krummel TM. Virtual reality: surgical application—challenge for the new millennium. J Am Coll Surg. 2001;192(2):372–84. https://doi.org/10.1016/S1072-7515(01)00769-4.
40. Muñoz-Saavedra L, Miró-Amarante L, Domínguez-Morales M. Augmented and virtual reality evolution and future tendency. Appl Sci. 2020;10(1):322. https://doi.org/10.3390/app10010322.
41. Pensieri C, Pennacchini M. Overview: virtual reality in medicine. J Virtual Worlds Res. 2014;7(1). Lantern Part 1. https://doi.org/10.4101/jvwr.v7i1.6364.
42. Pottle J. Virtual reality and the transformation of medical education. Future Healthc J. 2019;6(3):181. https://doi.org/10.7861/fhj.2019-0036.
43. Claudio P, Maddalena P. Overview: Virtual reality in medicine. Journal of Virtual Worlds Research. 2014;7(1).
44. Ruthenbeck GS, Reynolds KJ. Virtual reality for medical training: the state-of-the-art. J Simul. 2015;9:16–26. https://doi.org/10.1057/jos.2014.14.
45. Gallagher AG, Ritter EM, Champion H, Higgins G, Fried MP, Moses G, Smith CD, Satava RM. Virtual reality simulation for the operating room: proficiency-based training as a paradigm shift in surgical skills training. Ann Surg. 2005;241(2):364–72. https://doi.org/10.1097/01.sla.0000151982.85062.80.
46. Lungu AJ, Swinkels W, Claesen L, Tu P, Egger J, Chen X. A review of the applications of virtual reality, augmented reality and mixed reality in surgical simulation: an extension to different kinds of surgery. Expert Rev Med Devices. 2021;18(1):47–62. https://doi.org/10.1080/17434440.2021.1860750.
47. Pensieri C, Pennacchini M. Virtual reality in medicine. In: Handbook on 3D3C platforms: applications and tools for three dimensional systems for community, creation and commerce. Cham: Springer International Publishing; 2016. p. 353–401. https://doi.org/10.1007/978-3-319-22041-3.

48. Velazquez-Pimentel D, Hurkxkens T, Nehme J. Virtual reality for the digital surgeon. In: Digital surgery. Cham: Springer; 2021. p. 183–201. https://doi.org/10.1007/978-3-030-49100-0_14.
49. Westwood JD, editor. Medicine meets virtual reality 20: NextMed/MMVR20, vol. 184. Amsterdam: IOS Press; 2013.
50. Neuwirth LS, Ng S, Devors S, Lonjon N, Ros M. Comparing 4th-year medical student's procedural surgical skills learning outcomes between third-person point-of-view video lectures and first-person point-of-view immersive virtual reality: a pilot study surgically implanting a subdermal drain on cadavers. Front Virtual Real. 2024;5:1408092. https://doi.org/10.3389/frvir.2024.1408092.
51. Ros M, Neuwirth LS, Prasolova-Førland E. Editorial: Redefining the pedagogy in virtual and augmented reality in medical science education. Front. Virtual Real. Sec. Virtual Reality in Medicine, 2024;5:1533837. https://doi.org/10.3389/frvir.2024.1533837.
52. Cesanek E, Shivkumar S, Ingram JN, Wolpert DM. Ouvrai opens access to remote virtual reality studies of human behavioural neuroscience. Nat Hum Behav. 2024;8(6):1209–24. https://doi.org/10.1038/s41562-024-01834-7.
53. Chen W, Lyu Q, Ho MYM, Tan J, Teo WP, Chai SCA, Su G. The role of individual preparation on coordination in computer-supported collaborative learning: a neuroscience perspective on learners' inter-brain synchronization. In: 17th international conference on computer-supported collaborative learning (CSCL) 2024-CSCL proceedings, vol. 2024; 2024.
54. Diwakar S, Parasuram H, Medini C, Raman R, Nedungadi P, Wiertelak E, Srivastava S, Achuthan K, Nair B. Complementing neurophysiology education for developing countries via cost-effective virtual labs: case studies and classroom scenarios. J Undergrad Neurosci Educ. 2014;12(2):A130–9.
55. Isager PM, Lakens D, van Leeuwen T, van't Veer AE. Exploring a formal approach to selecting studies for replication: a feasibility study in social neuroscience. Cortex. 2024;171:330–46. https://doi.org/10.1016/j.cortex.2023.10.012.
56. Kannan R, Gendreau M, Hatch A, Free SK, Muriungi K, Garje YA, DeBoer J, Gage GJ, Harris CA. Building brains for robots: a hands-on approach to learning neuroscience in the classroom. bioRxiv. 2024. https://doi.org/10.1101/2024.05.15.594177.
57. Ramirez-Zamora A, Giordano J, Gunduz A, Alcantara J, Cagle JN, Cemera S, Difuntorum P, Eisinger RS, Gomez J, Long S, Parks B. Proceedings of the seventh annual deep brain stimulation think tank: advances in neurophysiology, adaptive DBS, virtual reality, neuroethics and technology. Front Hum Neurosci. 2020;14:54. https://doi.org/10.3389/fnhum.2020.00054.
58. Thompson PM, Jahanshad N, Ching CR, Salminen LE, Thomopoulos SI, Bright J, Baune BT, Bertolín S, Bralten J, Bruin WB, Bülow R. ENIGMA and global neuroscience: a decade of large-scale studies of the brain and health and disease across more than 40 countries. Transl Psychiatry. 2020;10(1):100. https://doi.org/10.1038/s41398-020-0705-1.
59. Wilson KD. Integrating non-invasive brain stimulation into the undergraduate psychology curriculum: an upper-level laboratory course using transcranial direct current stimulation. Teach Psychol. 2024;51(1):110–7. https://doi.org/10.1177/00986283211060324.
60. Coumans JV, Wark S. Impact of problem-based learning coaching and neuroeducation in the development of 21st century lifelong learners. Mind Brain Educ. 2024;18(1):35–42. https://doi.org/10.1111/mbe.12406.
61. Mehra RK. Evolutionary advantages of virtual reality in education. In: Transforming education with virtual reality. New Jersey: Wiley; 2024. p. 73–86. https://doi.org/10.1002/9781394200498.ch5.
62. Neuwirth LS, Jović S, Mukherji BR. Reimagining higher education during and post-COVID-19: challenges and opportunities. J Adult Contin Educ. 2020;27(2):1–16. https://doi.org/10.1177/1477971420947738.
63. Ros M, Neuwirth LS. Increasing global awareness of timely COVID-19 healthcare guidelines through FPV training tutorials: portable public health crises teaching method. Nurse Educ Today. 2020;91(104479):1–6. https://doi.org/10.1016/j.nedt.2020.104479.

64. Cardoso K, Zaro MA, da Silva PF. The evaluation of the virtual immersive learning laboratory in health and nursing (LIASE): a study in neuroeducation. Immersive Learning Research-Academic 2024:306–16. https://doi.org/10.56198/U6C0W3BAT.
65. Ghani N, Baker H, Huntsinger A, Chen T, Familara TD, Itorralba JY, Vanderford F, Zhuang X, Change CL, Vo V, Oh EC. Science education for the youth (SEFTY): a neuroscience outreach program for high school students in Southern Nevada during the COVID-19 pandemic. eNeuro. 2024;11(4).
66. Obrero-Gaitán E, Nieto-Escamez FA, Zagalaz-Anula N, Cortés-Pérez I. An innovative approach for online neuroanatomy and neurorehabilitation teaching based on 3D virtual anatomical models using leap motion controller during COVID-19 pandemic. Front Psychol. 2021;12:590196. https://doi.org/10.3389/fpsyg.2021.590196.
67. Sahu M, Gupta R, Ambasta RK, Kumar P. IoT-driven augmented reality and virtual reality systems in neurological sciences. Internet Things. 2024;25:101098. https://doi.org/10.1016/j.iot.2024.101098.
68. Singh R, Baby B, Singh R, Suri A. Role of virtual modules to supplement neurosurgery education during COVID-19. J Clin Neurosci. 2021;91:125–30. https://doi.org/10.1016/j.jocn.2021.06.039.
69. Zhang Z, Gao Y, Pan Y, Zhou J. Video education through the lens of educational neuroscience: a concise review. TechTrends. 2024;68(3):469–84. https://doi.org/10.1007/s11528-024-00946-1.
70. Winter L, Periquito J, Kolbitsch C, Pellicer-Guridi R, Nunes RG, Häuer M, Broche L, O'Reily T. Open-source magnetic resonance imaging: improving access, science, and education through global collaboration. NMR Biomed. 2024;37(7):e5052. https://doi.org/10.1002/nbm.5052.
71. Kashwani R, Kulkarni V, Salam S, Sharma S, Rathi P, Gupta S, Sinha P, Kumari A, Sharma A. Virtual vs. augmented reality in the field of dentistry. Commun Pract. 2024;21(3):597–603.
72. Sharma K, Aswal HS, Chauhan R, Rawat R, Gupta R. Neuroscience of virtual reality. In: 2024 international conference on intelligent and innovative technologies in computing, electrical and electronics (IITCEE). Piscataway: IEEE; 2024. p. 1–6.
73. Aridan N, Bernstein-Eliav M, Gamzo D, Schmeidler M, Tik N, Tavor I. Neuroanatomy in virtual reality: development and pedagogical evaluation of photogrammetry based 3D brain models. Anat Sci Educ. 2024;17(2):239–48. https://doi.org/10.1002/ase.2359.
74. Gurses ME, Gökalp E, Gecici NN, Gungor A, Berker M, Ivan ME, Komotar RJ, Cohen-Gadol AA, Türe U. Creating a neuroanatomy education model with augmented reality and virtual reality simulations of white matter tracts. J Neurosurg. 2024;141(3):865–74. https://doi.org/10.3171/2024.2.JNS2486.
75. Rodriguez-Amendariz AG, Saint-Germain MA, Khalafallah AM, Huq S, Almeida ND, Dornbos DL III, Graffeo CS, Sivakumar W, Mukherjee D, Johnson JN. The neurosurgery research & education foundation-young neurosurgeons committee webinar series: providing education and inspiration during the COVID-19 pandemic. J Clin Neurosci. 2024;120:221–8. https://doi.org/10.1016/j.jocn.2024.01.021.
76. Ros M, Neuwirth LS, Ng S, Debien B, Molinari N, Gatto F, Lanjon N. The effects of an immersive virtual reality application in first person point-of-view, video-based, on the learning and generalized performance of a lumbar puncture medical procedure. Educ Technol Res Dev. 2021;69:1529–56. https://doi.org/10.1007/s11423-021-10003-w.
77. Wach J, Winkler D, Vychopen M, Kropla F, Grunert R, Güresir E. Student's perspectives regarding needs for and opportunities with mixed-reality education in neurosurgery at German medical schools. Neurosurg Focus. 2024;56(1):E17. https://doi.org/10.3171/2023.10.FOCUS23628.
78. Xin Y, Wang R, Liang Y, Lu Q, Zhang N. A proposed framework of virtual reality system design for neuroscience education in mental health. In: International conference on human-computer interaction. Cham: Springer Nature Switzerland; 2024. p. 136–52. https://doi.org/10.1007/978-3-031-60012-8_9.
79. Brucker B, Pardi G, Uehlin F, Moosmann L, Lachmair M, Halfmann M, Gerjets P. How learners' visuospatial ability and different ways of changing the perspective influence learning

about movements in desktop and immersive virtual reality environments. Educ Psychol Rev. 2024;36(3):65. https://doi.org/10.1007/s10648-024-09895-w.
80. Neuwirth LS, Ros M. Comparisons between first person point-of-view 180 video virtual reality head mounted display and 3D video computer display in teaching undergraduate neuroscience students stereotaxic surgeries. Front Virtual Real. 2021;2:706653. https://doi.org/10.3389/frvir.2011.706653.
81. Bonini L, Rotunno C, Arcuri E, Gallese V. Mirror neurons 30-years later: implications and applications. Trends Cogn Sci. 2022;26(9):767–81. https://doi.org/10.1016/j.tics.2022.06.003.
82. Curry M. If I only had a brain: mirroring neurons, authentic relationship (regulate, relate, reason). In: The neural teaching guide. London: Routledge; 2024. p. 195–208.
83. Lauharatanahirun N, Won A, Hwang AHC. External and internal attribution in human-agent interaction: insights from neuroscience and virtual reality. Hum Mach Commun. 2024;8(1):119–39. https://doi.org/10.30658/hmc.8.6.
84. Bodet T. Neurobehavioral correlates of embodiment in immersive virtual-reality-assisted learning [Doctoral dissertation]; 2024. https://hdl.handle.net/10657/16154.
85. Laine TH, Suk HJ. Investigating user experience of an immersive virtual reality simulation based on gesture-based user interface. Appl Sci. 2024;14(11):4935. https://doi.org/10.3390/app14114935.
86. Ros M, Neuwirth LS. Immersion Dans Les Réalités Étendues—Outil De Formation VR Et Spécificités Du 180-FPV. Maitrise Orthopedique. 2023;328:1–12.
87. Ros M, Weaver L, Neuwirth LS. Virtual reality stereoscopic 180-degree video-based immersive environments: applications for training surgeons and other medical professionals. In: Stefaniak JE, editor. Cases on instructional design and performance outcomes in medical education. Hershey, PA: IGI Global; 2022. p. 92–119. https://www.igi-global.com/book/cases-instructional-design-performance-outcomes/244674.
88. Martins IF. Additive manufacturing in the development of low-cost immersive virtual reality solutions for education: a study with Google cardboard. Lond J Soc Sci. 2024;(8):39–51. https://doi.org/10.31039/ljiss.2024.8.209.
89. Rockcastle SF, Sollom-Brotherton Whalen G, Milgrim A, Davis R. The use of virtual screens in perceptual studies on electric light: a comparison between 2D, panoramic, cardboard, and head-mounted displays. LEUKOS. 2024;20(3):275–90. https://doi.org/10.1080/15502724.2023.2279517.
90. Bocanegra-Becerra JE, Sánchez JLA, Castilla-Encinas AM, Rioa-Garcia W, Mendieta CD, Quiroz-Marcelo DA, Alwaishel K, Aguilar-Zegarra L, Lopez-Gonzalez MA. Toward a frontierless collaboration in neurosurgery: a systematic review of remote augmented and virtual reality technologies. World Neurosurg. 2024;187:114–21. https://doi.org/10.1016/j.wneu.2024.04.048.

Batch number: 10732090

Printed by Printforce, the Netherlands